Creating Abundance:

Visionary Entrepreneurs of Agriculture

By the same author:

The Day of the Bonanza

The Challenge of the Prairie

Beyond the Furrow

Tomorrow's Harvest

Koochiching

Plowshares to Printouts

Taming the Wilderness

History of U.S. Agriculture
(*a.k.a.* Legacy of the Land)

Creating Abundance:

Visionary Entrepreneurs of Agriculture

Hiram M. Drache, Ph.D.

Concordia College
Moorhead, Minnesota

Interstate Publishers, Inc.
Danville, Illinois

Creating Abundance:
Visionary Entrepreneurs of Agriculture

Order from

Interstate Publishers, Inc.

510 North Vermilion Street
P.O. Box 50
Danville, IL 61834-0050

Phone: (800) 843-4774
Fax: (217) 446-9706
Email: info-ipp@IPPINC.com

World Wide Web: www.interstatepublishers.com

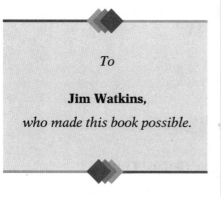

To

Jim Watkins,

who made this book possible.

AFTER GRADUATION from college in 1970, with majors in economics and fine arts, James D. Watkins joined Pillsbury Foods. Eight years later, his restless, entrepreneurial spirit caused him to move on. He founded Golden Valley Microwave Foods, Inc., where he developed Act II Microwaveable Popcorn. Then he pioneered Act II Microwaveable French Fries, which are extremely popular in the European market.

Mr. Watkins' success prompted him to venture again. This resulted in Golden Valley's acquiring 50 percent ownership of Lamb-Weston. In 1992 he sold his properties to ConAgra and was named president and chief operating officer of ConAgra Diversified Foods. Six years later he resigned to found J. D. Watkins Enterprises, Inc.

Since 1998, in addition to serving on various corporate boards, Mr. Watkins has been involved in rejuvenating Russian agriculture. He is joint venturing with Russians in the production of popcorn and potatoes, two crops not controlled by the government.

James Watkins' life typifies the visionary entrepreneurial spirit that, like the spirit of the farmers described in the following pages, is leading our nation's agriculture to new frontiers.

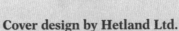

Cover design by Hetland Ltd.

In the face of a global economy, agriculture is coming of age. From the technological genetics (the circuit board) to creative marketing (signified by the numbers), the process of farming continues to change—not by the decade but *by the minute* (the global Internet). Its potential is as panoramic as its landscape, limited only by the constancy of the dedication, hard work, and entrepreneurial vision of its diverse people.

Foreword

IT IS IMPOSSIBLE to do justice to the unique contributions of the entrepreneurs of agriculture whose accomplishments Dr. Drache recounts in *Creating Abundance*. These leaders, together with their extended families, are continuing to change the total structure of the U.S. agribusiness system from genetic inputs to sophisticated, healthy, and nutritious food products to the development of functional food, nutraceuticals, pharmaceuticals, fiber, and energy.

In each case, the entrepreneur consistently found a more efficient way to produce commodities from the land, partner with the rest of the vertical value-added system, and become large enough to deal on a win/win basis with the consolidated retail food business and fast-food service industry that emerged.

The entrepreneurs were also mindful of their responsibilities to manage important land and water resources in a way that is responsive to the environmental, health, nutrition, food safety, and economic development needs of society.

As did the visionaries of the twentieth century before them, these entrepreneurs of the new century are providing ways to assure the consumer of consistency, quality, and safety by using expertise that can trace animals and crops from the gene to the final product.

These visionaries keep the U.S. agribusiness/agriceutical system competitive in a global marketplace. They are also leaders of integrity, grateful for the help they have received throughout their careers and always reaching out to the needs of others and society in general.

— Ray A. Goldberg, Ph.D.
Moffett Professor of Agriculture
and Business Emeritus
Harvard University

Preface

L ITTLE DID I REALIZE when I wrote *The Day of the Bonanza* in 1964 that it would be the beginning of a career writing about the industry that provided the foundation for America's rise as the world's leading industrial power. After *Bonanza* I wrote *The Challenge of the Prairie*, a book about homesteading in the almost identical area where the bonanza farmers broke sod. The homesteaders succeeded where the bonanza farmers failed, because they relied on the unpaid labor of family members to survive. The bonanza farms were far better managed than most of the homesteaded farms, but they had to pay their workers and could not compete in what was still a very labor-intensive industry.

My curiosity was whetted, and in the late 1960s I decided to continue research on agriculture and bring my study up to date. My goal was to seek the largest farmers in a five-state area and determine what they were doing that made them successful while so many others were failing. I learned that these farmers were the leaders in adopting technology. They maximized the benefits of the government programs. They were business managers. They used technology to reduce drudgery. They were risk takers and accepted the tenets of the capitalistic free-enterprise system. But, more important, they were entrepreneurs. They scanned for opportunities to change agriculture, which was the only sector of our society still not industrialized.

In 1975, speaking at the Bicentennial Symposium on Two Centuries of American Agriculture, I stated that 95 percent of our farmers did not grasp the rate of change that was taking place in their industry. At that point I vowed to continue my research. After finishing a college textbook, *History* of *U. S. Agriculture and Its Relevance to Today*, I felt it was time to write about the incredible visionaries who were not only industrializing agriculture but were taking it into the global era. These entrepreneurial producers, who generally started from scratch, willingly took risks that lesser-motivated individuals avoided. These people enjoyed the business of farming as they coped with the obstacles of a cyclical industry. At the same time, they faced criticism from family farm advocates, particularly politicians, church groups, and farm organizations.

In June 1990, I had my first interview for this book with 97-year-old Wheeler McMillen, a sage of American agriculture. In 1929 McMillen wrote *Too Many Farmers: The Story of What Is Here and Ahead in Agriculture.* Any farm in 1929 that grossed more than $30,000 was considered a large farm. The study revealed that 7,875 out of 6,288,648 farms, a mere $^1\!/_{10}$ of 1 percent of all farms, produced 4.5 percent of the total farm output. The study concluded that these farms were "more efficient than average farms as far as the use of land, labor, and capital." Consolidation continued, and today the top 4 percent of our farms produce 50 percent of all our agricultural products.

With one exception, in the chapters that follow is a geographical and enterprise sampling of individual entrepreneurs who today have placed their operations in the top rank of agricultural firms. Contrary to the common misconception about large farms, they are not "Wall Street–type corporate firms." All but one belong to an individual or a family. The owners of these large farms are the primary producers in a rapidly changing food chain with processors and retailers who are trying to satisfy the ever-demanding consumer with a greater variety of nutritious and safe food. They are providing the leadership that enables America to retain its position of having the best-fed society at the lowest percentage of income of any nation in the world.

These entrepreneurs are the "homesteaders" of the post-industrial global frontier. They are bringing the country to an era of even greater abundance than did the homesteaders of our first three centuries, whose task it was to break sod on our nation's frontier.

— **Hiram M. Drache**

Acknowledgments

FREE-ENTERPRISE, entrepreneurial-oriented farmers have always been my heroes, and the subjects of this volume deserve my first thanks. These interviewees generously gave of their time and that of their colleagues to provide me with essentially all the data used in this volume. This is oral history at its best, for who else could tell it better than the people who are making it happen and are setting the direction for tomorrow's harvest? Your cooperation was splendid. My wife said that every time I was off on another interview trip, I was as excited as a child. How could I be otherwise when I was interviewing such exciting people?

Upon the advice of Dr. Ray Goldberg, who wrote the Foreword, we did not pick 15 individuals as our "must" list. Instead, he wisely guided me through the maze of selecting the potential interviewees and let "the network" play its course. I knew, or had heard of, or had read about several of the persons chosen, but the others were not known by me prior to my first contact. Our chief concern was that we have both an enterprise and a geographical cross section of the very best in their respective fields of endeavor. Thanks, Ray, for the time you devoted to the project. I was also assisted by an expert team of individuals who are so knowledgeable about who the leading producers in American agriculture are. This group was made up of Gene Johnston, Dan Looker, John Majors, Dr. Porter Martin, Bob Moraczewski, Randall Pope, Don Senechal, and Murray Wise. Your help made the task of picking the very best easy.

Jim Watkins, to whom this book is dedicated, has devoted his life to agriculture. Jim has a real entrepreneurial spirit. He liked the theme of this project and offered to fund it. Jim, your financial input, moral support, and suggestions are much appreciated.

Ada, my wife of nearly 53 years, originally told me not to waste the effort because there was no interest in a book of this type. But I was determined because I sensed that great changes were taking place in agriculture that had to be recorded. After she saw the first chapters on the computer screen, she realized that each of these individuals had a real story to tell that was far different from the traditional information about farming. She deserves the credit for getting another manuscript in top form for Ronald L. McDaniel, Vice President–Editorial at Interstate Publishers, Inc., the best editor a writer could have. Thanks, Ronnie, for your very thorough editing and helpful encouragement. It is always comfortable to call Kim Romine, Interstate's desktop publishing expert, knowing that no matter how challenging the problem, she never loses her cool. She has helped the Draches out of many difficult snags. Kim, you are terrific! How fortunate we are to have such a gifted editorial team.

The library staff at Concordia College provided its usual wholehearted cooperation. It is always dangerous to single out individuals, but Linda Swanson and Jim Hewitt were most often called on to answer my queries. Thanks, Linda and Jim, but also the entire library staff. Thanks also to Concordia College for providing an office with access to all the library facilities, plus the Internet—an author's dream.

— **Hiram M. Drache**

About the Author

HIRAM M. DRACHE'S roots are in rural Minnesota. He delivered milk to customers in the village of Meriden in the 1930s, first with a sled or a bicycle with a sidecar and finally a 1927 Studebaker. While working on an uncle's farm, he milked by hand, shocked grain beside his aunt, and did other chores. His first field work was driving a single-row two-horse cultivator. During two threshing seasons he drove a bundle team. With those experiences behind him, he has never understood the strong feelings many people have about labor-intensive farming of the past.

Drache's college career was interrupted by service in the Air Force during World War II, after which he received his B.A. from Gustavus Adolphus College, his M.A. from the University of Minnesota, and his Ph.D. from the University of North Dakota. He taught high school, worked in various businesses, and purchased his first farm in 1950. From 1952 through 1991 he was a professor of history at Concordia College, Moorhead, Minnesota, where he specialized in European and economic history. Since then he has been historian-in-residence. Because of several innovations, his farming operation was the subject of many articles in regional and national farm periodicals. In 1962 he developed a computer record system for the farm and his cattle feeding business. After 31 years of farming, the Draches have rented out their farms.

In the past four decades Dr. Drache has spoken to hundreds of audiences in 36 states, 6 provinces of Canada, Australia, and Germany.

This is his ninth book. In addition, he has been a contributing author to 7 books and has written more than 50 articles on contemporary agriculture and/or agricultural history.

Contents

(In alphabetical order)

The Baileys

"They Learn How to Make Change Work for Them"

ONCE Americans were assured of a food supply by raising crops that were adapted to the area, it did not take them long to start experimenting with other plants to enrich their lives further. In 1728 a botanical garden was established in Philadelphia, where many imported varieties of plants were cultivated. Two years later an orchard of fruit trees was established in Maine, and the first commercial nursery was developed on Long Island.

As commercial activity and urban centers grew, the United States started the inevitable shift from being virtually a rural agrarian nation, with 95 percent of the people living on farms, to being primarily an industrialized nation, with less than 1 percent of the people engaged in production agriculture. In the process, home production of fruits, vegetables, and flowers shifted to commercial truck gardening and greenhouses.

Except for the Civil War years, commercial production grew rapidly during the nineteenth century. In 1885 the Society of American Florists held its first meeting. By then 24 million roses and 120 million carnations were produced annually by commercial nurseries. Only four years later the industry was made up of more than 10,000 firms, with 619 total acres under glass and total sales of more than $13.5 million. The green sector—greenhouse, floriculture, and nursery businesses—had become a major industry.

As the nation became more urbanized, the green sector expanded, but because of the lack of adequate transportation, it was always careful to stay near centers of population. Not until the 1950s, with the advent of the interstate highway program, which some say legislated success for the green sector, did the industry become more disbursed. At the same time, a home-building boom commenced, and the combination of the two caused the green sector to mushroom.

During the 1970s all phases of the industry grew at the rate of 10 percent per annum. By the late 1980s the green sector was one of the fastest-growing alternative enterprises in U.S. agriculture. In the 1990s the green sector outpaced all other major agricultural segments. In 1990 cash farm gate sales of nursery stock of $8.1 billion were 10 percent of all farm cash crop sales. In 1996 the industry was made up of 5,500 firms, employing 45,000 full-time and 105,000 part-time workers. Its $10.9 billion in sales made it the seventh-ranking sector of agriculture.

An interviewee for this chapter stated, "It is fun to see so many people interested in nursery goods—it satisfies their attachment to agriculture." This is a most appropriate statement at a time when nostalgia about agriculture as it once was still exists.

THE OPERATORS of Bailey Nurseries, Inc., can trace their roots in the United States back 10 generations to Levi Bailey, born in New England in 1742. In 1881 John Vincent Henry Bailey, along with his wife, Isabella, and their six children, moved to Red Rock (now part of Newport), Minnesota. By 1897 John Bailey was a well-known farmer and fruit producer in the area. His farm was about 2½ hours by team and wagon from St. Paul, enabling him to deliver his produce to that market.

John Vincent (J. V.) Bailey, Jr., was born February 10, 1873, while the family lived near Hastings, in Goodhue County. Like all farm children, he was assigned regular chores. He liked nursery work and majored in horticulture at the State School of Agriculture—the St. Paul campus of the University of Minnesota. After his graduation in 1896 he invested $10, his total monetary assets, into a partnership and rented 15 acres of land to grow produce. That venture ended in failure.

In 1897 J. V., Jr., working by himself, tried again. He rented 20 acres of nitrogen-deficient soil on a stony hillside near Red Rock. This time his college agricultural education paid off. He knew that to get a good growth on the 17 acres of muskmelons he had planted he would have to treat the soil with nitrogen. He made three applications of nitrate soda (a.k.a. sodium nitrate). Then, he purchased twenty-three thousand 5-by-7-inch glass negatives from local photographers and constructed a small hot house around each plant. This was done to speed up growth, for he knew that the first produce on the market each season captured a premium price. He received $3,000 for his melons, a virtual bonanza.

On December 16, 1898, J. V., Jr., paid $2,790.75 for 80 acres of land that was so badly infested with quack grass and wild oats that the former owner had virtually given up farming it. But that farm answered the dream of the Bailey father-and-son team, for it was on top of a hill overlooking the Mississippi River Valley. J. V., Jr., fenced the land and stocked it with about 600 western sheep, purchased from the nearby So. St. Paul Stockyards, to graze off the quack and wild oats. After the quack and wild oats were consumed, the sheep were finished for market with hay and oats. They yielded a net profit of $1,100.

Next, he covered the land with manure, which presumably came from the stockyards, and then plowed it under. He immediately seeded the land to hay, which he plowed down in the fall. In the spring of 1900 he re-plowed the land and then disked and harrowed it several times to expose the quack roots to the sun until mid-June, when he plowed again. Next, he planted Holland cabbage transplants and Miller's Cream Hybrid, California Citron, and Hackensack melons on the fertile mellow soil.

After a couple years he realized that he had to change his cropping pattern because the soil was washing off the hillsides. He remembered from his college courses that asparagus was a good crop to slow erosion. He planted five acres along the slopes. Much of the first crop of asparagus was placed in cold storage until the price improved. The profit from that crop was used to purchase another 35 acres, increasing the farm to 115 acres.

O N MARCH 26, 1902, J. V., Jr., married Elizabeth Anna Biery, who had just completed her course in home economics from the St. Paul campus. Late in 1902 the young couple moved into their new home on the farm at Newport. This stately structure served the family until 1966, when it became company headquarters.

In 1905 J. V. Bailey Nurseries was established. A company ink blotter stated, "We grow a complete line of fruits and ornamentals." Melons, apples, plums, and 2,000 currant bushes were grown in addition to vegetables. During harvest season Bailey loaded a wagon with produce at the end of each day. The next morning he rose at 3:00 A.M. in preparation for making the 2¹/₂-hour trip of 9 miles to the St. Paul market, which opened at 6:00 A.M.

In 1911 Bailey purchased an automobile, which he modified to haul produce. This enabled him to make several trips a day. He also used the auto to transport workers from the city to the fields and to return them to their homes in the evening. By 1911 the nursery crops included flowers, ornamental shrubs, and evergreens. Bailey products also were sold through a catalog distributed in the Twin City area.

When the Bailey children were old enough, they were expected to do field work. They received a penny a quart for picking berries. A good picker could pick as many as 100 quarts a day, but the children generally did far less. About 1911 Bailey established a roadside stand at the foot of Bailey Hill, where the children sold melons, berries, apples, and plums as the produce became available.

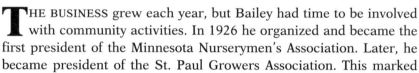

THE BUSINESS grew each year, but Bailey had time to be involved with community activities. In 1926 he organized and became the first president of the Minnesota Nurserymen's Association. Later, he became president of the St. Paul Growers Association. This marked the beginning of a position of leadership in state and national nursery associations that family members and company employees have continued to this day.

By 1930 the company had 4 full-time employees and 30 seasonal workers. The business grew too large to be run from the house, so a building was erected for retail sales, an office, and quarters for workers. On October 22, 1930, because of the downturn in the economy, John and Elizabeth Bailey were forced to mortgage some of their holdings to secure operating funds. They had installed their first irrigation system during the 1920s and, with the dry years, realized that they had to expand that system to ensure their chances of getting a crop.

By 1933 they had recovered financially and were able to purchase the nearby Danberg and Grey Cloud Island farms. This brought their total holdings to 300 acres, which was more than double the average-sized farm in Minnesota at that time, a real feat considering nearly all their acres were planted to labor-intensive crops. More storage space was added to include vaults equipped with forced ventilation for the

fruits and shrubs being held for the marketing season. A dairy herd of 30 purebred Holsteins was maintained to consume the forage and grain raised in rotation with the truck garden and orchard crops.

J. V. BAILEY, JR., was 60 years old in 1933. Although not in the best of health, he was far from ready to turn over the reins to Vincent, born in 1904; Gordon, born in 1907; or Beth, born in 1909. By then they all had graduated from college, were married, and intended to become involved in the business. Vincent worked in production, Gordon in management, and Beth as the secretary and bookkeeper. Two other daughters, Virginia and Margaret, chose not to enter the business.

J. V. Bailey, Jr., saw the addition of three children to the business as an opportunity to step out and expand. Fortunately, the three children were capable and, just as important, they willingly worked together, which is not the standard for most family businesses. Three decades later family relations changed, and Gordon Bailey became the dominant figure in the firm.

ON JUNE 17, 1933, Gordon Bailey and Margaret M. Fritz were married. Margaret and her family were neighbors and friends of the Baileys. The Fritzes had a truck garden on which they raised vegetables, berries, and some orchard crops. Margaret grew up working in the fields, helping harvest asparagus, strawberries, raspberries, muskmelons, apples, and other crops as they ripened during the season. Her big treat was getting up at 4:00 A.M. to ride to the St. Paul market in her family's Kissel truck. Margaret said, "The best way to explain the roads is dust or mud." When the truck was not used to haul produce, three rows of seats could be installed, making the Kissel resemble a station wagon without sides.

After graduating from high school in 1927, Margaret worked for the *St. Paul Daily News*. By then she and Gordon Bailey were courting. Gordon used the family Model A Ford coupe with a rumble seat but had to pay for each mile he drove it. This meant that he always looked for the shortest route, which was not necessarily the best. In 1928 Gordon and Margaret were dressed to attend a formal dance at the university, but they became so stuck on the muddy road that they never got there.

In 1929, after graduating from college, Gordon started working at the nursery and lived at home. After their marriage in 1933, Margaret and Gordon rented a bungalow in Newport for $25 a month. Gordon was paid $100 a month, which he thought was "good pay." He became the first road salesman for the company and traveled by train and/or bus to St. Louis, Chicago, North and South Dakota, and major Minnesota cities. Vincent also spent time on the road selling in other territories. Gordon and Vincent rotated being on the road so one of them was always on hand to help their father with the business. During the growing season, while Vincent worked in production, Gordon primarily managed the business. At harvest this involved receiving orders and preparing produce for delivery each day. This meant late hours, "at least 11:00 P.M. each evening."

In 1937 the first of several glass greenhouses was erected, and Clarence Seefert was employed to supervise the work there. Securing Seefert was a boon for the company, for he not only was an excellent propagator and production manager, but he also was an outstanding teacher for the future generation of Baileys.

Margaret recalled that after the greenhouse was erected, a bell was installed in the Baileys' house. Whenever there was a problem at the greenhouse, the alarm woke the entire family. The office telephone was also wired to ring in their home during nonbusiness hours.

From 1935 to 1948 Margaret gave birth to five children but continued to help at the nursery. This was especially so after 1938, when the family moved from Newport to one of the Bailey houses on the farm.

Apples were stored in the warehouse, and on fall weekends business was so heavy that all the family members had to work serving customers. Margaret was also in charge of the berry-picking crews to make sure they did not leave berries on the bushes. She commented that by the late 1940s the pay for picking berries had increased to 2¢ a quart, double what she had received as a girl.

Expansion continued, and by 1940 the company had 12 full-time employees and 50 seasonal workers. In 1944 Gordon, Margaret, and their children moved to the "nursery home," which they shared with Gordon's mother, Elizabeth, whose husband, J. V., Jr., had died the previous year. The alarm bell and off-duty telephone calls continued, but Margaret never took part in office activities even though her home was on the same yard as the office. Beginning with the labor shortage of World War II, she cooked for the 18 migrant Mexicans who stayed in the dormitory on the farm from March through October. Her helper was Helen Franek (later Chiarella), who started working for the Baileys at age 13, first as a babysitter and then as a hired girl. Helen said that her day started at 7:00 A.M. and ended at 6:00 P.M. or whenever the supper dishes were washed. For working those hours six days a week, she received $15, which she said was good pay. She added, "I never felt that I was overworked. I worked there all four years of high school."

Margaret recalled that the men liked her food but said it was not hot enough, so she purchased gallons of peppers "to spice it up to suit them." During the war years she had difficulty securing the ration coupons necessary to purchase food for the workers. She added that neither she nor the Mexican workers could speak or understand the other's language, but somehow they managed.

Margaret cooked for seasonal workers from the early 1940s until 1966, when Gordon and she moved into their new home across the street. She noted that many of those Mexican workers returned year after year and eventually moved into the area and became permanent employees. After Gordon and Margaret left the big main house, Elizabeth shared the home with two office girls and the Lance Peterson family. Peterson was a field foreman.

VINCENT BAILEY, Gordon Bailey, and Beth Bailey Fritz each had about a decade of experience in the business when their father died in 1943. They inherited a solidly financed and well-run business, and all that had to be done was to make some typical generational changes. Wages had risen during the war years, and it was necessary to mechanize to reduce costs. One of the first changes was the switch from using six- and eight-horse digging teams to tractors. Next, Carl Johnson, the chief mechanic, used his skills to build a tractor-mounted cutter/lifter/digger that cut the roots of shade trees and lifted them out of the ground. This eliminated heavy lifting and speeded up harvest while using fewer workers.

Vincent Bailey was recognized in the industry for his ability as a propagator and production manager. In 1941 he was elected president of the Minnesota Nurserymen's Association. Industry leaders commented that he was almost a genius as a nursery production person, but within the company he was considered abrasive and very difficult to work for. On the other hand, Gordon was a most astute business manager and one who liked, and was liked by, virtually everyone. Beth apparently worked well with her brothers, and the trio led the business into an exciting decade.

When Helen Franek graduated from high school in 1947, she worked for Minnesota Mining and Manufacturing (3M). Gordon Bailey knew her work ethics and asked her to work at Bailey's. In July 1947 she started for $30 a week, which was $4 a week more than she had received at 3M. Helen was the only office employee among the 24 full-time employees. Initially, she did all the office work except the bookkeeping, which Gordon Bailey or Beth continued to do. She was supposed to work 40 hours per week, but for two months each spring during the heavy shipping season, she worked longer hours, including Saturdays. During the rest of the year, if there was no work she could take off.

Vincent and Gordon Bailey and Clarence Seefert shared the other room in the office. The office was built over the warehouse and had a cement floor, which was always cold. The space was heated by a free-standing oil stove, so Helen turned the thermostat up to get more heat. Whenever Vincent walked in, usually from the warehouse, he automatically turned the thermostat down. Nothing was ever said, but the thermostat received its daily workout. To communicate to the warehouse crew, Helen used a canister attached to a small rope line that was connected to pulleys and ran through an opening in the floor. She wiggled the line to gain the attention of people on the other end, and they pulled on the line to get the message.

The Baileys were not left out of the rapid increase in the landscaping business that started in the 1950s. The company grew to the point that the brothers decided they needed to devote full time to management and gave up being on the road selling. In 1950 Joe Whelan became their first full-time salesperson. Whelan, a career nurseryman, had a real insight into the business from the retail sales side. His marketing knowledge was particularly valuable to the fast-growing business, especially to members of the third generation who soon joined the ranks.

THE BUSINESS CONTINUED TO PROSPER during the 1950s, even though it was operating under the John Vincent Bailey trust, which meant that it had to pay a higher income tax than corporations or other forms of businesses. This made it more difficult to accumulate capital than would otherwise have been the case. The burden of the higher tax rate intensified the problem between family members as they matured and developed different goals. Two daughters of John and Elizabeth Bailey, Virginia and Margaret, were not involved in the business but held stock in it. In 1953 Beth and Vincent were 44 and 50, respectively, had no children, and did not agree with Gordon, whose children probably were interested in making a career with the nursery.

Four of the siblings were interested in taking cash from the business, while one wanted to build equity. From 1953 to 1962 the company was in court over a family disagreement regarding their father's trust. Stan Wolkoff, who started as an external auditor for the company in 1953, noted that Gordon, Vincent, and Beth were very dedicated to the business, and the two men particularly were "very close" when it came to running the business. Wolkoff commented that if John Bailey had realized the future problems with his trust, he might have set it up otherwise. Wolkoff noted from his long experience, "Any time there are outside family members, the business is going to have a problem." But in this case even the "insiders" disagreed.

The problem was made more complicated with the death of Beth in 1959, and her sister, Margaret, in 1961. Virginia, who was still single and not working in the firm, acquired the stock of her sisters, neither of whom had children. However, according to company auditors, the trust stated that the stock of any child who died without issue should revert to the company. In any case, Virginia and Vincent then held two-thirds of the stock and voting power. The court concluded that "the will authorized the trustees to distribute income to the heirs only after the investment of earnings in the business." The court interpreted that John Vincent Bailey, Jr., wanted the company to continue through his bloodline.

DESPITE THE ABOVE PROBLEMS, J. V. Bailey Nurseries prospered. In 1954 the company adopted a profit-sharing plan for its permanent employees—reputedly the first in the industry. This proved very important in retaining employees.

In 1956 the company discontinued its retail sales business because it did not want to compete with its wholesale customers. At that time it was doing sales and service in 12 states. Joe Whelan was promoted to sales manager, and Cliff Ecklund, who had handled retail sales, was moved to the sales department.

When the company left the retail market, it phased out of the orchard business. The orchard trees were removed and replaced with pines and shade trees. In the process a nine-year crop-rotation plan was adopted: six years for nursery stock, followed by an oats-alfalfa-brome seed down. During the second and third years, the alfalfa-brome was chopped. It was then left on the field until at the end of the third year, when it was plowed down as a green-manure crop. In addition, several thousand tons of manure were purchased from the So. St. Paul Stockyards at 50¢ a ton and spread on the land at the rate of 40 tons per acre. This was a continuation of the soil-building program that J. V. Bailey, Jr., had started in the late 1800s.

In 1957 Catherine Franek (later Lighten), whose sister Helen had worked in the office since 1947, started her employment with the company. Gordon Bailey had gone to Newport High and asked to see Catherine's records and hired her on the spot. She started work as a bookkeeper the Monday after graduation. The two Franek sisters recalled that each Christmas Vincent and his wife invited the office girls to their home for treats and then to the Lexington Hotel for a dinner. This was Vincent's sociable side, which thoroughly impressed the staff. Helen and Catherine both commented that the Baileys were known for their good pay and benefits and that they preferred working at J. V. Bailey Nurseries instead of in the Twin Cities.

In 1957 Gordon Bailey, Jr., finished college and immediately joined the company. He initially worked with his Uncle Vincent, who was in poor health and needed help in production. This was difficult for Gordon, Jr., because he was caught in the middle of the family struggle. He said that he came to the company to replace his Uncle Vincent "who did not always agree with Dad. Vincent was difficult to work with. He had no people skills. It was he who decided to sell out."

The good side of Gordon, Jr's. involvement at that time was that "it made him sensitive to the real problem and he was able to work with his father in developing a long-range strategic plan" involving family transition. Gordon, Jr.'s brother, Rod, majored in horticulture and after graduation in 1959 started in the production end of the business. Gordon, Jr., was moved to the office and worked in many capacities.

He established the human relations department. He also handled insurance matters. Gordon, Jr., and Rod worked very well together, which enabled the company to prosper for the next several decades. Even though Rod worked in production, he was completely informed of all management decisions.

Despite Vincent's shortcomings, his talents in propagation, production, soil management, soil conservation, and promotion of winter hardiness were well recognized in the nursery industry. He was a "genius in propagation and production." His election as president of the American Association of Nurserymen in 1960 can be attributed in part to the fact that J. V. Bailey Nurseries was recognized for its leadership and innovation in the industry. It had a reputation for delivering quality products and standing behind them.

Larry Bachman, of Bachman Nurseries, stated that there was probably only one other nursery in the country that was as good at backing its products. Bill Reid, who has known the Baileys since the 1930s and been a customer since 1950, commented that he could count on doing business with Gordon, Sr., any time from 6:00 A.M. to 10:00 P.M., and Bailey's always got its orders out immediately. "If we ever got a poor product, all it took was a phone call, and we got credit or a replacement. That is what built their business."

Reid added that Gordon, Sr., was known as a real people person. "He really cared about people, and he treated everybody the same. He felt that employees were the real key to his success. I am sure that when he was honored for the nursery hall of fame, it came because of his innovations regarding the work force."

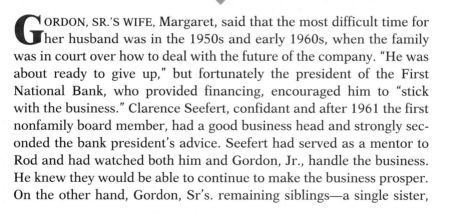

GORDON, SR.'S WIFE, Margaret, said that the most difficult time for her husband was in the 1950s and early 1960s, when the family was in court over how to deal with the future of the company. "He was about ready to give up," but fortunately the president of the First National Bank, who provided financing, encouraged him to "stick with the business." Clarence Seefert, confidant and after 1961 the first nonfamily board member, had a good business head and strongly seconded the bank president's advice. Seefert had served as a mentor to Rod and had watched both him and Gordon, Jr., handle the business. He knew they would be able to continue to make the business prosper. On the other hand, Gordon, Sr's. remaining siblings—a single sister,

Virginia, who did not marry until late in life, and his brother, Vincent, who was married but had no children—were not interested in expansion, and both wanted to withdraw funds from the company.

Gordon, Sr., encouraged by the good job his sons were doing and by the fact that they got along so well, decided to stand firm with his program of continued expansion. Margaret, who sat in on most meetings, commented that her husband never had trouble with financing, even though she often wondered why he kept buying farms or expanding whenever they had a little extra money. She added, "It always turned out for the best."

In 1962, after the dispute over the J. V. Bailey trust was settled by the Minnesota Supreme Court, Gordon, Sr., pushed ahead. In 1965 he engaged attorney Thomas Campbell, a son-in-law, to investigate the titles on 30 parcels of land held by the trust so they could be transferred. Campbell added his professional insight: "Family businesses have the inescapable conflict between family members who want dividends because they are outside versus the inside members who want to reinvest the profits." Company auditor Stan Wolkoff, who had observed the struggle over the trust, noted that the Bailey situation was no different from other agricultural enterprises he had worked with. He said that the good result was that it made Gordon, Jr., very sensitive to the real problem and has helped him to guide the transition to the fourth generation.

By 1962 the company had grown to 26 full-time production people, 10 office staff, and 100 seasonal workers. The farm was now 600 acres, of which 450 were in nursery stock. Bailey's led the industry with the installation of an automatic humidity and refrigeration system in its huge storage facility. The company no longer needed shredded cedar (a.k.a. shingle tow) as packing material, which was a great saving.

JOE GARCIA grew up in a typical Mexican migrant labor family that traveled from job to job with the planting and harvesting seasons. Each spring the Garcia family left its home in Mission, Texas, and headed for the sugar beet fields in the Red River Valley of the North. After the Garcias finished with beets, they moved to Ohio to harvest tomatoes and then to Lubbock, Texas, to harvest cotton, before returning to Mission for the winter. In 1959, while making their annual circuit, the Garcias started seasonal work at Bailey's. They liked the work, and the pay was good. By becoming a full-time employee, one could receive profit sharing and later a pension. So, in 1964, Joe's father took a job with Bailey's, and the Garcia family became permanent residents of the Newport area. Joe was the oldest child, and because his father did not permit Mrs. Garcia to work outside the home, Joe had to start work full-time after he finished the eighth grade. This enabled his younger siblings to obtain more schooling.

Joe did seasonal work at Bailey's, but after he became of age, he left for Texas. He had a permanent job in a carrot cannery in Texas and at age 23 was married. His parents wanted Joe to rejoin the family and come to work for Bailey's on a permanent basis. Joe and his wife wanted their child to be born in Texas, so they did not want to leave until after that event. The family persisted, and in April 1965, Joe became a permanent employee at Bailey's. He received $2.03 an hour, which was 80¢ more than he was earning in Texas and 50¢ more than he could get at "most other places."

Joe's father was made a field foreman during his final six years at Bailey's. He did not want to be a foreman because he did not think his English was good enough, but Rod Bailey kept encouraging him until he finally agreed. In 1979 he retired with a pension and social security to a life he could never have enjoyed if he had continued as a migrant worker.

When Joe returned permanently, there were only slightly more than 100 field workers. Joe worked under field foreman Gust Schultz doing ball and burlap. This meant taking a tree from the ground with the clump of soil (the ball) and then wrapping the ball in burlap for moisture preservation while storing or shipping. The biggest trees were about 8 feet tall and weighed as much as 300 pounds. As many as four people were needed to handle them. A tractor was used when working with some of the shade trees, which were taller and weighed as much as 1,000 pounds.

Schultz retired after a few years, and Joe was promoted to field foreman. Joe proudly stated that he remembers several fourth-generation Baileys in diapers. He later supervised some of them when they did field work. He has more than 34 years with Bailey's and could retire with "a very good living because of profit sharing and pension," but he prefers to work because his wife has a day-care service. He "cannot stand the noise, but she likes it." Joe made sure that his children received the education he was denied. He and his wife have four daughters, all of whom have college degrees and one has become a medical doctor.

A 1992 photo of Gordon Bailey, Sr. (center), who entered the family business in 1929 and provided entrepreneurial leadership into the 1970s. On the left is Rod Bailey, third-generation member, who joined the company in 1959. On the right is his brother, Gordon, Jr., who started in the business in 1957.

Oᴎ JANUARY 3, 1967, Gordon Bailey, Sr., purchased the J. V. Bailey Nurseries and reorganized it as Bailey Nurseries, Inc. He became president; Gordon, Jr., secretary-treasurer; and Rod, vice president. That year the business grossed $1.8 million in sales, 10 percent of Minnesota's $18 million volume in the nursery business. Bailey's was recognized as the largest grower of general nursery stock in the state. The nursery contained 2,100 acres, of which 600 were in nursery stock and the remaining acres in soil preparation crops. About 80 percent of the company products were grown there, but roses were propagated near Bakersfield, California, and maple and birch trees were started in Oregon.

The bylaws were written to minimize future family disputes by creating preferred nonvoting stock for family members not working in the company. The shares were written so they could not be pledged as collateral for a loan or be transferred to anyone outside the company. This assured that voting control would remain with those persons active in the firm.

Gordon, Sr.'s second major action reflected his well-known empathy for the needs of others. He established a pension plan for all permanent employees. This was in addition to profit sharing, which averaged about 10 percent of the employees' annual wages. Those plans, plus a good medical program, proved to be a positive factor in securing and retaining quality employees, which he believed to be the key to having a successful business. This provided a broad benefit program that was virtually unheard of in agriculture at that time. The employees have reciprocated the loyalty, if their longevity is any indication about how they feel.

The company was on a roll, growing more than 10 percent a year, with gross sales of over $2 million. In 1969 Bailey's moved into new office quarters, only the second time since leaving the family home built in 1902. The firm had 45 year-round employees, and during peak season, its employment rose to 150 people. The nursery contained more than 40 miles of terraces, which slowed erosion on the rather hilly terrain where most of the farms are located. Storage had grown to 1.25 million cubic feet. During the 1960s the company became a leader in the industry, with storage facilities totaling 30,000 square feet of climate-controlled space for maintaining dormant nursery stock in the winter season. In 1970 the controlled storage was increased to 90,000 square feet.

Bailey's was not satisfied with the poured cement buildings, so the company searched for a better structure. Rod learned about Behlen buildings, which were not only post-free but also virtually maintenance-free and could be insulated as much as necessary for controlled atmosphere. Bailey's erected a 100-by-300-foot structure and was very pleased with its performance. That building increased the refrigerated storage space to 125,000 square feet, nearly three acres. Later, Bailey's heard about fabcon cement walls, which had insulation sandwiched between layers of cement. With roofs by Behlen, the company now had insulated post-free storage. By 1999 the refrigerated area had increased to eight acres, and another building was planned for 2000. The adoption of steel pallet racking and the use of forklifts

Center is the headquarters building built in 1902, with the addition completed in 1989. The building at the far right is the main shop. The five large buildings at the left cover 310,000 square feet and consist of more than 6 million cubic feet, nearly all of which is refrigerated and automatically humidified. (Photo by Bordner Aerials)

have greatly increased the storage capacity of the existing buildings as well as speeded up loading from warehouse to trucks, and vice versa. The company probably has more refrigerated storage space than any other nursery in the world.

By the early 1970s the office staff had grown to six, but business was increasing so rapidly that the firm was often 10 days behind in billing during the rush period. Gordon, Sr., pushed for computers to make the bookkeeping and office procedure more efficient. Once the computers were online, bills were processed and ready to mail by noon of the day that the goods were shipped. This was extremely important financially, for a major portion of shipping was done during two months each spring. Helen Chiarella was put in charge of the computer department.

IN SOME RESPECTS Bailey Nurseries is unique because it was fortunate to have had two entrepreneurs, J. V. Bailey, Jr., and Gordon Bailey, Sr. For practical purposes their active business lives spanned from 1898 to 1987. In both cases the next generation had come on board during their predecessor's tenure. Gordon, Jr., and Rod, third-generation family members, have carried on in a noble manner but clearly state that they are not the entrepreneurs their father and grandfather were.

With two seasoned sons in their thirties, 63-year-old Gordon, Sr., took every opportunity to expand. Herb Dalglish had come to know Rod from calling on the company as a road salesman for an automotive parts supplier. Dalglish liked his well-paid position and had no thoughts about changing, but Rod persisted. Dalglish was really pleased when the company requested that his wife also be interviewed, because his job would entail considerable time on the road and the Baileys were concerned about her feelings.

Dalglish spent several years learning every job in production and learning about plant material before he was sent to his territory. He soon discovered how constant the flood of new products was, how quality- and price-conscious the Baileys were, and how good they were about replacing any defective product. The company employs a former state inspector to check on all facilities and land to see that everything conforms with guidelines, including field practices.

Bailey sales personnel are paid a salary and not a commission because the company wants to be sure that salespeople do not push products just to make a sale. The nursery business is probably affected even more than the food industry by fickle consumers who change their likes rapidly. Rod said, "The customers tell us what they want, and that dictates what we grow. For example, recently we had orders for hops. What we don't sell we burn, and there is never a year when we sell out of every product." It is common practice to burn most products that are in surplus. "If we do not have more than 10 percent surplus, we are lucky. In extreme years we may burn up to 30 percent of production on some stock. This is in addition to the culling that comes when pruning." Burning assures better quality by having fresher products to sell, and it prevents any temptation to cut prices. Bailey's is the price setter because it is the leader in the production of bare roots (roots that are devoid of any soil).

When asked for some highlights of his 30 years with the company, Dalglish told about a shipment to Japan, which, like most countries, insists on disease-free plants. The bare rootstock that had been ordered was placed in cedar shavings to keep the bugs out and hold moisture. The plants were air-shipped to Japan and arrived in excellent condition. Dalglish was quick to add that in his 30 years in the business, his greatest satisfaction still comes from the challenge of teaching a young person how to be successful. "It is very satisfying."

◆

WEATHER is important to the nursery business, just as it is to any other sector of agriculture. During the 1960s Bailey's suffered three severe hailstorms. One year a farm suffered two hailstorms that completely wiped out that farm's crop. Fortunately, individual farms are scattered in the Woodbury, Cottage Grove, Newport, and Hastings areas. This provides reasonable protection against natural disasters. Bailey's has expanded its irrigation system regularly since first installing it in the 1920s. The company's land also is well tiled, but that did not protect against a prolonged rainy spell in 1977. To make up for lost time, Bailey's installed lights on all its equipment and added a night shift to complete harvest. Starting October 19 and continuing for the next 14 days, Bailey's ran two shifts, each working 10 hours a day, to complete harvest before fall freeze-up.

◆

IT BOTHERED GORDON, SR., that the company was paying out so much for western nursery stock, which it used as foundation material. He felt that owning a nursery in Oregon would help to insure them against natural disasters as well as improve the quality of some of the foundation stock. He reasoned that the milder maritime climate of western Oregon fostered better and faster tree development. In 1976 the company purchased 350 acres of land near Yamhill, Oregon. The initial intent was to concentrate on more hardy and faster-growing roses and Norway maples.

◆

DON POND changed his major in college to horticulture after a field trip to Bailey's. He worked for the company during his final college years and for a period after graduation. But his dream was to work into a partnership with his father and brother on the family dairy farm. That dream withered after about 18 months: "It was too much grief and drudgery for the return." He had enjoyed his previous work with the Baileys, so in 1972 he wrote to them about a job and received a letter by return mail. After an intensive interview, he joined them as a general field worker. Pond knew that the company's 12 field foremen had an average of more than 15 years of service and were excellent supervisors, so he sensed that there would be good opportunities for learning production and for being promoted.

It was Pond's impression that except for Vincent, the Baileys, from Gordon, Sr., down, were very people-oriented. He remembered how firm Vincent was and how Gordon, Sr., "would always cover for you. Vincent was so firm that even the veteran foremen were afraid to try something new. He was such a contrast to Gordon, Sr., who let you make mistakes so you could learn." Pond marveled how well Gordon, Sr.'s strategy worked with the large, well-organized crews of Mexicans, most of whom came back every year and many of whom became permanent employees.

Pond learned rapidly working under the seasoned foremen, and new opportunities came even sooner than he had anticipated. In 1974 he was made a field foreman and thoroughly enjoyed working with the well-trained crews. In November 1976 Gordon, Jr., and Rod came to the storage building where he was working and asked him if he wanted to go to Yamhill, Oregon. He was not enthusiastic about the offer, because he wanted to stay in Minnesota and he enjoyed what he was doing. He commented: "A few days later I told Linda [his wife]. Surprisingly, she was excited about the idea, or I would never have agreed to consider the offer."

After a trip to Oregon to survey the operation, which had been used as a truck garden farm, Pond decided the job would be a challenge, and Bailey's named him manager of the Yamhill farm. For the first four years, the previous owners did the tillage, planting, and harvesting. After that, Bailey's had enough land planted to trees and shrubs to justify its own equipment and permanent labor. The company started the operation slowly, planting only 38 acres to trees and shrubs to determine which varieties would do the best. In 1977 the four winter months, normally the wet season, saw only an inch of rain. Pond said, "That was a good learning experience."

THE NURSERY BUSINESS is very labor intensive, but because the demand for specialized nursery equipment is so limited, manufacturers do not find that market attractive. In addition, many of the smaller nurseries rely primarily on family labor and, like many small farm operations, are not concerned about its cost. But for large commercial nurseries, such as Bailey's, labor is a major expense. Therefore, they must seek ways to improve efficiency.

One of the first mechanical challenges Bailey's faced was to develop a tractor-pulled cutter that could eliminate the need for a man with a spade to cut roots around trees and shrubs when harvesting them. The old method of ball and burlap was hard on the land because, if practiced continuously for 20 years, it removed an entire foot of soil, which had to be replaced or at least restructured. Ball and burlap also required more labor and, because of the weight involved, made the trees and shrubs difficult to handle and costly to transport. Bailey's adapted a blade to a crawler tractor for cutting and lifting trees, which was an improvement. Next, the firm secured a tractor with an engine over each drive wheel to provide more clearance. This enabled the equipment to travel over the trees and lift them straight up.

The next step was to rebuild the crawler tractor so it had a 5-foot clearance. At the same time, the tractor was modified to harvest two rows of trees at a time. Once that was perfected, hydraulic shakers were adapted to get rid of the dirt around the roots. The machine cut 16 inches deep, lifted a tree or shrub, and shook the dirt from the roots while traveling 1 mile an hour. The end product was bare rootstock that was about half as costly to harvest, store, and transport as the ball-and-burlap trees.

VERN BLACK grew up on a small diversified farm and did not like the life, but he wanted a career in which he could work out of doors. When he received a flyer from a nearby technical college explaining its horticulture program, he decided that was the answer to his dream. While attending college he had a summer internship at Bailey's, which was interesting and paid $2.79 an hour instead of the 50¢ an hour he would have made baling hay for neighboring farmers. His life changed, and in 1972, immediately after graduation, he began his career with Bailey's. A year later he became an assistant field foreman.

At first he experienced the hard work of moving irrigation pipes, which had to be done repeatedly during the season. Reputedly, the Baileys were pioneers in using center pivot irrigation. This first self-movable system did not have wheels. It "was a ratchet and push system, which used water pressure to lift the entire rig and move it ahead." Black commented about what a relief it was for the field workers when they no longer had to move pipes by hand in the wet, muddy fields, usually on hot days.

The next improvement was to put time clocks on the center pivot systems, which not only saved labor, but also improved the timing of water application. Water is extremely critical in maintaining young seedlings. For example, soft wood cuttings can be lost within less than an hour without water on hot, windy days. To safeguard against such losses, Bailey's has invested in 30 wells. All are connected so that if any of them should fail, there will still be water. Fortunately, Dakota County has ample underground water.

The company has always done whatever is necessary to maintain its good reputation as a propagator, which Vincent Bailey established. As head of production Rod Bailey felt that it was necessary to make regular rounds during planting, propagating, and harvesting times to observe the progress or spot problems. When he relied on a pickup, it took an entire day to visit each of the crews twice. After the firm purchased two-way radios, he was able to continue his work in the office, traveling to the fields only when called by a field foreman. The radio was far less costly and stressful than operating a pickup and freed time for more management. By the late 1970s two-way radios were furnished to all field foremen and higher management and were placed in 25 tractors and in 19 cars, busses, and trucks.

Using technology to replace labor enabled the Baileys to greatly expand their business while keeping the number of workers and the labor cost reasonable. Their equipment yard contains a massive array of specialized and costly machines and a very large shop. Bailey's has trucks for local delivery only and relies on common carriers for over-the-road work. The company prides itself on quick delivery, but frequently small orders may be delayed as much as a week. In such cases a semitrailer rig may contain orders from as many as 20 dealers that may be spread over a considerable distance.

ON APRIL 6, 1978, Gordon, Sr., 71, became chairman of the board; Gordon, Jr., president and manager of administration; and Rod, secretary-treasurer and manager of production. The permanent staff had grown to 75, and seasonal staff to 250. The payroll of $2 million equaled the total sales of just seven years earlier. The company kept pace as the leader in the industry with 1.75 million cubic feet of refrigerated storage.

Gordon, Jr., and Rod, ages 43 and 41, respectively, with about 20 years of hands-on experience, were well prepared to take over the leadership. The transition from second to third generation was far more peaceful than the previous generational transfer. Those close to the insiders marveled how smoothly the family functioned. The Baileys were capable of having hot sessions about business matters, but when they left the meeting room, there was no evidence of any disagreement. Few family businesses are that fortunate.

Gordon, Jr., commented that his parents had never pressured him or Rod to enter the business. He had considered majoring in psychology. "When you are raised in a business, it is part of a family life. You cannot separate business and family. We always had the office phone in the house. Someone always had to be there to answer it." When he entered college he decided upon a degree in business administration, but before he finished he took some courses in horticulture "just in case" he decided to enter the business.

When asked to what he attributed his success, his instant reply was, "I think it was hard work. I'm not smarter than other people, and my vision is not greater than theirs, but I have been able to hire and retain good people." The company writes thorough job descriptions for key employees so they know what is expected of them. Whenever possible, Bailey's likes to interview an applicant's spouse to acquaint that person with the company and the people with whom his or her partner will be working. "We want someone to fit into the culture of the company." Gordon, Jr.'s final comment was, "We are a family business that carries with it both challenges and opportunities."

Each generation of Baileys has been very concerned about passing down the work ethic and values to its children. The children are expected to work on the summer field crews and to follow directions from the field supervisors, who are comfortable in keeping the pressure on. Local youth, college students, and Mexican migrants all work the same pace; there are no exceptions. Gordon, Jr., and Rod started in field work when they were 11 or 12 years of age. Unfortunately, since the 1970s fewer youth have sought field work because they apparently do not want to do manual labor. This keeps constant the pressure to find ways to mechanize.

BECAUSE the company is constantly growing, it has had an ongoing program of restructuring. Don Selinger joined Bailey's in 1973 and, after a thorough training program in production, became a road sales representative. In 1979 he took over production planning, in addition to some marketing functions, customer service, and the search for new varieties.

Production planning entailed deciding the quantity of each variety to plant and determining the rotation on the land. Sometimes, after certain crops are removed, the soil is restructured. Computers are used to track the history of each field, which determines the above practices. Bailey's works steadily at realigning fields to increase size in an effort to gain economies of scale. This is just as important to the nursery industry as it is to any other sector of agriculture.

Selinger's chief job in customer service was handling complaints. If Bailey's determines that a complaint is about a common problem, it will replace. If the problem is about quality, it replaces about 90 percent of the time. Bailey's uses its own personnel in loading all trucks to avoid breakage in filling the trucks as well as in shipment. This is a major task, because from Newport alone, more than a thousand trucks are loaded out from February to June of each season.

Bailey's pioneered the picture tag service. Each plant is identified with a tag resembling an artist's pallette. The tag provides the generic and common names as well as other information about the plant. With more than a thousand varieties listed in the company's catalog, tagging is a major task but a great sales aid.

When not busy with the above, Selinger visits universities, arboretums, and competitors to learn about new plants. The nursery industry is unique, for from a competition standpoint it is much more friendly than most other sectors. When Bailey's finds new plants it would like to sell, the firm sometimes will purchase them from the producer, and other times it will do its own propagating. Royalty is generally paid at the time of sale. Royalty per plant varies but can be as much as $2 on some flowers. Selinger's final task is pricing the plants, which will be listed in the annual catalog.

GOOD PROPAGATION is necessary to produce quality stock, and fortunately the Baileys have good propagators. In 1981, after the usual search for a new propagator, Rod persuaded Vern Black to take the position. Black quickly proved his propagation skills. Bailey's was short on land and greenhouse space for more propagating, so Black used some blacktopped parking space. Sand was dumped on the blacktop and then misted to get the plants growing. Up to that time, propagation had always been done under controlled systems, but the outside lots worked well.

Next, the nearby 127-acre Nord farm was purchased specifically for propagation because it had an ideal soil structure. With that purchase, farm land increased to 2,128 acres, of which 728 were in nursery stock. Black created a 10-year plan of platting the farm for propagation. Twenty-nine 30-by-150-foot poly-covered Quonset hoop-framed greenhouses with traveling irrigators timed to mist the plants every three minutes were purchased. Those 29 houses were to have been adequate for 10 years, but the system worked so well that the seedbed operation on the sand land near Hastings was moved in, and within 3 years there were 50 hoop houses. They were superior to the traditional glass greenhouses in some respects and far less costly.

Almost simultaneously with the above, it was discovered that propagators were more successful by taking cuttings directly from plants in the field and leaving the foliage attached when planting them in the seedbeds. The leaves provided nutrients for the new plants, enabling them to grow faster, and labor costs were cut by half. Formerly Bailey's had used stool blocks, which were special plots set aside with several varieties of plants just for production of foundation stock. Because stool blocks were small areas and were away from the rest of the farm, they were expensive to maintain.

About the same time, the salespeople suggested that the company produce bedding plants (annual flowering plants) for retailers. A change in the seeding by adopting the plug system enabled more consistent seeding and quality and put Bailey's on the cutting edge of the bedding plant sector. All the above changes could not have been more timely, because the company had just put additional sales personnel

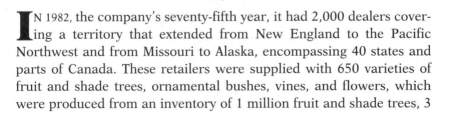

Approximately 4 million seedlings per year are raised in the outdoor beds. The 16 acres of poly-covered greenhouses enable double-cropping of bedding plants from January through June and a crop of softwood shrub cuttings for propagation through November. Bailey's propagates 8 million rooted cuttings and 250,000 flats of annuals and perennials in three facilities. (Photo by Bordner Aerials)

on the road and demand was phenomenal. The number of poly-covered Quonsets increased until they covered 15 acres and just kept up with the propagation needs at the St. Paul division.

◆

IN 1982, the company's seventy-fifth year, it had 2,000 dealers covering a territory that extended from New England to the Pacific Northwest and from Missouri to Alaska, encompassing 40 states and parts of Canada. These retailers were supplied with 650 varieties of fruit and shade trees, ornamental bushes, vines, and flowers, which were produced from an inventory of 1 million fruit and shade trees, 3 million ornamental shrubs, and 1 million potted plants. Much of the inventory was conditioned in the 2.5 million cubic feet of refrigerated storage covering nearly six acres.

Bailey's was the sixteenth largest wholesale nursery in the nation, with sales of $8.5 million. Its staff consisted of 115 permanent personnel and more than 175 seasonal workers, plus an additional 90 employees in Oregon. The company grew more than 10 percent in 1982 and climbed to fifteenth ranking among the nation's wholesale nurseries. Its growth continued.

IN 1884 the Sherman Nursery Company was founded in Charles City, Iowa, and grew steadily until the 1920s, when it was considered one of the largest growers of roses in the world and the largest producer of evergreens. Then, Mr. Sherman became involved with a herd of purebred cattle, and reputedly funds and management time were diverted to that enterprise, which weakened the nursery operation. In 1931, with the downturn of the economy, the nursery fell into the hands of a holding company, which operated the business until 1947, when M. R. Cashman, a nationally recognized nurseryman, purchased Sherman's. Cashman rapidly expanded the business from 500 acres to 954 acres and had 150,000 square feet of glass greenhouse plus 1.8 million cubic feet of controlled storage. Sherman's was one of the largest wholesale nurseries in North America, selling to more than 3,500 retail centers and other wholesalers.

On October 10, 1983, after nearly a year of negotiation, Bailey's purchased the Sherman Nursery. Dale Siems had joined Sherman's in 1955 as credit manager and in 1975 was made president and general manager. The Baileys were impressed with the Sherman operation, as well as with Siems' management, and asked Siems to remain in his position. Sherman's was made a free-standing subsidiary in part because Gordon, Sr., felt that it would be good competition for Bailey Nurseries.

In 1983 the Sherman Nursery had 800 acres, whereas Bailey's had 2,500 acres at Newport and 900 acres at Yamhill, Oregon. Sherman's employed 50 people. Siems was given a free hand in the business, but expansion and renovation expenses had to come out of Sherman's profits. In 1986 Siems began an extensive renovation program on the existing facilities. Next came the construction of four new green-

houses, propagation structures, and an order processing/shipping lay-
out capable of handing five semitrailer trucks at once. Under Siems'
leadership Sherman's prospered and by 1995 was able to reward the
parent company with a healthy return on its investment. By 1999 the
Charles City subsidiary, with its 56 year-round and 20 seasonal
employees, generated $11 million in sales and ranked forty-ninth
among the nation's wholesale nurseries.

WHEN Bailey Nurseries was incorporated in 1967, Gordon
Bailey, Sr., yielded many of his formal duties to his two sons.
Gordon, Jr., said that at the time it appeared his father was quite willing to shift responsibilities. However, he later felt he had given up too
much, so in 1984, at age 77, "he purchased a nursery in Oregon on his
own. Dad was always young in his thinking and always progressive."
In 1982 Gordon, Sr., had been asked by a financial institution to make
an appraisal of Carleton Plants near Dayton, Oregon. That nursery
had 515 acres of owned land plus 450 acres of leased land, with 60
year-round and 30 part-time employees.

Gordon, Sr., was impressed with Carleton Plants, and when the
financial institution put it on the market to be sold, he went to the
Bailey board and recommended that the company purchase it. The
board objected, because Bailey's had just purchased Sherman Nursery in 1983 and because it felt there was a problem of ethics since
Gordon, Sr., had been a consultant for the bank involved. His wife,
Margaret, said:

> I was opposed to it because we had just purchased a cabin
> on Snow Bank Lake near Ely, Minnesota, and I thought at age
> 77 he should slow down. But he just loved Oregon. It had a lon-
> ger growing season, and the plants generally grew faster there.
> He spent about five days there each month. He figured we
> would be better off if we produced our own products there
> instead of buying stock from another nursery.

Annette Mullen, who was secretary to Gordon, Sr., and both sons,
said, "When Senior purchased Carleton Plants, he did it against the
wisdom of the board and family members. He showed me a million-
dollar check that he was taking out there. He was so happy." Margaret

added that at age 77, he found Oregon to be a challenge, and it was good for him. "During our 63 years of married life, he usually talked to me about the business when we were trying something new. This time I did not agree with him, but it worked out for the best." Carleton Plants proved to be a very successful business, and Jon Bailey Bartch, son of Virginia Bailey Bartch, daughter of Gordon, Sr., and Margaret, purchased it from his grandfather's estate.

ALTHOUGH the board and the family disagreed about the purchase of Carleton Plants, they solidly backed the operation at Yamhill, Oregon. In 1980 Bailey's purchased a second farm there; another in 1982; two in 1983; and 320 more acres in 1992. In the meantime it also bought a 600-acre block on Sauvie Island, west of Portland, and a desert location near Sunnyside, Washington. The Sunnyside land was purchased because it was virtually "virus-free and was a great seed-growing area." It was good for ornamental and fruit products, including prunus, a plum that can be dried without spoiling. Those three operations were all under the management of Don Pond.

Pond related that some of the land at Yamhill is under long-term lease by Bailey's "because you never have enough good ground. We have three large reservoirs filled with liquid gold [water]." Some water comes from the Yamhill River, and the remainder comes from small creeks. Pond has to ration water about 3 out of 10 years because of the intense heat or a lack of snow water. Water rights are a must, and new rights cannot be purchased. On the other hand, the rainy season is during the winter, and most of the harvesting is done from October through January, so the heavy clay soil can present a real problem if it stays wet during harvest. Even though drought is the major concern during the growing season, much of the land has to be tiled to drain excessive winter rains. Fortunately, some of the tiled water can be funneled into the reservoirs.

The Oregon division has grown so that it represents about 40 percent of the company's production, not including the Sherman Nursery. About 275 semitrailer loads of bare-root stock and another 200 loads of container stock are shipped to Newport each year, as well as about 50 loads to Sherman's. Shipping expenses on those loads total nearly a million dollars annually, but the security of having production under a more controlled climate offsets the cost factor. Deregulation of the trucking industry has proven a real boon, because rates have remained very competitive since the late 1970s.

Container-grown shade trees at Yamhill, Oregon. Left to right: Sergio Ortiz, Container Production Manager; Juan Flores, Container Tree Production Foreman; Marlin Brethower, Inventory Manager, West Coast; and fourth-generation member, Mark Bailey, Traffic Manager.

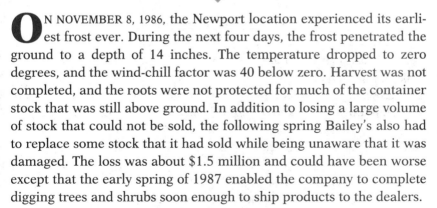

ON NOVEMBER 8, 1986, the Newport location experienced its earliest frost ever. During the next four days, the frost penetrated the ground to a depth of 14 inches. The temperature dropped to zero degrees, and the wind-chill factor was 40 below zero. Harvest was not completed, and the roots were not protected for much of the container stock that was still above ground. In addition to losing a large volume of stock that could not be sold, the following spring Bailey's also had to replace some stock that it had sold while being unaware that it was damaged. The loss was about $1.5 million and could have been worse except that the early spring of 1987 enabled the company to complete digging trees and shrubs soon enough to ship products to the dealers.

The weather at Newport did not present any further serious challenges until October 31, 1991, and the following week, when two storms deposited 36 inches of snow on the company's unharvested crop. Fortunately, the snow cover prevented the ground from freezing, so Bailey's was able to continue harvest even though muddy fields made it difficult for both workers and machinery. Vern Black said that a snow blower went ahead of the harvest crew to remove snow from ground around the plants. It was slow work, and extra tractors were needed to pull the machines. A crew of 180 worked all the daylight hours seven days a week for two weeks to save the crop and won. The Bailey wives and children made brownies, pizza, sandwiches, and hot chocolate, which they took to the fields for the workers. That was a "real treat for those struggling to rescue the crop; they really appreciated it. It meant more to them than just money." In addition to the above problems, the company also lost eight greenhouses in those storms.

Despite such weather adversities, Bailey's believes that the Newport location has been one of the keys to its success because the company has good land, has a good water supply, has excellent transportation, and has been able to maintain a good labor force. Also, Minnesota business is 30 percent of the company's market. However, Bailey's takes comfort in its Iowa, Oregon, and Washington locations, which provide it with additional security against most weather contingencies. Larry Bachman, who represents one of the firm's largest customers, noted that Bailey's grows a much broader variety of products than any other northern nursery and that it has made a real effort to specialize in hardy, northern-grown stock. He said that is probably what has made Bailey's the largest northern nursery.

A T BAILEY'S, direct field labor is 30 percent of production costs, and the total labor cost is 48 percent of all costs. This compares with total labor costs of 2 to 3 percent for grains and 10 percent or more for potato production and dairying. Thus, labor management and productivity is the key to survival. The Minnesota division has 45 salaried and 210 hourly year-round employees and as many as 550 seasonal workers. The West Coast locations have 13 salaried and 122 hourly full-time employees and up to 320 seasonal workers. Even during off-season, both locations have need for up to 100 part-time workers. The Minnesota location has 82 individuals with four-year college degrees and 16 with associate degrees, while the West Coast sites have 16 and

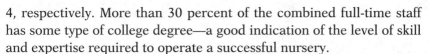

4, respectively. More than 30 percent of the combined full-time staff has some type of college degree—a good indication of the level of skill and expertise required to operate a successful nursery.

Bailey's has a long history of hiring minorities, many of whom, like Joe Garcia, have made careers with the company. But with each passing year, it has become increasingly difficult to employ people, including school-age youth, for manual, outdoor, agricultural work. There is some indication that more college-age females are becoming interested in the nursery field. Bailey's is pleased that in recent years more women have taken horticultural training. In 1988 the company established a scholarship program of up to $12,000 annually at Woodbury Park and Hastings High Schools for students wanting to go into horticulture. The program provides $1,500 per year for each student's college education. A student may continue receiving assistance until graduation.

The search for good workers is historic and will remain a major concern of most businesses, especially those in agriculture, which has traditionally relied on unskilled and immigrant workers. In that respect Bailey's appears to fare better than most agricultural employers. Don Pond, in charge of the West Coast operations, says that every year when migrants return from Mexico, they always have new friends or relatives along. This greatly eases the task of seeking new workers, but government regulations are always a major concern when employing migrant workers. At Newport, Joe Bailey is in charge of seasonal hiring, safety, and Immigration and Naturalization Service (INS) problems and paper work. He speaks fluent Spanish and travels to Mexico each year to recruit workers.

Bailey's employs an INS practicing attorney, who supervises all the paper work on the documentation of persons applying for jobs. This is the company's best means of protecting itself against any potential violations. Despite all the precautions, Annette Mullen stated that from her vantage the most severe challenge the company has faced came on October 29, 1996, when federal INS agents raided the nursery. Annette said that helicopters swooped down to watch that no one got away. "I felt like we were in an armed camp. The INS had people stationed at every door."

The local newspaper reported that the INS raided the 14 Bailey farms in the area and targeted 190 individuals, of whom 138 were apprehended. The INS had checked nursery records in May, but suspicions arose when local police received anonymous tips and arrested nursery employees for offenses. The following September the Baileys were notified that some Mexican workers had used falsified documents or assumed identities of others to get jobs. Despite the nursery's cooperation, it was to be fined for failure to complete the necessary paper work for its workers. Some of the Mexicans left voluntarily, and others faced deportation. Along with most sectors of agriculture, the national association of nursery people is trying to get immigration laws changed because of the need for all levels of workers.

Bailey's has had a program of educating its employees about safety. Dan Schneeman, the company's long-time insurance agent, noted that the education program has been so effective that workers compensation insurance rates for Bailey's have dropped to "almost unbelievable lows. It reflects the true concern of Bailey's for its employees. The nursery has very few claims with field equipment and has not had a serious accident since 1989." Its rates have dropped so low that "it would not pay Bailey's to self-insure." This comes from the firm's outstanding success with getting people back to work after an illness or an accident.

I N 1986 Bailey Nurseries' sales reached $15 million, and the company, exclusive of Sherman's, was ranked the fourteenth largest nursery in the nation. The business was handled by a sales staff of 8, an office staff of 11, 18 field supervisors, 105 full-time workers, and several hundred seasonal workers. More than 400 varieties of fruit and shade trees, ornamental shrubs, hedge plants, bedding plants, evergreens, roses, and perennials were sold to 2,000 dealers. The controlled-storage facility had grown to 2.7 million cubic feet, and plastic greenhouses covered 240,000 square feet (more than 5.5 acres) of land. Fifty tractors and 40 trucks and cars were required to handle the production from 1,200 acres of nursery stock at the Newport location.

The employee handbook stressed the importance of being on the job. A weekly bonus of $5 was paid quarterly for each week of perfect attendance. Also, each Christmas a bonus of 1 percent of wages times number of years of service up to a maximum of $900 was given. The above benefits were in addition to a funded disability, health, life, and dental program, a profit-sharing and pension plan, tuition for approved courses, rainy-day coverage, jury duty compensation, and generous paid vacations.

Interior view of bare-root dormant shrub storage. A Behlen free-span insulated structure, with plants in pallets and steel racks, graded, counted, and bundled for shipment. Storage temperature is held at 37 degrees and humidity near 100 percent.

The company employed every possible labor-saving device to
reduce labor expenses. When horses were still used for power and
before the day of herbicides and pesticides, not much could be done to
cut labor expenses. During the 1940s a shop was specifically established to repair and modify equipment. Len Bondeson started working
at Bailey's as a teenager and continued there while he went to high
school and college. He became a teacher and enjoyed the work but
was dissatisfied with the pay, so he applied for a position at Bailey's.
He was offered 50 percent more than he made as a high school
teacher. Bondeson understood that he had to work more hours but
was happy to be back with the company. At first he "filled voids wherever there was a need for an equipment operator." He was able to
observe all the ways that machinery was used in the nursery.

Because he had formal training in mechanics and had a knack as a
mechanic, Bondeson eventually was assigned to work in the shop,
where his brother Terry served as maintenance supervisor. Terry
headed the shop for several years but decided to go elsewhere. When
he left, Len was named superintendent of maintenance. One of the
first challenges was to perfect the shakers on the tree lifters so they

would not shake the machinery apart while shaking the tree. One of the newest shakers on the market was purchased, and "it had to be modified or it would have self-destructed." Such a piece of equipment was essential for an easier and faster tree harvest. The problem was solved by applying an orbit hydraulic motor to do the vibrating with minimum wear on the equipment and by greatly refining the shaking process. Soon the Bailey foundry was busy making more lifter-shakers and improving them. Prior to the improved machine, 12 people were needed to dig trees. Now only 4 were required, and they did the job faster, easier, and with less expense.

A crew of 10 could plant 6,000 trees an hour, which was quite efficient but required hand-planting and "lots of bending." Mechanical planters were made so the workers could sit while placing the tiny plants in the soil. The process was slower, but bending was eliminated and the work was far less tiring. The planter-trencher had to be modified to avoid packing the soil around the plant and to reduce the resistance so the roots could grow in all directions instead of following the trench line as formerly.

Although not all the above are the direct workings of the shop, when mechanical modifications are needed, it is the challenge of the shop staff to come up with the answers. Of the staff of 14, 4 work at fabricating equipment, 5 repair, 1 supervises, 2 take care of electrical and refrigeration, 1 washes and cleans the equipment, and 1 moves machines to and from their place of use. John Bailey, who is in charge of purchasing, maintains that the company must continue to mechanize to reduce costs.

In 1993, as part of its effort to find labor-saving devices for tasks that are highly repetitive, the company started using robotic transplanters. The search for enhanced robots for the industry is currently being conducted by the Carnegie-Mellon Foundation, in cooperation with the National Aeronautics and Space Administration (NASA). The big challenge will be to standardize robots, making them usable for enough different tasks that the market will be sufficient to keep the cost down.

One of the greatest technological boons to agriculture has been the development of herbicides and pesticides to control weeds and insects. In 1978 the company decided that it was time to engage a specialist in entomology and agronomy. Don Peterson had worked at the nursery in his school days and, after a very successful career in sales with a large international company, decided he was ready for a change of pace. He said he never realized that coming back to his roots would be the best 19 years of his career. His task was to find an herbicide that would control weeds but not harm the nursery plants.

The Baileys cooperated with chemical companies to find the proper chemicals for their various needs. Goal, Post, and Devrinol were the first chemicals that had a major impact on the nursery business. They controlled the weeds, were safe around the nursery stock, and greatly reduced the labor bill. Peterson provided illustrations showing that in 1987 the savings from using herbicides on shrubs was $230 an acre. The per acre labor savings with containers was even greater. With the cost of an herbicide and its application at $64 per acre, the cost of production dropped from $910 to $190, a savings of $720. The chemicals for insect control were helpful in preventing plant damage from pests but did not have the same impact on labor cost as the herbicides.

PREVIOUSLY it was mentioned how Bailey's has diversified its land holdings as a precaution against loss from natural disasters. The company realizes that larger fields are as necessary for gaining economy of scale in nursery crops as in small grains and most row crops. But nurseries face staggering input costs that producers of small grains and row crops do not. The average nursery crop takes about 3½ years from first seeding. Each year about 10 to 12 million plants are propagated and then transplanted to fields or containers. In either case they must be treated against disease and insects.

Containerization (potting) was a persistent bottleneck in the production and sale of nursery-grown plants. Started in the 1950s, it grew slowly until Gordon, Sr., learned about using No. 10 tin cans, and then the problem was minimized. However, the real breakthrough came in the 1970s, when plastic pots became available. Potted plants soon became very popular. That proved a boon to the nursery industry because it extended both the planting and the marketing seasons from early spring to late fall. Container stock is grown from one to three years before it goes into a container, and then it is grown another one to three years before it is marketed.

In the case of field-grown evergreens, it takes three years before they are transplanted, and then it can be another three to six years before they are harvested for sale. All is very labor intensive. For container stock, as much as $15,000 can be invested per acre, in addition to land cost. That does not include the inventory cost. Banks generally do not like to lend on inventory because of too many uncertainties. Gordon, Jr., pointed out that the capital needed to get started has kept many farmers in other lines from looking to nursery crops as alternative crops.

In the nursery business, virtually every plant is sold individually and, therefore, must pass closer scrutiny than most other agricultural products. Rod Bailey previously explained that the company destroys 10 to 30 percent of its production annually because of imperfections or surplus. Pat Bailey, Gordon, Jr.'s son, related how he and his siblings were given an acre of land to plant to pumpkins and gourds and after harvest were allowed to sell them at a roadside stand. His father came by, spotted a pumpkin that was not very good, and said, "If you wouldn't want that pumpkin, you should throw it out." Pat said that struck him as proof that the company stood for quality even though it adds to the cost.

In 1989 Bill Velch joined the company after 30 years in the nursery business as marketing manager for another firm. Each year members of the marketing staff tour garden centers throughout the United States, Canada, and the United Kingdom to watch the market. They also attend at least 40 trade shows annually. Velch said, "It is important to know what the customer is looking for. We regularly propagate new plants that come from Europe." He hopes that in the near future, wholesalers will go into pre-pricing and adopt bar coding to help the retailers.

By 1991 Bailey's was marketing to about 4,000 nurserymen, landscapers, and garden centers in 47 states and Canada, with $35 million in sales. It had become the nation's fourth largest wholesale nursery and had the world's largest cold-storage facility, capable of holding 75 million plants.

Gordon Bailey, Sr., had been named to the Hall of Fame of the American Association of Nurserymen for his "compassion, his foresightedness, and his good old-fashioned business acumen." Previously he was named to the Minnesota Nurserymen's Association Hall of Fame. In 1980 he and Margaret gave a million dollars to establish a chair of horticulture at the University of Minnesota, the first in the nation. Later the company, the Bailey Foundation, and family members pledged funds to construct and endow the Gordon Bailey, Sr.,

Shrub Walk at the School of Agriculture Campus, University of Minnesota, a fitting tribute to a pioneer entrepreneur nurseryman.

Jim and Gen McCarthy were managers of the Minnesota Nursery and Landscape Association. They credit much of the success of that association to Bailey Nurseries, which was one of the progressive leaders in the industry and gave so much support from an "altruistic and quality standpoint," real testimony of the true character of the company, its founders and their successors.

REFERENCE was made that each year nursery stock producers destroy some of their surplus production to keep supply in line with demand. So the question was posed, "Is the nursery industry less subject to cycles than other agricultural industries?" The consensus was that the economy is so diverse today that some sectors will always need nursery products. Gardening is the most popular leisure-time activity—more popular than eating out, baseball, football, golf, entertainment, or other hobbies. It is not nearly as costly as most of the other activities. Even in slow economic times, people try to enhance the appearance of their homes, so they garden or otherwise have plants.

Larry Bachman recalled that in the 1930s Bachman Nurseries and Bailey Nurseries were saved by their truck garden products. He felt that both were more diversified today than they were then, so "that should give them some protection." Dick Buell, who for many years has concentrated on landscaping for developers, thinks that nurseries ride the crest of the economy. He justifies that opinion by the fact that as long as there is new building, nursery products will sell, because the cost of landscaping is figured in with the home mortgage. "A million-dollar home can have a $50,000 to $100,000 landscaping investment. Some developers plan $10,000 for landscaping into the price of a lot in medium-priced developments."

On the other hand, the nursery sector is consolidating just as rapidly as all other sectors of agriculture and our economy. Because of the large amount of capital required, the number of nurseries keeps declining and the survivors keep getting larger. Marketing is also changing because "big box stores, such as Home Depot, Wal-Mart, and Menards," sell on price alone. They give very little service but control an ever-increasing share of the total sales. They prefer to buy their stock from the larger producers. The current unknown is how the Internet will impact the industry. Will it be an influence to educate the public on the value and use of nursery products? Home and gardening television is rapidly changing the demographics of gardening. Nursery people believe that in relation to other areas, such as Europe, we have only scratched the surface as far as potential sales.

A S CITED EARLIER, the Bailey family spent nine years in court to determine how proceeds from the business should be allocated among the heirs. Once the court decided, Gordon Bailey, Sr., concentrated his entrepreneurial skills on creating a successful and nationally recognized business. He set up the new family corporation in such a manner that in the following generations those who are active in the business will have the voting stock and will determine the course of operation. Gordon, Jr., Rod, and their father "had roaring disagreements but always left the room as if nothing had happened." They understood how to settle matters in such a manner that the business continued on the track that their father established, and it flourished. The three sisters of the third generation understood and accepted the dictates of the company's bylaws. To guard against future challenges, Gordon, Jr., said, "We do not give voting stock to family members until they have proved their commitment to the business. If they leave they must redeem their stock. Nonbusiness family members hold only nonvoting stock."

Starting in 1983, the first of the fourth generation joined the company, and at the advent of the twenty-first century, eight members of that generation were actively involved. Insiders, including both family and nonfamily individuals, are first to admit that the biggest single challenge facing the company is "the ability and willingness [of family members] to get along. There is no way around the family issue." Fortunately, the family has long had a tradition of having successful outside individuals on the board, which is a positive influence. A long-time employee spoke about the "strong family culture" but was not sure how the fourth generation would carry on. However, the consensus of most nonfamily individuals interviewed is that the members of the fourth generation will find their strengths and weaknesses and will iron out their own problems.

In December 1997 a management team was organized composed of John, Pat, and Terri, of the fourth generation, and Vern Black and Don Pond, each of whom has more than 30 years with Bailey's. The company's long-time auditors have observed that team at work and feel that it has done an excellent job of planning ahead. One of them emphasized, "I haven't seen any family do such a good job of long-range planning. The mechanics of transition are in place. The company has the routine challenges of any agricultural enterprise, but it is far better diversified than most farm firms. The Baileys will find a solution. They learn how to make change work for them."

Bibliography

Interviews

Arneson, Jerry. Newport, MN, November 24, 1999.

Bachman, Lawrence W. Newport, MN, November 22, 1999.

Bailey, Gordon, Jr. Newport, MN, November 18, 1999.

Bailey, John. Newport, MN, November 23, 1999.

Bailey, Margaret M. (Mrs. Gordon, Sr.). Newport, MN, November 17, 1999.

Bailey, Pat. Newport, MN, November 18, 1999.

Bailey, Rodney (Rod). Newport, MN, November 24, 1999.

Bayless, Ben. Newport, MN, November 19, 1999.

Black, Vern. Newport, MN, November 22, 1999.

Bondeson, Len. Newport, MN, November 24, 1999.

Buell, Richard W. Newport, MN, November 24, 1999.

Campbell, Thomas J. Newport, MN, November 24, 1999.

Chiarella, Helen Franek. Newport, MN, November 23, 1999.

Dalglish, Herb. Newport, MN, November 18, 1999.

Garcia, Joe. Newport, MN, November 22, 1999.

Lighten, Catherine Franek. Newport, MN, November 23, 1999.

McCarthy, Jim and Gen. Newport, MN, November, 19, 1999.

McEnaney, Theresa "Terri" Bailey. Newport, MN, November 22, 1999.

Morlock, Paul. Newport, MN, November 22, 1999.

Mullen, Annette. Newport, MN, November 23, 1999.

Orloff, Arnie. Newport, MN, November 23, 1999.

Peterson, Don L. Newport, MN, November 18, 1999.

Pond, Don. Yamhill, OR. Telephone interview, December 7, 1999.

Power, Tim. Newport, MN, November 22, 1999.

Reid, Ed. Newport, MN, November 22, 1999.

Sargent, Forrest. Newport, MN, November 19, 1999.

Schneeman, Dan. Newport, MN, November 24, 1999.

Selinger, Don. Newport, MN, November 18, 1999.
Siems, Dale M. Newport, MN, November 19, 1999.
Velch, Bill. Newport, MN, November 23, 1999.
Wolkoff, Stan. Newport, MN, November 23, 1999.

Miscellaneous

Bailey Nurseries' printout of its labor force as of January 7, 2000.

Bailey Nurseries' scrapbook of clippings and mementos.

"Beauty Grows with a Family: The Story of Bailey Nurseries Inc. and Its Founding Family." Dan Moriarty Associates, 1989. A company-sponsored publication.

Grissey, Forrest. *Country Gentleman*, June 1911.

Minneapolis Tribune, August 23, 1982, October 29, 1986.

Nollette, Evon. *St. Paul Pioneer Press*, February 2, 1933.

"Spring Stirs in Nurseries," *Minneapolis Star*, March 24, 1967.

St. Paul Pioneer Press, June 21, August 23, 1990; May 23, 1998.

"Story of a Successful Book Farmer," *The Farmer*, August 15, 1908, 536–537.

Washington County Bulletin, February 11, 1971, April 6, 1978, August 13, 1987.

Woodbury Review, November 5, 1996.

A. Duda & Sons, Inc.

*"Entrepreneurship Is What It Takes
to Keep Agriculture Alive"*

IN the 1500s, when the Spanish came to Florida, they ignored the agricultural potential and concentrated on looking for gold. By contrast, in the 1600s, when English colonists arrived in Virginia, which was far less endowed agriculturally than Florida, they immediately set to work tilling the land. So, by the 1800s, when the federal government first turned its attention to developing Florida, Virginia was well developed, and much of its soil fertility was already on the decline. Florida, however, had few settlers and little cultivated land.

In 1845 the federal government granted Florida 500,000 acres in addition to the customary grants received by all states to hasten internal improvements. By 1850 Congress realized more needed to be done and passed the Swamp and Overflowed Lands Act, giving Florida 20 million more acres to be sold to finance drainage and reclamation projects. The following year Florida established the Internal Improvement Fund. In 1881, to reduce the debt of the Internal Improvement Fund, Congress sold 4 million acres of land near Lake Okeechobee and the Kissimmee River to Hamilton Disston, of Philadelphia. Disston formed the Glades Atlantic and Gulf Coast Canal and Okeechobee Land Company. He brought in a wood-burning dredge to start canal construction. Congress then decreed that Disston was entitled to half of any other land he drained.

After the freezes of the late 1800s, citrus growers moved south to "the ridge" in Polk County and to the East Coast in Indian River County. At the same time, growers planted thousands of acres of new citrus in the Caloosahatchee Region, centered around Hendry County. But it was not until after the freeze of the 1960s, which hurt the groves in central Florida, that citrus production in southwestern Florida boomed and Hendry became Florida's leading citrus county.

In the early 1920s it became apparent that the Everglades drainage system, which lowered water tables during dry periods, caused the developed area to flood during wet periods. The flooding resulted in heavy abandonment during the 1920s by homesteaders who were assessed with a drainage tax that was greater than they could afford.

BEFORE southern Florida became attractive to settlers, three technological breakthroughs were necessary: drainage, mosquito control, and air conditioning. The development of all three took place shortly after World War II. Before then, much of the area was in range land for cattle. However, during the war large numbers of troops were stationed in the state, and after the war many veterans decided to settle there. Until 1949 the open range law was in effect, which meant that farmers had to erect fences to protect their crop land from cattle.

Things changed rapidly starting in the 1950s, and agriculture became the state's leading industry. Only in recent years has tourism surpassed it. As pointed out in the chapter about Louis E. "Red" Larson, dairying suffers from a climatic disadvantage in Florida. A wide variety of fruits and vegetables, however, thrive under the same conditions. Florida's nearly 35,000 farms contain 3.8 million acres of land, of which nearly 1.8 million are irrigated. Those farms provide work for more than 80,000 people and generate $18 billion in economic activity. The entire food chain, from farm gate through supermarket, generates more than $45 billion annually. The state produces 75 percent of the nation's citrus, 10 percent of its vegetables, and 25 percent of the domestic sugar supply. The total value of all agricultural products sold is slightly more than $6 billion, making Florida the ninth-ranking state. Florida ranks second only to California in the production of sweet corn and other vegetables, melons, fruits, nuts, and berries.

Following the national trend of farm consolidation, Florida's largest 5.5 percent of all farms gross an average of more than $500,000 in sales annually and account for 82 percent of all agricultural output. The smallest 79 percent of all farms gross an average of less than $50,000 in sales annually and generate less than 3.7 percent of agricultural sales. Most of the smaller farmers have off-farm income, whereas the largest producers are concerned that they may not be big enough to meet the demands of the large retailers.

THE ANDREW DUDA FAMILY lived in the community of Velcice, Slovakia, a part of the Austro-Hungarian Empire. Andrew worked in a French-owned veneer factory that made furniture. His wife, Katarina, a skilled seamstress, was frequently employed by Austro-Hungarian royalty. They had four children: Anna, born in 1900; John, 1904; Andrew, Jr., 1906; and Ferdinand, 1909.

Andrew was involved in a labor dispute with his employer and because of unsettled conditions in Europe decided to leave Slovakia for the United States. He left his family in June 1909, only six weeks after his youngest son was born, and went directly to Cleveland, Ohio, where Katarina's sister and her husband lived.

Andrew secured a job at the Kuntz Brothers furniture factory and joined Holy Trinity Lutheran Church, Slovak Synod, where the services were conducted in the Slovak language. The Slovak Synod later merged with The Lutheran Church—Missouri Synod, which played a major role in the lives of the Duda family. Each month from 1909 through 1911, Andrew sent part of his earnings to Katarina, who used the money to buy a small plot of land. She supported herself and the four children from her seamstress activities.

In 1911, when Florida was experiencing a boom, 12 Slovak members of Holy Trinity and their families migrated there. They purchased stock in the newly formed Slavia Colony Company, which used the funds to acquire about 2,300 acres near Oviedo. Then, they established the community of Slavia, where Slovak was the chosen language. Andrew arrived in Slavia in December 1911 and exchanged his stock as a partial payment of the $800 cost for 40 acres of land along Mikler Road. He then notified Katarina that she should emigrate. She sold the land in Slovakia to purchase passage for herself, Anna, Andrew, Jr., John, Ferdinand, and Andrew, Sr.'s brother George. They landed in New York and took the train to Oviedo, arriving there in June 1912. On March 17, 1912, Andrew Duda and five other members of the Slavia Community organized the St. Luke's Congregation. For the next 22 years, Andrew and other senior members conducted services because they could not afford a minister.

The Dudas experienced the typical hardships of frontier life. It took six men to plow the community garden plot—five pulled the plow and one guided it. In addition to clearing his plot of ground, Andrew worked in a citrus packing shed and in the swamp cutting cypress trees that were hauled to the mill to be made into shingles. For many years he was prohibited from clearing more than 12 of his 40 acres because the pine trees on the land were under lease for turpentine production.

Anna was 12 when the family arrived in Slavia and wrote about the family's experience: "We didn't like it here—too many mosquitoes, stinky water, not a window in the house, and we had to work on the farm." The windowless home was a former sawmill shack. Everyone slept under mosquito netting and constantly had to sift bugs out of the flour, which was kept in a wooden barrel. But the Dudas endured. Later, the family had a cow, and Anna commented, "Between the cow [the milk], sweet potatoes, and the flour, we had it made. It was even better after we had pigs for pork."

Except for the father, none of the family knew a word of English. Anna attended first grade for two years just to learn the language. For the first two years, the three children of school age walked 1¹/₂ miles to school. In the third year, Anna drove a horse and wagon, and her brothers rode with her. The Dudas were so poor they had no shoes for the Christmas play, so the teachers, headed by Miss Mary Aulin, chipped in to buy some. The Duda children all agreed that the other children were always kind to them and they were not teased when they made mistakes with the language.

IT WAS DIFFICULT getting the 12 acres to produce enough to make payments on the mortgage and to support the family. So, in April 1916, after four years of struggling with farming in Florida, the family returned to Cleveland, where Andrew renewed his job in the furniture factory. Anna remained in Slavia and married George Jakubcin. The Dudas lived in the Cleveland suburb of Lakewood for two years. Then, Andrew located a farm at Olmsted, 12 miles west of Cleveland, where they lived for the next eight years. When they moved to the farm, the boys ranged in age from 9 to 14 and were able to help with the crops.

The soil on the Olmsted farm reportedly was exhausted, but Andrew was not deterred. He hauled manure from the local stockyards and livery barns to replenish the land. He purchased a cow, some pigs, and some chickens, which not only helped fill the family larder but also provided additional manure for the land. The Dudas raised carrots, beets, onions, cabbage, cauliflower, tomatoes, green beans, peas, lettuce, and turnips—all labor-intensive but high-return crops. Andrew continued his job at the furniture factory and continued to make mortgage payments on the Florida farm. John quit school after the seventh grade, but Andrew, Jr., and Ferdinand went through the ninth grade. Andrew rented a second farm, which had a small orchard of apple and plum trees. Then, the Dudas started to raise sweet corn, raspberries, and strawberries, which lengthened the marketing season. This enabled the family to save money after making the mortgage payments.

Katarina was bothered by rheumatism and spent the winter of 1925–26 with Anna and George in Florida. She felt much better in the warmer climate. John, age 21, accompanied his mother. The turpentine lease had expired, so he plowed the entire 40 acres. John learned from Mr. Mikler that the economy in Florida had improved. The rented land in Olmsted was sold, so in November 1926, after selling all the produce, the Dudas headed south. John and Ferdinand, accompanied by the family cat and dog, rode in the Model-T Ford truck loaded with furniture, while Andrew, Katarina, and Andrew, Jr., took the Sterling Knight auto. Mr. and Mrs. Martin Stanko, with their Sterling Knight, accompanied them. The remaining furniture and farm equipment were sent by rail. The 1,055-mile trip from Olmsted to Florida took from Monday morning to the following Sunday noon. On the way the travelers stayed at homes that had signs for lodging.

THE DUDAS made the last payment on the farm at Slavia in 1925 and had about $2,000 in savings. But they conserved the cash and lived in a small structure made of old barn lumber. It had just two rooms—a kitchen and a bedroom. Several years later they added three rooms and a garage to the original structure. The father and three sons went right to work on the farm, where every job was done by hand except the plowing, which was now powered by a mule. Things did not go well, and after the 1928 crop season, the Dudas were right back where they had been in 1916—they owed more than their assets were worth. Fifty years later Ferdinand reminisced: "That's when we really found out the value of the dollar. But we never asked for help and were never given anything."

For many years all they farmed were the original 40 acres, which were not all cropped. When faced with opportunities to go in different directions, the father and the brothers held fast to the belief that they could do much better if they stuck together. They saw that firms in other industries were forming partnerships so they could operate on a larger scale. They reasoned that maybe they could do the same. The father probably was hoping this would happen, but he does not seem to have been heavily involved from this point on, though he was only 52 years old. The sons developed a three-way entrepreneurial venture. Those who knew them best say that probably "none of the three could have made it on his own. [But] the brothers functioned under the premise that the Lord was the guiding influence at all times."

Ferdinand was the visionary. Andrew, Jr. (hereafter no longer referred to as junior but only as Andrew), was the organizer and the business disciplinarian. He was the front man and had the money instincts. John was the inventor, but he was also recognized as the one who could size up people. The word "patriarchal" was never used, but the Duda wives never became involved. They were expected to take care of their families and be active in the church. The three brothers also played active roles in the church, with Andrew and Ferdinand each teaching Sunday school for 40 years.

Once the three decided to work in partnership and identified their roles, they never looked back but always looked to the future. They often said: "Looking back only causes us trouble. Those who do [look to the past] don't want to change." In their spare time they traveled throughout the country-side to check on how the competition was doing and pick up ideas

This 1970s photo shows (left to right) Ferdinand, John, and Andrew Duda in the yard of the Lutheran Church in Slavia, in which the family had a very active role.

from their observations. From early on they had an intense desire, almost an obsession, to own land, which kept them short of cash but was a key to company success.

In 1928 drainage ditches were constructed in the Slavia-Oviedo area. Originally, all of Seminole County was tiled and subtiled, initially for drainage and later for irrigation. Once the land was drained, the brothers saw that land east of Highway 426 had no cypress roots. By experimentation they learned that the soil tested low in copper and manganese. It was easier to correct for the deficiencies than it was to cope with cypress roots, so when opportunity came to buy, they knew where to look for land.

Even before they recovered from the financial setback of 1928, they had a chance to buy land. An experienced farmer from Granite City, Illinois, had attempted to farm in the Florida community and did not succeed. He offered generous terms on his 10 acres, and the Dudas acquired their second farm, which nearly doubled their tillable acres. The difficulties of the 1930s presented more opportunities as members of the original colony sold their small farms. In 1932 the Dudas planted celery on one 10-acre purchase, but the quality was so poor that the proceeds did not pay either for the fertilizer or for the crates for the celery. They could not borrow money against their land for living expenses because the fertilizer dealer had filed a lien on it. It took them nearly two years to overcome that setback.

In 1935 they secured 140 acres of what was characterized as Mitchell Hammock muck soil. This land was totally forested, so they cleared it by cutting off a ring of bark around the trunk of each tree and letting the trees stand for three or four years, after which they were much easier to remove. This soil proved to be very good for late celery because it was porous and the celery endured the heat better there.

The Duda brothers purchased these plots for about $100 an acre with a small down payment and with a clause in the contract that if they did not have enough money to make the annual payment, they could pay only interest and not be foreclosed on. By 1948 the Dudas owned 140 acres of land from the original colony members and a total of 500 acres in the Slavia-Oviedo area. Unlike most farmers, they farmed in an area where people wanted to live, which gave them an early opportunity to profit from land appreciation. They agreed that whenever they sold land, they would immediately invest in other land so that they would not erode their farming base.

A 1935 photo of Andrew Duda, Sr. (right), with a fertilizer salesman in the company's celery seed beds near Oviedo.

The combined strength of the three brothers and a larger-than-average land base gave them the synergies they needed for continued growth. By the early 1930s, after their celery production grew beyond that of the smaller farmers, they became disturbed about the commission they paid to the American Fruit Growers, which operated as the Blue Goose Cooperative. As a result, they started direct selling. It was Andrew's idea to do their own marketing because he felt that they could do a better job than the cooperative. At the same time, they concluded that they could save money if they asked for bids on their fertilizer purchases. Once they learned how much they saved on competitive bids for fertilizer, they asked for bids on all their inputs.

◆

LIKE MOST PROGRESSIVE FARMERS, the Dudas were anxious to get away from the tedious, backbreaking, labor-intensive tasks required in truck farming and from the use of inefficient animal power. In the 1920s they purchased their first tractor, a Fordson, quickly followed by a second. In 1934 they purchased their third tractor, one of the first rubber-tired Olivers. Flotation was important in the muck soil, and rubber tires were a real asset in that respect. To increase flotation they attached lugs to the wheel rim that extended beyond the rubber-tired wheel, a forerunner of dual- and triple-tired attachments. Plowing was the first task the tractors were used for, but modifications were made so they could do other jobs. John built a drawbar wide enough to pull three four-row horse-drawn fertilizer spreaders. This was a major time and labor saver because twice each year they spread four tons of 5-6-5 analysis fertilizer per acre on land used for celery. One person could do the job of three at a much faster rate, and no animals had to be cared for.

Probably the most laborious never-ending task in celery production was removing the tops of the plant so that sunlight could penetrate to the heart to enhance growth. A person walked the row with a butcher knife and cut the tops. To do the job more efficiently, John built a two-bladed rotary mechanical topper that was tractor drawn.

Insects, always prevalent in a warm, moist climate, were a constant threat to crop production. To control worms in celery, workers carrying knapsack sprayers on their backs walked down the rows and sprayed the celery with either arsenate of lead or Paris green. Similarly, red spiders were controlled by dusting with sulfur, and aphids by using nicotine dust. John's next major machinery innovation was a horse-drawn but engine-powered sprayer capable of doing several rows on each pass through the field. This was before the farm had electricity, so all the holes in the steel framework had to be drilled by hand-held brace and bit. This was a major task, but the Dudas were unable to get anyone to manufacture such a machine because of its limited market potential.

ONE OF THE MOST FORTUNATE CIRCUMSTANCES in the Duda story is that the three sons of Andrew, Sr., got along so well. This is generally not the case when family members must work closely together in business. But they were a unique blend, and the synergies were so strong that the brothers were motivated to keep on. They were in their thirties, were well seasoned, and were primed for expansion.

John was always on the lookout for ways to make a machine do the work. His innovations proved to be great labor savers but were also the key to enabling the Dudas to pay better wages than most farm workers received. This was important to their long-range success. Ferdinand was the farmer. He liked to clear the forest, drain the land,

and see the soil produce crops. Andrew was the businessperson, the salesperson, and the banker.

All three were sticklers for detail and kept meticulous records. However, in some respects they were almost haphazard in their approach to expansion. If they had a good year, they looked for more land to buy; if they had a poor year, they tightened their belts. This was relatively easy, for neither they nor their wives sought a high life style. They were true risk takers, and when they got "out on a limb," they knuckled down and solved the problem.

As they expanded they realized that they could not do everything themselves and early in their careers learned to delegate authority. They had a natural compassion toward others and worked at securing and retaining good employees. This meant paying as well as possible in their labor-intensive business. When asked how they solved problems, they readily admitted that they had their share of arguments and discussions. But early on they took the position that if the three could not agree 100 percent on an idea, they would drop it. When interviewed late in life about that decision, Ferdinand said, "There were plenty of opportunities for things we could agree on."

BY the mid-1930s the Dudas had overcome the early reversals, and from then on into the 1950s they expanded rapidly. In 1939 they made their first large purchase when they secured the 500-acre Lake Hart Farm south of Orlando. The price was right, $40 an acre, but they soon ran into difficulties. Cattle were being pastured on the land, and as the brothers walked looking for cattle, they traversed a canal that ran through the property and realized they were on muck land. They purchased their first dragline to widen the canal and improve the drainage, but the county stopped them. The canal was both a positive and a negative factor. It was needed for drainage. However, when Lake Hart rose, the canal flooded the farm. The soil was highly acidic and did not drain well, but it raised celery, cabbage, sweet corn and peppers. Unfortunately, production levels were not good, and after a few years the Dudas returned the land to cattle grazing.

In the 1950s the Lake Hart Farm was traded to a neighbor who wanted it for cattle production. The Dudas received a farm at Clermont, west of Orlando. Forty years later the authorities widened the canal on the Lake Hart property, which they had stopped the

Dudas from doing in the 1940s. In 1941 the brothers acquired land at Zellwood, near the village of Lake Jem, northwest of Orlando. This land was along the McDonald Canal, within the Zellwood Drainage District. It was under water and had to be surveyed from a row boat. The owner wanted to build a dike to keep the water out, but he was in financial trouble and was not considered a bona fide farmer. Therefore, he did not qualify for the $300,000 government loan necessary to cover the cost. The agricultural agent of the Seaboard & East Coast Railroad knew that the Dudas were looking for land and would meet all the criteria for government approval and support. As they rowed over the alligator- and snake-infested water, they probed the lake bottom and came up with "good muck." They also observed saw grass that was over their heads. This convinced them to buy the 3,000 acres at $10 per acre, but it took about 25 years to develop the acreage into good crop land. This was in 1941, and the federal government was concerned about having an ample supply of food for the impending crisis, so it gave the Dudas permission and funds to pump off the water. By doing the drainage with their own equipment, they were not obligated to pay drainage taxes.

The lake was spring fed, so draining it was a slow process. Before the water level became too low, they used a boat to stake out the land into 25 fields of 120 acres each. After most of the water was pumped out, they used a small HD 9 crawler tractor with extra-wide cleats for toting material to build a dike. After the dike was finished, the crawler was used to make mole drains (drainage tubes) that carried the water to the larger lateral drains so crops could be raised.

Once the debris was removed, the Dudas planted potatoes, but that crop was a disappointment because it was scabby. The land would have been good for celery, but the government would not buy or allocate fuel to raise it, so the Dudas produced string beans, cabbage, and carrots, which the government wanted and allocated fuel for. Those crops were sold to Campbell Soup and H. J. Heinz companies, which processed for the military. The Zellwood farm was very productive and gradually was increased to 4,750 acres before it was sold to the government in 1997 and was included in the Lake Apopha restoration plan.

DESPITE the Zellwood land's high productivity, it was too far north to permit more than spring and early fall crops. The Dudas also needed winter cropping to reach their goal as a year-round supplier of vegetables. Thus, they had to farm farther south. Their next acquisition was not far enough south to offer year-round production, but it was land the three brothers could not resist. They had seen a newspaper article about Cocoa and Rockledge ranches, which totaled 43,000 acres, being offered by a realtor for $5.25 an acre. Those properties contained livestock, so the Dudas rationalized that probably they should diversify into cattle. They were also intrigued by the thought of having a vast expanse for hunting.

Before they could act, their attorney reported that those properties were being foreclosed on and would be sold at the Titusville courthouse. They appeared at the courthouse at the appointed hour, but no one was there. The owner had turned the properties over to the insurance company with a quitclaim deed, and a third party had purchased 24,880 acres for $50,000 and wanted to sell them for a $6,000 profit. The Dudas did not quibble, because they had been interested in buying the land for $5.25 an acre and the new price was only $2.25. They purchased the balance of the 43,000-acre holding at $3.00 an acre. At the same time, they acquired several other small parcels nearby at Turtle Mound.

The Cocoa purchase included 350 head of cows and calves for $45 per head. This was their first sizeable cattle herd, and it yielded some return until they could determine the land's potential for cropping. They acquired this property during the dry season, but the neighbors had warned them that the land was too wet for cropping. However, their surveyors informed them that the land had a 17-foot drop from one end to the other and it had a low water table. This meant it could be drained. They tested the soil and learned that it was low in cobalt, but correcting that deficiency was no challenge.

Owning swamp land meant that they would be plagued by too much water in the wet season. Regular pumps were used initially to drain the land. Later, a special pump called the mud sucker was brought in and had to be submerged deep enough so it could completely drain the land. It had so much suction that it sucked up everything in the drainage ditches, including pieces of metal. It also sucked out the muck, which had to be returned to the land. Because the mud sucker had to be operated while submerged, there was no way to lubricate it. The sand and other materials it pumped were very hard on the bearings and the propeller, so the machine had to be repaired at least two or three times each year. That was a costly process, but it was the only way to make the submerged muck land available for production.

At the same time, the Dudas put their dragline to work digging ditches two miles apart through the property to drain off the surface water. They developed a pumping system along with their drainage ditches so they could provide irrigation water during the dry season. All this work made Cocoa a good sand-land ranch. They constructed a similar system at Zellwood. By 1985 population pressure around Cocoa was such that "raising houses" became more profitable than raising crops, so they started selling some of that property.

THE DUDAS continued their southward movement. As soon as they had the Zellwood and Cocoa properties in full production, they took the next big step in geographical diversification and went into the Everglades Agricultural Area (EAA). In 1946 John and Ferdinand scouted for land near Belle Glade, which lies southeast of Lake Okeechobee, and bought two parcels totaling two sections at a cost of $25 per acre. The soil was muck, which they described as peat with high organic matter. The land was laid out in section-sized fields instead of the traditional 50- and 120-acre fields they were farming. This purchase was a stroke of luck, for it seldom freezes in that area, and the swamp land there was being reclaimed. Two years later they secured two more sections from U.S. Sugar Corporation, followed by another acquisition of eight sections. Howard Surellis, the owner of those eight sections, financed the sale 100 percent. He had a total of 30,000 acres he wanted the Dudas to purchase, but, much to their later regret, they declined that opportunity. In any case, the 12 sections, plus a large block of leased land, gave them a substantial unit at Belle Glade.

Some of the property was contained within larger holdings and had only a dirt road to it, so they brought in their dragline and dug a ditch along the road to start draining. Most of the land not being cultivated was covered with saw grass, which did not present a clearing problem. The soil needed to be disked several times, and the usual ditches had to be dug. Cattle were raised on the land until it was needed for sugar cane or vegetable production.

This land was found deficient in several minerals, especially copper and manganese. Testing was not as advanced then as it is today, so the Dudas made block checks to determine how much to apply for crop needs. Initially, they grew a large acreage of sweet corn and some leaf vegetables. About the same time, they started large-scale radish production there and decided they had to speed up harvesting. No mechanical harvesters were available, so John met the challenge and built a six-row unit.

After the soil deficiencies were corrected, Belle Glade produced beautiful celery, much greener than what grew on the sand land in the north. But once again the brothers ran into a problem. They relied on outsiders to sell their products and prepare them for shipment. That practice changed after they sold 15 carloads of celery that Lake Charm Food Company had not properly cooled and they were not only unable to collect for the celery but were also stuck for the commission and the freight. They realized that they had to process and sell their own products as they had done at Oviedo and Zellwood. This meant building their own cooler at Belle Glade. The cooler had to be large enough to handle what they anticipated to be greatly increased production. This necessitated substantial borrowing. They were heavily leveraged at the time, so the Atlantic National Bank would lend only half of what they needed. Andrew was desperate. He made an appointment with the president of the Seaboard & East Coast Railroad to ask for money. He had his minister accompany him to bolster his courage, but even that did not help get all the funds he needed.

John and George Jakubcin, Anna's husband, were responsible for planning and overseeing the construction of a pre-cooler and a packing house along the railroad tracks. That set the stage for a rapid increase in production, and the Dudas plunged ahead on blind optimism. They netted $97,000 the first year, which they considered "lots of money for that day." They doubled their car loadings that year and tripled their loadings the second. In two years they had paid off the total loan on the cooler.

When they built the pre-cooler and the packing house, they initially ran the vegetables through cold water on a submerged chain to take out the heat. One day one of the flumes had no water in it, and the chain with vegetables riding on it passed through with cold water being sprayed on it. The Dudas noted that the spray seemed to penetrate the produce and leave it just as cold as, or even colder than, produce that had been submerged. They changed their practice, which reduced the amount of water needed, saved cost, and speeded up the cooling. Later, they adopted vacuum cooling, and that process was even more efficient. Everything worked well, and Belle Glade was soon recognized as the gem of the Duda farming operation.

CELERY was the major crop in Belle Glade, along with some leaf vegetables and sweet corn. Worms were a problem in the corn. To control them a worker walked down the rows carrying on his back a container filled with mineral oil. Attached to the container was a bicycle inner tube that had a hypodermic needle at the end. The worker squirted a couple drops of oil into the silk of each ear to stop the worms. If too much mineral oil was put in the silk, it retarded ear growth, so the worker had to be precise.

Another challenge came in harvesting the celery. The Dudas operated with the following philosophy: "To harvest a perishable crop, you do not have much choice. You either *get it* [on time] or *forget it.*" Once again John came to the rescue. He developed a rig called the "mule train." This consisted of a power unit to which a conveyor equipped with a moving belt was attached that stretched across 12 rows of celery. The rig was so heavy that they had to keep increasing the axles

A 12-row modern-day mule train used in harvesting lettuce in southwestern Arizona.

until they had ample flotation. Four axles, each equipped with two sets of dual wheels, proved to be enough. The unit moved through the field at 3 feet per minute. The celery was cut manually, stripped, and placed on the belt, where it was topped, washed, and graded and from which it was packed into boxes to be taken to the warehouse.

Yields are generally better if celery can be hand-harvested, but in some years securing the necessary labor is difficult. Mechanical harvesting damages the outside ribs of the celery. The damage results in an increase in the cost per package. It also discourages the consumer, who prefers to purchase a product that looks perfect.

Even though Duda's is the world's largest producer of celery, much of its Dandy brand grown in Florida, Texas, and California is still hand-harvested and packed fresh in the fields.

Later, the "mule train" was adapted to harvest corn. The rig traveled very slowly through the field, giving the workers time to snap off the corn ears and place them on the belt, which moved them to where they were boxed. The company received a patent on the "mule train," and some units were manufactured for other producers.

I N the 1940s and 1950s the three brothers spearheaded a tremendous expansion, after which they lost some of their youthful exuberance. They had a sizeable estate, and it was time to refine their management practices. True to their faith, they did not forget their blessings, and in 1948 the Dudas made a major commitment. The family "bankrolled Lutheran Haven, a combination retirement center and nursing care facility" located on 60 acres just west of the original Duda farm in Slavia. Andrew Duda, Sr., the company founder, died in 1956, and in 1958 the Duda Foundation was established. Each year a percentage of the corporation's profit is given to Lutheran and other Christian organizations.

In 1953 the firm was incorporated as A. Duda & Sons, Inc., and made continued expansion its mission. Prior to 1953 the Duda brothers had invited a University of Florida professor of economics on a regular basis to teach them business analysis from an MBA's perspective. At this time they established a policy of including nonfamily members in their top management. Shortly after the reorganization, they set out to employ experienced senior executives from other large firms. This gave them a broader perspective of the world of business and the leadership ability to steer one of the top agricultural producing and marketing companies of the world. The Dudas openly shared information and profits with their professional managers. The groundwork was laid for the firm to become a year-round full-line supplier of vegetables and citrus. The Duda management realized that the company had to continue to grow in order to compete in the global market.

In 1957 Duda's established the Oviedo Tractor Company as a separate, free-standing division, which became a major dealer for four-wheel-drive Steiger tractors, the trend setters in that field. In the summer of 1999, Duda's sold the business to a local equipment company. This was due in part to changing trends in the machinery industry that enabled large farmers to buy direct just as cheaply as dealers could. The family wanted to focus on its core business and other enterprises that had more profit potential than a machinery dealership.

D UDA'S has attracted considerable negative publicity in recent decades because it is a large agricultural firm with labor-intensive crops and because it has employed a gradually increasing number of minorities. Those two factors have made Duda's an ideal target for the media, which have tended to regard large-scale farms as "bad farms" and small family farms as "good farms" and to imply that agricultural firms with a large percentage of minority employees have hired them to work at lower wages and under poorer conditions than other laborers would accept. However, agriculture has historically relied on immigrant labor, and since World War II Caucasians have shunned agricultural jobs. In November 1960 the company's reputation was severely challenged when Edward R. Murrow narrated a CBS-TV documentary, "Harvest of Shame." This distressed the three brothers, who had always worked side by side with their employees and felt they had treated them well and generously. They were well known for being benefactors of their community and their church and felt that the program was not a balanced portrayal of the situation.

This was a critical time in the life of the company, for it was just when members of the third generation were joining the business. The second generation admonished the newcomers to abide by their faith and work together, because the combined knowledge of all was far greater than the singular knowledge of any individual. Today, the word "synergy" would apply, but that was not a commonly used word in the 1960s and not in the vocabulary of the three elders with their limited education. However, they knew from experience that the best way to succeed was to use their combined talents, and they delivered a long dialog on the virtues of sticking together.

When Ferdinand Duda, Jr., started with the company in 1961, some of his cousins were already employed there; his brother Joseph came soon after. All the cousins "had never thought of working any place other than the family business." The third generation all received "basic training" by working in the fields beside the other employees on an equal basis. The field managers were given instructions that any Duda "had to cut the mustard." The other field hands knew the culture of the family and generally were proud to be working in a row of vegetables next to a family member who someday might be managing the company.

With the prospect of several members of the third generation joining the company, it was only natural to look for opportunities to expand. The Belle Glade farm south of Lake Okeechobee proved to be very satisfactory for the kind of crops the Dudas wanted to produce. They searched for more land in the area and in 1962 located the 7,040-acre Circle Bar Ranch, owned by Asa Townsend, west of La

Belle in Hendry County. Next, they secured the 20,000-acre Wellhouse estate. Both units were purchased for $100 per acre. The original cost by today's prices seems low, but the cost of preparing this raw marsh land for eventual production far exceeded the initial price.

Once again the company dragline was put to good use constructing the extensive Townsend Canal, which included a combination drainage and irrigation system that permitted recirculating water that is not used. That involved building a costly lift station to take water from the Caloosahatchee River during the dry period. The five lift pumps are capable of pumping 135,000 gallons per minute to a head of 28 feet. That system, devised by John Duda, has been studied by many but has not been duplicated. The motto he used to explain the system was "Trifles make perfection, but perfection is no trifle." To protect the land from overland flooding from the southeast, the firm built a 2-mile dike on the eastern border, joined by a 7½-mile dike on the southern border.

To eventually farm the nearly 28,000 acres, Duda's constructed more than 150 miles of canals. Much of this work involved blasting with dynamite, for the ground was underlain alternately with hard and soft layers of rock. Again, the canals were built for drainage and for irrigation. Even with all the canals, not everything was safe; for after one big rain, the entire place was flooded. One pump did not start, and another failed. Then, Duda's established a policy whereby the entire system would be tested every month and every three years each pump would be overhauled. The company finished the entire project by building high dikes around the place—something that would not be permitted today. It has not been flooded since.

The Dudas' intentions at La Belle were to maintain cattle there until such time as the muck land became depleted and they could plant citrus trees. In the meantime, they planted tomatoes, watermelons, cabbage, sweet corn, potatoes, and other vegetables, but none were really successful. In 1963 they turned to their original goal and planted several hundred acres of citrus, and in 1966 they harvested their first crop of nearly 7,000 boxes.

International events played a part in what took place next at La Belle. The Cuban Crisis of the early 1960s led to opening Florida for sugar cane production. The Dudas first started raising sugar cane at Belle Glade, then added acreage at La Belle, at Moore Haven northeast of La Belle, and, in the 1990s, in the Rio Grande Valley in Texas. Sugar cane came at a time when the returns on vegetables were not particularly attractive, and from 1965 to 1999 it proved to be the most consistently stable and profitable crop for Duda's. Ferdinand, Jr., and Joseph believe that Florida provides the most economical conditions for cane sugar production, and by 1999 the firm was growing sugar cane on 40,000 acres. The company's goal is to be the largest independent grower in both Florida and Texas by 2004. Duda's is a major investor in the local cooperative, where some of its cane is ground and refined. A side benefit of the elaborate canal and pumping system is that with rains of 3 inches or greater, some of the excess water can be diverted to the cane fields.

DESPITE the criticism brought on by the media presentation, the members of both generations, including the younger ones with college degrees and newer perspectives, decided to "keep doing the best they could and continue on." The three brothers were too centralized in their management style, which sometimes led to difficulty with expansion plans. In 1964 they made a major change in policy after Andrew, then in his 60s, had attended a seminar on business management and heard a speech by Keith Louden, of the American Management Association. Andrew was so impressed with Louden that he invited him to survey the Duda operation and asked him to "strip away the homespun qualities of the family business and bring it into the modern age by giving it a more 'collegial' management with everything from long-range planning to succession management."

Louden implemented a system that called for long-range planning to replace the previous style of operation and take the company to the next tier of professional management. This essentially refined the firm's business practices. At that time (1964) the three members of the second generation declared their full-time retirement dates—age 68 as president and 72 as chairman of the board. This meant John would retire from the positions in 1972 and 1976, respectively; Andrew in 1974 and 1978; and Ferdinand in 1977 and 1981.

Outside talent was needed and employed, and these individuals, along with Duda family members in the company, were sent to seminars of the American Management Association and the Stanford Research Institute and to Dale Carnegie courses. Auditors from Chase Manhattan Bank were called in to review the finances. A full-time economist was employed to follow world economic and political situations. This was in anticipation that the company would become involved in international business. When all of the suggestions were completed, Louden was asked to join the board of directors as the first nonfamily member. Board activities became much more formal than previously.

ALTHOUGH the three senior members had announced their retirement dates, they were far from giving up. Almost as if to prove this, they entered into a far-flung international venture. Ferdinand, Sr., said in an interview later in life: "We felt that freedom in agriculture was disappearing in the United States and that it might be good to have some holdings in Australia in case things became unbearable here." In the 1960s, independent of the company, the brothers purchased leases on land in Australia's Cape York Peninsula, bought the United Cattle Station in the northeastern part of the country, near Townsville, and acquired a 6,500-acre freehold, all of which were used for a cow/calf operation. They visited the property, and later other family members did also, but they were not involved in onsite management.

They had learned about Australia from their attorney, Cushman Redibaugh, who was active in the National Cattlemen's Association. Years later, when asked if they had it to do over whether they would enter into the Australian venture again, Ferdinand, Sr., replied: "When food gets more scarce as population increases, Australia will have a market. If family members did not have an interest in Australia, Duda's would have to learn to work with management companies. Australia's potential for citrus and vegetables was just as good as Florida's, and its potential for sugar cane was better." In 1982 the company purchased the Australian property from the three seniors and managed it through its U.S. cattle division. In 1999 it sold those holdings.

IN THE 1960S the company established its Redi Foods division at Lake Jem, northwest of Orlando, near where the Zellwood farm was located. To make use of the celery produced there, it developed processes to freeze and to can diced celery for institutional users.

The mile-long rows of weedless celery in the fertile muck soil are part of the reason for the success of Duda's Redi Foods division.

Today, Redi Foods is the nation's leading supplier of this product. It also provides fresh-cut celery to food-service customers and food manufacturers. Redi Foods is aggressively expanding, with the intention of adding more commodities to its processed line.

IN 1966 Duda's made its most southerly acquisition in Florida when it purchased what the brothers referred to as the Naples Sand Farm. The original intention was to grow "dry" produce, such as potatoes, peppers, and tomatoes, as opposed to "wet" produce, such as leafy vegetables. The company quickly discovered that although it could raise peppers and tomatoes there, it could not grow potatoes. However, flowers did very well, and there was a lively market for them. Soon Duda's was double-cropping on the sand land, using plastic to speed up the production time. The first crop was tomatoes, followed by cucumbers or watermelons. The Naples unit was successful and was enlarged to 6,000 acres. The company had considered pre-sprouting the seed for the Naples farm because that was being done in other countries, but it sold the property without initiating that step. Duda's did not fully integrate as a producer/marketer of its tomato production at Naples as it had done with other products. When Duda's purchased the land at Naples, it realized that the acquisition would become valuable development property. However, when I-75 followed "alligator alley," the potential development value of the land was diminished. In the 1980s that farm was sold to a competitive tomato grower.

AFTER La Belle was in production, Duda's wanted to get into processing there. Connecticut General Life Insurance Company had repossessed some land in the area and was looking for a partner. Duda's, Connecticut Mutual, and the Collier and Turner corporations joint ventured to build a citrus packing and concentrate plant, which became the Citrus Belle facility at La Belle. Duda's supplies about half the total produce used, and the facility has proven to be a real asset to the company. This fit in well with long-range plans for continued diversification. When the last of the second generation retired as president, the company was well diversified in citrus, vegetables, sugar cane, sod, and cattle.

Members of the second generation were very much in favor of expanding internationally. Besides their personal venture in Australia, they had tried to get involved in Morocco, but the culture of the region was not compatible with their way of doing business. They also spoke to leading people in Egypt, but the biggest problem there, as other American firms have experienced, was the unstable government. The same situation was true in many other countries. However, none of the proposed international undertakings would have been any greater proportionately than the company's taking on Belle Glade had been in 1946 when its resources were smaller.

Duda's was located in Florida and by the 1970s in California. In the 1980s it was laying the groundwork for moves first into Texas and then into Arizona and Mexico. After computer specialist Alan Newton was employed in 1972, each division was connected to the home office at Oviedo. This greatly facilitated management of the extensive business.

In 1973 the sod division was started at Oviedo by Walter Duda, Sr., of the third generation. In 1974 sod operations expanded to Cocoa and La Belle, and soon after to Fort Lonesome. In the 1990s branches of the sod division were opened at Hastings and Lake Placid. Today, the sod division encompasses 6,000 acres with 150 employees and annually sells enough sod to cover 150 million square feet, nearly 3,500 acres. That division has been a steady profit producer.

The La Belle farm is a good example of diversification in a single operation. Those nearly 28,000 acres contain a citrus nursery farm, a 1,800-head cow/calf enterprise, a 480-acre sod farm, and several thousand acres each of grapefruit, oranges and sugar cane.

IN 1976 Duda's entered California when it purchased Southland Produce Company, a marketing firm that gave access to fruits and vegetables in the area. That acquisition was followed in 1979 by the establishment of the 12,000-acre Gene Jackson Farms, Inc., named for the person who set up the West Coast division. This unit had operations in the communities of Salinas, Reedley, Lodi, Oceanside, and King Island. It quickly became the state's leading celery producer and a large grower of iceberg and romaine lettuce, broccoli, and asparagus. It also had its own seed farm. In addition, the unit was responsible for working with contract growers and for farming in Arizona and Mexico. At the same time, Duda's joint ventured in South Carolina and Ohio and contracted with producers in Missouri for apples and in Idaho for potatoes.

With the acquisitions in 1979, the goal of being able to supply customers 12 months a year was achieved. By 1982 the company had expanded to the point that it was growing 30 varieties of fruits and vegetables and had access to 20 more for its suppliers. During the 1970s, annual gross sales increased from $20 to $200 million. At the end of the decade, Duda's controlled more than a half million acres from Florida to California to Australia.

THE MEMBERS of the second generation always had a liking for cattle, especially the cow herd. They understood the value of cattle as a liquid asset that could be easily turned into cash. In that respect, cat-

tle were an ace in the hole. By grazing cattle, the Dudas could keep a property classified as agricultural. This fit in well with their desire to own lots of land. Cattle were part of a holding pattern for maintaining land until they had the funds or the need to develop it for cropping. Grazing is possible 12 months a year, and roughage can be harvested in most months. Citrus pulp is readily available, but most of the time Europeans were willing to pay more than local feeders could justify for having pulp as part of their ration. The company never used citrus pulp in its cattle ration. Like all agricultural commodities, cattle were subject to cycling, but the third generation felt that, contrary to the cycles of most vegetables, the beef cycle was predictable, and they profited from "going counter-cyclical."

The cow herd expanded along with the other enterprises. It peaked in the 1970s when the company owned 30,000 cows. At that point conversations were held with Paul Engler, whose career is recorded in the following chapter, about establishing a totally integrated livestock business from range to feedlot to packing house. But Engler felt so strongly about the advantages of feeding cattle in Texas that he decided against coming to Florida. Duda's tried feeding cattle on its own without success and turned to Horace Fulford, who did very well in the business. Fulford helped Duda's in its second effort at feeding cattle and training personnel at its lot in Ocala. Today in Florida the company feeds out its own locally grown calves only. However, it purchases feeders for the Texas operations. Duda's considers cows that are 25 percent Brahman to be ideal. A lesser percentage interferes with the ability to maintain a good breeding pattern. A greater percentage brings a packer discount.

IN 1977, when Ferdinand, Sr., retired from the presidency, it was time for the members of the third generation to "step to the plate." Keith Louden once again entered the picture. He was still the only nonfamily board member but served as a mentor to the eight Dudas who were now in the company. He assembled them in the board room for the purpose of determining whom their new leader was to be. Edward was selected to be the president. Ferdinand, Sr., was still chairman of the board, so restructuring the board was delayed until he retired in 1981. All eight of the new leadership were college educated and brought a much broader vision to the company, but they were still guided by the basic principles that had motivated the founders. They were not content with their own views and during the next decade created a seven-member board with four nonfamily members. They learned from their fathers and from management courses that if they wanted to succeed they had to demonstrate that they could work together.

All members of the third generation had started working on the farm at age 12 and during the summers had been rotated to different jobs, so they understood all facets of the business. That practice was continued with the fourth generation, and in 1999 a 16-year-old member of the fifth generation worked part time during his summer break. As the family grew, more of the members were no longer involved with the farm. The attitude of those individuals toward the business was much different from that of the insiders. This problem intensified as time passed.

I N 1976 the company was again targeted by the media, this time in a series of articles about its migrant labor camps by columnist Jack Anderson. Duda's had been upgrading its camps for the past several years but apparently not enough to satisfy everyone. When the articles were written, the Naples farm already had a day-care center and the company offered a full health and insurance program and some retirement benefits. This probably put it in the front ranks among agricultural enterprises. Public relations personnel were added to handle communications with the public and combat the negative programs of the media. Don McAllister was engaged as director of public relations and in the course of his work recorded much of the company history.

V ICTOR BEERS has been a contract grower for Duda's at the Citrus Belle plant for nearly 30 years. Before coming to Florida in 1965, he had 24 years of experience in the nursery business, so he was well prepared to manage a citrus grove. Within two years he was operating a 5,000-acre grove. From 1965 until the freeze of 1976, fruit prices were low, and the Dudas had a real challenge paying down the debt on their business. Soon after Beers arrived in Florida, he came to know the Duda second generation. He was impressed by their character and by the fact that they knew what it was to lose money but had the faith to persist.

Following a freeze in 1962, prices rose sharply. Then, as generally happens in agriculture, production increased faster than the fresh market could absorb, and prices declined. In the 1970s, production exploded, causing a very sharp drop in prices for the fresh market product. Contracting with a processor became imperative, but even with a reliable contract, the prices were so low that it was still a struggle.

The freeze of 1976 caused the next change in direction. Half of Beers' crop suffered frost damage, but the price per pound of solids on the remaining portion rose from $0.60 to $1.75. One can grasp the potential of this price change by understanding that a 90-pound box of oranges yields from 6 to 7 pounds of solids and that 500 boxes are considered a normal yield per acre.

Contracting worked very well from 1976 until about 1992, when prices again "turned south," as the expression goes. Starting in the late 1990s, many financially pressed growers contracted for cash on delivery at a price set in advance of the harvest. This has forced the processor to purchase futures contracts. Then, the processor contracts with the growers at a price that will cover their expenses based on the futures contracts. Growers in better financial standing use participation contracts with the processor. They do not get paid at time of delivery but instead receive payments each month as the plant sells part of its inventory and pays the growers on a pro-rata basis. This is a slower payout, but traditionally the price improves during the year, thus giving growers a chance to receive better returns. Duda's supplies about 45 percent of the total needs of Citrus Belle and is a participation grower.

Charles "Chuck" Harvey, plant manager for Citrus Belle, related that many growers are struggling under the above-described market prices. The market for concentrates takes the largest portion of the orange crop, but because of imports the market stays flooded with supply. Harvey maintains that U.S. growers can outproduce foreign farmers on a per-acre basis, can produce for less because of our technology, and have the best safety standards and the biggest domestic market. But because foreign growers and processors do not have the regulations that U.S. growers have to contend with, they can sell cheaper in our market. Harvey contends that much of the problem arises because agriculture is "used as a tradeoff to get exports for [our] high-tech products." He is particularly disturbed that Citrus Belle must furnish houses at cost for its workers but that only 70 percent of the structures are occupied. In his opinion, such costs, combined with the cost of regulations, keep American agricultural processors at a disadvantage when, on an even playing field, they could compete with the best.

HISTORICALLY, homesteaders probably made more money from appreciation of land than they ever netted from farming. Rising land prices because of growing population have been a boon to landowners. This is especially true for farmers in states where urbanization has taken place most rapidly. In this respect, one of the biggest surprises Duda's has experienced is that much of its property has been in the path of urbanization. Florida was a late bloomer, but since the 1950s it has had a steadily increasing population.

In 1981, Edward also became chief executive officer and chairman of the board. At that time the company began to envision the potential of converting some of its 125,000 acres in Florida into commercial and residential development. In the early 1980s, in response to that vision, the Viera Company, a wholly owned real estate subsidiary, was created. Viera, whose name means faith in Slovak, operates in Seminole, Brevard, and St. Lucie counties in the central part and on the eastern shoreline of the state. The second generation had the foresight to develop the policy that whenever they were in a position where it was more advantageous to sell or develop land than to farm it, they would immediately buy acreage elsewhere. If they had not adopted that philosophy, the Dudas might not be farmers today. Early in their farming career they were forced into the position of developing raw land into productive crop land because they were moving into the agricultural frontier of Florida. In a sense, they have been developers from the start. Draglines, mud suckers, and earth movers have been part of their progression. Developing commercial and residential centers was an inevitable step.

The largest Duda development to date is the master-planned community of Viera. Located in Brevard County, the community was started in the late 1980s. A master blueprint identifies areas for homes, businesses, shopping, services, schools, etc. All roads and necessary infrastructure were planned to accommodate growth and to provide the highest level of convenience for the residents. By 1999 that 9,000-acre site had become home to 6,000 people living in 2,500 homes. They have a K–6 grade school with a capacity of 945 students, a company-built fire station, medical services, and a retail complex. One of the several neighborhoods is an 80-acre site for a restricted-age community that has room for 300 homes. The Viera Company is headed by Joseph Duda, president of the land, cattle, and citrus divisions. In Seminole County, near the site of the original Duda homestead and corporate offices at Oviedo Crossing, the Viera Company is developing a regional mall, in addition to retail and food outlets, medical services, and other commercial tenants.

In 1997 the state of Florida, through Florida's Preservation 2000 Program and the USDA's Wetland Reserve Program, purchased 14,000 acres of the 36,000-acre Cocoa Ranch at $1,764.29 an acre. The sale did not have a major impact on either the cattle or the sod operations. It would have taken several decades of farming to yield that return, whereas the income was now immediately available for expansion. The company has leased the land back for grazing, and by controlling growth through mowing to offset part of the lease cost, it has economical cattle pasture.

In 1997 the state of Florida, in conjunction with the St. John's River Water Management District, purchased Zellwood Farm. The facilities were closed in 1999, and most of the employees were transferred to other locations. Zellwood was one of the first expansion units and a major production center. Duda's had hoped to retain the operation there, but increased pressure from state and environmental groups led to a decision to sell the property.

In June 1999 the company divested itself of land leases and cattle at Corse O' Gowrie and Old Glenray Cattle Stations in Australia. Jenny Alfor, the Australian manager, was relocated to a U.S. cattle operation.

AN ARTICLE written in 1996 about family transition made it clear that A. Duda & Sons, Inc., was still a closely held corporation with about 20 family members taking an active part in the operation and with all 101 shareholders receiving dividends. At one time 46 descendants of Andrew Duda worked for the company, but in 2000 only 13 are employed there. The second generation made it clear that future generations were welcome into the company "if they have the desire. [But] they should be interested in contributing, and [their] talent will show up."

When Ferdinand and Joseph, the two senior officers of the company today, were asked what is the biggest challenge facing them in managing the business, they almost in unison were quick to reply: "The family problem is far more difficult than any farming problem we have had. How can you handle it? You are always trying to figure out what the family members are thinking. Fortunately, the company has preferred and common stock [so the active members on the board have a stronger voice]."

Duda's faces the age-old struggle of outside family stockholders wanting to receive the highest possible dividends and those within the operation wanting to retain earnings to keep the company strong and expanding. Ferdinand and Joseph commented on how the second generation had many down years, but they had less-demanding family interests to cope with. They were driven to own land and did what they had to in order to accomplish that goal.

Tom Duda, a member of the fourth generation, vice president of the citrus division, had a slightly different reply when he was asked what is the company's greatest challenge. He said, "Having a vision that is consistent and properly focused. Knowing your place with your competitors and customers. [In today's world] you have to have constant global communication." He related how the second generation had such a passion for work and knew how to get the job done, even though they were not as concerned about the formalities of running the business. Tom concluded his thoughts: "Entrepreneurship is what it takes to keep agriculture alive."

After agreeing that family and/or succession appears to be a major issue for most businesses regardless of whether children are involved, it was appropriate to discuss the problems in the business of farming. The biggest challenge, as Ferdinand and Joseph saw it, is trying to determine how to stabilize income. They elaborated with a comment that would shock many smaller-scale farmers: "We are diversified into several businesses, and that evens out our income. In addition to enterprise diversification, we are also in four states but need to go further." They were clearly thinking global but, at the same time, were concerned about keeping the company a viable family business for the next generation. They discounted any possibility of going public to raise funds because, in their opinion, agriculture generates income too slowly due to the fact that it is land based. That is partly why it is so capital intensive.

WHEN asked to comment on the success the Dudas had in labor relations in operating a large labor-intensive business, Stuart Longworth, long-time vice president of human resources, replied:

> The second generation were remarkable people without trying to be remarkable. They had a very egalitarian attitude. They were humble and were very kind toward their people. The three decided not to hire job-service people and made heavy capital investments, about three years' profits, to make it [a good labor program] happen. The second generation did what was right even if it strained the budget. In that matter they did not worry about returns. [But] the workers have repaid the Dudas far beyond their expectations.

Late in his career Ferdinand, Sr., of the second generation, commented: "Lots worked with, not for, us [and worked] as hard as they would have on their own. All wanted to do a better job." Today, almost a century after it was founded, the company has nearly 1,000 full-time and 2,500 seasonal part-time employees and displays the same dedication toward its colleagues as previous generations did. Direct labor cost on individual farm units runs 20 to 25 percent of expenses. Part of the investment Longworth spoke of involved building living quarters that, at one time, provided space for 1,000 individuals. No rent has ever been requested; however, a nominal charge is made to cover utilities.

Longworth maintains that Dudas are considerably ahead of most agricultural growers. "For full-time employees, we will rank up with most of the Fortune 500 benefit packages." Professionals can earn up to 35 percent of their salaries in incentives, based 10 percent on personal achievement and up to 25 percent on company performance. The benefit package for the part-time employees includes a no-waiting period and low deductibles for medical insurance, free onsite day care, vacation and holiday pay, retirement plans, supplemental 401(k), and a mobile Head Start program. A range of benefit programs is offered, but most workers choose not to pick the more costly plans. Part timers generally work seven to eight months annually. When not at Duda's, some work in other businesses. Others have homes in south Texas and collect unemployment compensation during the off-season. The extensive benefit program has helped to stabilize the work force. More than 70 percent of the seasonal workers return year after year. Some have returned for as many as 40 years, proof that seasonal workers do respond to the benefits provided. In 1993 the *Personnel Journal* presented the Dudas with the Optimas Award in recognition for their efforts in labor relations.

After Jack Anderson's 1976 very slanted account of labor conditions at Duda's, the media left the company alone until 1994, when "CBS Reports" called wanting to do a follow-up to "Harvest of Shame." The CBS crew spent several months on the farms interviewing farm managers and seasonal workers and even followed some workers back to Mexico to make sure they were speaking freely.

Dan Rather conducted the final interview of the hour-long special, "Legacy of Shame," with Ferdinand and Ed Duda. When the report aired in 1995, the final three minutes portrayed Duda's as an example of "how to do it right." But it was quite explicit in pointing out that the firm was an exception in agriculture, which gave the wrong impression about the industry.

HUGH ENGLISH, an employee since 1965 and eventually vice president of the citrus division, can testify to the change in the makeup of the labor force. When he was first on the scene, workers were a cross section of Caucasians, some African Americans, some Puerto Ricans, and a few Hispanics. In 1999 there were only a few elderly African Americans left among the majority of Hispanics. English related how technology had changed citrus production in his time from nearly all manual labor or early crude equipment to a highly mechanized system. However, because of the nature of the product, much hands-on work was still required. The evolution in agricultural chemicals and machinery made possible an increase in the number of trees per acre from 48 or 50 in the 1960s to 150 in the late 1990s. This has helped to reduce labor cost per unit of production. English was well aware that the evolution has to continue if abundant low-cost food is to be available.

English spoke of the challenge faced by American growers:

Because of the use of chemicals, equipment, and other technologies, the company is very competitive from a production standpoint. [However], when it comes to harvest, it is at a competitive disadvantage with other countries. Nobody wants to pick fruit [or do other hand-harvest] in the U.S., but in other countries laborers fight for a chance to work in the harvest.

To gain maximum efficiencies, the company contracts with custom operators for such tasks as hedging and topping trees, applying chemicals, and performing similar specialized jobs. At one time Duda's was more mechanized, but it reverted to hand-harvesting because the company "ends up with a greater portion of top-quality crop, which is essential for fresh-pack products, including celery."

J ACINTO ALVAREZ was only four years old when his father, who was a U.S. citizen, died. For the next nine years, Jacinto lived with his mother's family in Mexico. In 1967, at age 13, he returned to the United States and finished his last two years of elementary school. As soon as he graduated from the eighth grade, he took a job in a fruit and vegetable packing house at Oxnard, California. In 1979, after nearly a decade at his first job, he joined the Gene Jackson Farms, a division of Duda's, and was placed in on-the-job training as a cutter and loader of lettuce. His leadership talents were quickly observed, and within two months he became foreman of the celery harvesting crew. Four years later he was promoted to supervisor of the celery harvesting crews at Oxnard and Salinas and then to assistant harvest manager of both farms. Under his leadership those farms had the best celery yields in the Oxnard-Salinas area.

In 1987 Jacinto became harvest manager for the entire Duda California division. He has as many as 2,850 workers under his supervision at peak season and never less than 200 in the slack period. When harvesting, each crew works with a "hump"—a mobile packing station. As it travels through the field, three cutters cut the celery and place it on a bed. Then, three other workers trim the celery into one of six sizes and put it into the cartons designated for the respective size. When the cartons are filled, they are set on the ground. A closer follows, stapling each carton. Then, the cartons are picked up by a forklift and loaded onto a truck. At times there can be as many as 150 pack choices, which involves a great deal of selecting. When the cartons arrive at the cooler, a hydrovac is used to bring the temperature of the celery in the cartons down 35 degrees before the cartons are placed in the cooler. At any one time, Jacinto may have as many as 50 orders of different-sized celery, ranging from 40 to 1,200 cartons per order, being handled by 100 or more crews. It is his responsibility to select the fields from which those orders are filled.

When Jacinto was asked to describe the most difficult time in his 21 years at Duda's, he replied that he never had trouble on the job, "but when I got a divorce, that was difficult." His oldest daughter has graduated from college and is a bank manager. His oldest son is employed at the refrigeration facility at Duda's in Salinas. His youngest son is in technical school taking a course to become a career fire fighter. Jacinto's wife does not work outside the home because "she stays at home to raise the [two youngest] children," who are still in elementary school. Jacinto says he is happy with his job and does not want to change work because "there are so many different things to know about and do with celery." He supervises "crews that are 100 percent Mexican."

Jacinto is another example of a person who came "across the border" to pursue the American dream. Besides being an excellent employee and citizen, he has established a family that is making a good accounting for itself. Without people like Jacinto, a major portion of American agriculture would not be able to operate.

WHEN the Dudas were asked what they thought the future holds for labor-intensive agriculture, they replied that if there is no relief in restrictions against immigration, crops that cannot be mechanized are doomed for further U.S. production. "We don't know how long that will take." They are in hopes that the bill to legalize the status of the illegal immigrants will pass. Agriculture is finally getting some support from unions, which have opposed the bill but have changed their stance because they cannot organize illegal aliens. The situation is very unstable because so many illegals have come into the country, but the government has not enforced the laws against them. When asked about shifting production to Mexico, where labor supposedly is less costly, the Dudas pointed out that the Mexican system has many built-in inefficiencies that destroy the economics of going there. For

their part, they would prefer to continue production in the United States.

IT IS OBVIOUS that Duda's is larger than the average-sized family farm, and when Ferdinand and Joseph were asked what they thought the future has in store regarding farm size, their answers were surprising but perceptive. They said that life style in the United States is dependent on cheap food. "If we were all small family farmers, consumers would be paying far more for their food than they are today. So there is not much chance that will happen [go back to small farms]. If we really had free trade, we could lick anyone in the world." While on the topic of low-cost food, they commented about organic production. It was clear to them that organic was a "niche market and is here to stay. But if we went completely organic, we would have to triple our acreage. This would be a great polluter, and society would complain [about that]. Going organic is not realistic."

Even though Duda's is large when compared with the average-sized grower, its leaders believe the company may not be large enough. It is their opinion that within the next decade the retail food industry will be controlled by five chains. "We are making alliances to protect our future. We would like to align with a few chains. Wal-Mart currently is our biggest customer, and we like that, but are we big enough to [continue to] have its attention?" To maintain the interest of the large chain buyers, Duda's has made alliances in two directions. It contracts with smaller-scale producers to secure their production and help it fulfill the Wal-Mart contract, and it aligns itself with other grower/shippers, such as Riverfront Groves, a large packing company. In both cases Duda's is aligning itself to generate a critical mass sufficient to meet the needs of the largest chains.

Because the large buyers understand the need for the producers to make a profit, they stress quality, just-in-time delivery, and a steady year-round supply over cost. The end price is immaterial. This is important to the growers, because a slight change in the supply of perishable commodities can make a big change in price. Thus, when Wal-Mart and Duda's agree on a price for an entire year, it enables Duda's to contract with smaller-scale growers at a firm price. This stabilizes its income. Because Duda's is a grower, it knows what it costs to produce a given commodity as well as what quality to expect from its growers. The Dudas are meticulous farmers and set high standards for those who grow for them. That is essential so that the retailer can meet the demands of the consumers who are accustomed to expecting the best.

Duda's is the world's largest producer of celery. It produces and has it available every month of the year, so it has no difficulty meeting the demand for that commodity. But retailers want to deal with providers who can supply a wide variety of produce on a year-round basis. This means that Duda's has to align itself with growers of as many different fruits and vegetables as possible. Its alliance with Riverfront Groves and other major citrus growers gave it the ability to supply Wal-Mart with 1.9 million cartons of produce in the 1998–99 season. In addition to the above, the company has aligned with other grower/packing houses to make every possible connection with the large buyers.

Duda's understands that the larger the food retailer is, the larger it expects its suppliers to be. A chain of 2,500 stores does not want to buy from a large number of small growers because of the inefficiencies in assembling the commodities, the variables in quality, and the inability of securing a year-round supply. All of the above would add greatly to the acquisition cost, which would have to be passed down to the suppliers so the retailer could remain competitive. It wants to do business with suppliers who can fulfill its expectations because it wants food retailing to be a service industry. This is good for the producers if they can meet the retailer's criteria.

Dan Richey, who heads Riverfront Groves, related that Duda's is the only producer in Florida that has the ability and the willingness to meet the three objectives: supplying a wide variety of commodities, maintaining consistent quality, and assuring year-round availability. "It is a sleeping giant that is about to awaken." This is good for the larger growers who want to stay in business, but it virtually forces the smaller producers to contract with packers or larger growers like Duda's. Generally, the smaller growers prefer to focus on production and not on all the external facets of farming, such as regulations, inspections, and marketing. Many of them do not want to take the risk of being exposed to the market for one or two months a year, so they contract to protect themselves. The changing market trends have made it almost impossible for the inners and outers to survive, for the big buyers want to deal with growers who have known track records.

Duda's is very comfortable farming in Florida, but its move to California, Texas, and Arizona has proved to be a real boon. Sammy Duda, general manager for the Duda West Coast operations and a member of the fourth generation, realizes that California dominates American agriculture because its climate is "so predictable that it is truly superior to any other location. We know what we can produce, but we do not know the price. With this reliability, we know how much we can contract to produce." He defended his comments by pointing out the vagaries of the weather in other parts of the nation. California dominates because crops can be produced 12 months a year there.

Some growers are quite large but want to focus strictly on production, so contracting is their only out. Ferdinand Duda feels that is their only choice unless they want to turn to vertical integration, such as packing, processing, branding, and selling their product. The other way of diversifying is by joining with others in a totally integrated cooperative. Duda's feels that a stronger alliance with the larger retailers would possibly provide the necessary backing needed for expansion. The company wants to contract at a higher level. It feels that it has an edge because it is a customer-oriented, low-cost producer. Duda's appreciates large buyers because they tend to be the most reliable payers.

Company computers are online with Wal-Mart distribution centers, where inventory is controlled for sweet corn and other vegetables. The distribution centers are restocked automatically with just-in-time delivery on a pre-priced basis. In 1998 Duda's won the Wal-Mart Supplier of the Year Award and earned the same honor in 2000, making it the only Wal-Mart supplier to win the award twice. In April 2000 the Citrus Belle division won the Supplier of the Year Award from Wal-Mart in the retailer's private label division. Citrus is processed, blended, and packaged under Wal-Mart's private label at Citrus Belle. In May 2000 Duda's won the Red Book–Mixtec Produce Excellence Award. The award honors firms that display "leadership in product development, innovation, customer service, management, and other areas that drive progress in the industry." Looking to the future, the consensus of Duda's is "There is a big market out there."

DUDA'S has given thought to getting involved in global production and marketing since the 1950s. In the 1960s the second generation invested as private individuals in Australia and saw potential there. In recent years the company has also been active in Latin America, Europe, and the Pacific Rim nations. It is well aware of the impact of the global market, for it is buying and selling around the world. This has helped to ensure year-round delivery, but it has opened the United States to competition from nations with lower regulatory standards and, hence, lower production cost. At the turn of the century, the company experienced a downturn in exports to Europe because of pressure from other nations. It then turned to the Pacific Rim countries and has increased exports of fruits and vegetables there. Ironically, while the produce industry nationally lost export market during the downturn in the economy in the Pacific Rim countries in the late 1980s and 1990s, Duda's actually had steady to stronger exports there because it had developed good relationships with the families that own firms in the Japanese fruit business.

One of the real surprises in the export market took place starting about 1995. A decline in per capita consumption of celery, causing a serious drop in the domestic demand, was offset by a sharp rise in exports of that commodity to China. Sammy Duda commented that the Chinese want to buy more than Duda's has available. The Chinese want top-quality celery, which is exported in refrigerated containers and arrives in China in less than a week. The Chinese, like the people of all the Pacific Rim nations, want to buy from the United States because "They think of us as a quality producer." Duda's deliberately labels all its shipments in English, rather than in the language of the country of destination, for even if the people cannot read what the printing says, they know it is American. The Dudas all agree: "We are going overseas because we are in a global economy." People of all nations want to enjoy a year-round supply. This gives Duda's a chance to export to others but also enables it to supply the American consumer every month of the year. Tommy Duda said: "Computers and the Internet may help in global business, but we have built that business on trust, which makes it fun and probably more reliable. People and our relationship with our overseas customers and providers are the core of our marketing information."

WITH gross farm revenues of $250 million in 1999 from nearly 100,000 acres in Florida, not counting real estate development,

Duda's seems poised for the future. It is well diversified in vegetables, citrus, sugar cane, sod, cattle, and real estate development. It operates solely in four states, joint ventures in several more, and has sales around the world requiring 120 employees in the home office to track the daily progress and take the company into the future. The Citrus Belle citrus concentrate and frozen concentrate plant built in 1973 was expanded in 1999 to increase the capacity from 3 million to 7 million cartons annually and can load a tanker semi with juice in 40 minutes. It has a 145-foot freeze tunnel capable of freezing 25,000 cans of concentrated orange juice every 90 minutes.

All of the above would make Andrew Duda, who left Europe in 1909 because he feared the future there, proud of what his family has accomplished. His sons were quick to recognize the need to employ the best available management help, give them a free hand, and share the profits with them. They stuck by their motto "Don't be afraid to admit that you made a mistake and be quick to change."

Bibliography

Interviews

Alvarez, Jacinto R. Salinas, CA. Telephone interview, August 9, 2000.

Beers, Victor. La Belle, FL, March 15, 2000.

Duda, Ferdinand S. Oviedo, FL, March 14, 2000.

Duda, Joseph. Oviedo, FL, March 14, 2000.

Duda, Samuel D. (Sammy). Salinas, CA. Telephone interview, August 9, 2000.

Duda, Thomas L. (Tommy). La Belle, FL, March 15, 2000.

English, Hugh M. La Belle, FL, March 16, 2000.

Harvey, Charles H. La Belle, FL, March 15, 2000.

Howard, Susan. Oviedo, FL, March 13, 2000.

Howeller, Michael E. Oviedo, FL, March 13, 2000.

Longworth, Stuart W. Oviedo, FL. Telephone interview, July 17, 2000.

Richey, Dan. Vero Beach, FL. Telephone interview, July 12, 2000.

Miscellaneous

"The Caloosahatchee River and Its Watershed, A Historical Overview." Florida Gulf Coast University Library Service, April 1998.

A. Duda & Sons, Inc., Company Fact Sheet, December 1999, and company brochures and video.

Florida Agricultural Facts. Florida Department of Agriculture & Consumer Services, Tallahassee, 1998.

Growing. A company publication. Various issues.

Dunn, John M. "Marketing Chain Lighting," *Florida Trend,* January 2000, 94–96.

Manuel, Frank. Typed manuscript of an interview with Andrew, Jr., Ferdinand, and John Duda for the Orlando Public Library, June 10, 1975.

McAllister, Don. Typed manuscripts of interviews with Andrew, Jr., Ferdinand, and John Duda, October 18, November 7, December 12, 1978; June 1, 1979. In company files.

Morey, Lesa. "A. Duda & Sons: Bigger Is Better," *Florida Growers,* May 2000, 14–18.

Neff, Ernie. "Innovation and Relationships at Riverfront Groves," *Citrus Industry,* April 1998, 21–22.

Patrico, Jim. "Passing the Torch," *Top Producer,* Mid-February 1996, 14–16.

Progressive Farmer, December 1993.

Stavro, Barry. "How a New Crop of Managers Took Root at A. Duda & Sons," *Florida Trend Magazine of Business and Finance,* January 1982.

Welsh, Joanne. "Caring Counts: Motivating Workers Comes Down to Simple Respect," *Top Producer,* December 1997, A11–A13.

Zaslow, Jeffrey. "The Duda Clan," *Inside Florida,* Vol. 28, No. 49, October 25, 1981.

W. D. Farr
and Paul F. Engler

Two Commercial Cattle Feeders

This chapter varies in format from others in this volume because it is about two entrepreneurs and their businesses. Both individuals played unique roles in transforming cattle feeding from an adjunct farm venture tied to corn production to a commercial business. They worked in different locations, both remote from the traditional Midwest Corn Belt, and relied on professional management and economies of scale to overcome the assumed built-in advantages of the farmer-feeder. W. D. Farr was a pioneer commercial cattle feeder in Colorado from 1930 to 1995. Paul Engler began his career in cattle feeding in 1960 in Texas and at the advent of the twenty-first century led the nation in that business.

— Hiram M. Drache

W. D. Farr

"I Was Constantly Opposed by People Who Did Not Want to Change"

W. D. FARR and Warren Monfort are generally considered the two individuals who developed large-scale commercial cattle feeding. To put things in perspective, Farr was born in 1910, and Monfort

W. D. Farr in 1998 at age 88. A pioneer commercial cattle feeder.

was about 20 years his senior. Unfortunately, Mr. Monfort is no longer living and his son Kenneth, who carried on the business, was unable to be interviewed. Warren Monfort and Farr both farmed at Lucerne, north of Greeley, Weld County, Colorado. Monfort started feeding before Farr, but their careers were sufficiently overlapping that they jointly gave rise to Weld County as the birthplace of large-scale commercial cattle feeding.

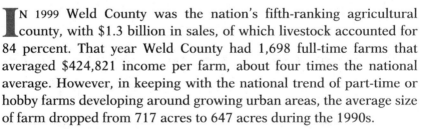

IN 1999 Weld County was the nation's fifth-ranking agricultural county, with $1.3 billion in sales, of which livestock accounted for 84 percent. That year Weld County had 1,698 full-time farms that averaged $424,821 income per farm, about four times the national average. However, in keeping with the national trend of part-time or hobby farms developing around growing urban areas, the average size of farm dropped from 717 acres to 647 acres during the 1990s.

In 1867 the Union Pacific Railroad was constructed through Weld County, and soon after, Horace Greeley, editor of the New York *Tribune*, founded a colony there and promoted the idea of irrigated farming. In 1877 the Desert Land Act was passed to encourage irrigation, and the first ditch for that purpose was dug in Weld County. That was the year William H. Farr, grandfather of W. D. Farr, arrived in Greeley from St. Thomas, Ontario.

In 1905 the Great Western Beet Sugar Company built a plant in Greeley. The sugar beet industry provided a good cash crop for farmers. It also served the livestock industry because beet crowns and tops were fodder for both cattle and sheep. Lambs produced on the western ranges were purchased each fall after harvest and fattened on beet tops, beet crowns, alfalfa, and stripped cornfields (the fields after the corn had been picked). Most farmers preferred sheep because one person could handle up to 1,500 head while the animals were grazing the fields and then pen them up for oats and water each evening. Farmers who did not have livestock were receptive to letting their neighbors graze livestock on their land because that was a way of getting the manure for fertilizer. Because of its fit with sugar beets, lamb feeding was king from the early 1900s to the 1940s. But the sugar beet industry also provided an opportunity to feed western range cattle, for after the sugar was removed from the beets, wet pulp was available, and the best use for it was as cattle feed.

HARRY W. FARR, W. D.'s father, was a banker in Greeley, but he was heavily involved in irrigated farming and sheep feeding. Harry Farr much preferred sheep to cattle because they had been more consistent money makers. W. D. rode to the farms with his father and, like his father, did whatever had to be done on those farms. When his father was busy at the bank, W. D. (hereafter Farr) often biked out to the farms to help. While still in high school he spent two summers working on a cattle ranch. He learned two things during those summers: "I learned about cattle, and I learned to love them. . . . That is where I got the feel of cattle, and after the second summer I knew I wanted to be in cattle. That is why I went to Wisconsin to learn about livestock." Even though he thoroughly enjoyed life on the range, he realized that he did not want to get into ranching, because he did not think it provided real money-making opportunities. His father had often told him that he "never saw many ranchers who made a profit other than through inflation." However, in later years Farr owned a ranch that served as a buffer for his feedlot. If cattle were cheap but the feedlot was full, he could still buy cattle and carry them on grass until there was room in the lot.

In 1928 Farr went to the University of Wisconsin intent on majoring in animal science. He was there only one semester when he became ill and returned to Colorado. However, while he was in Wisconsin he had observed cattle feeding and experienced part of a midwestern winter.

In the spring of 1929 Farr returned to Greeley and went to work on the family's farms. That fall he was put in charge of feeding 14,000 lambs that his father had contracted for in the spring for 12¢ a pound. Those lambs were fed during the fall and winter of 1929–30, and in the spring of 1930 they were delivered to Chicago and sold for 7¢ a pound. This was Farr's first full-fledged feeding experience, and it resulted in a loss of $70,000. This was a lesson that he never forgot and may have been a reason he preferred to feed cattle. His father had to mortgage his land to cover the loss, but he still preferred sheep to cattle.

A FTER Warren Monfort graduated from college, he taught country school in Illinois, where he lived with a cattle feeder. He observed that only corn and hay were fed and that hogs followed the cattle to consume the undigested corn. This was the traditional way that midwestern corn farmers marketed their crop. When Monfort's parents died, he returned home convinced that the Colorado climate was more conducive to feeding cattle and that he could feed them better than Illinois farmers.

Monfort was already feeding cattle at Lucerne when Farr returned from college in 1929. The Monfort and Farr farms were close together, and despite the 20-year age spread, the two men became good friends. Both realized that Weld County had a better climate for feeding cattle than the Corn Belt. They also knew that with the prospect of an increasing water supply, there was hope that crop production in the area would increase, which would be a plus for cattle feeding.

Both learned that World War I was the turning point at which the national demand for beef started to increase while the market for sheep began to decline. The packers sensed the shift in demand and encouraged the conversion. Corn Belt farmers raised corn and fed hogs and some cattle. If frost hurt the corn crop or if the corn crop was larger than normal, they purchased more cattle to utilize their immature or surplus corn. Both men realized that Midwesterners were not "professional" feeders. Their primary interest was in raising a crop and selling it, chiefly through hogs but also through beef.

In the late 1930s Monfort and Farr, together with other farmers, formed the Weld County T-Bone Club, which met twice monthly to exchange ideas and to learn from packer representatives and commission buyers. They had observed that soon after they sold their fat lambs and cattle each spring or early each summer, the price of cattle rose steadily until early December. They were encouraged by industry people to alter their feeding program to take advantage of the summer and fall prices. They also learned that they could buy cattle from the western ranges and ship them in transit to Greeley. When the cattle were finished, they could be sent to Denver or the midwestern markets at the transit rate. Monfort, Farr, and Bert Avery soon learned to stagger their shipments during the summer to keep the supply of finished cattle limited on any one day. "These summer-fed light heifers made us real money."

WHEN other Weld County farmers realized how well the T-Bone Club members were doing, some of them shifted from feeding sheep to feeding cattle. This caused two problems. The local supply of corn was insufficient to provide the feeders with their total needs, forcing them to purchase grain from Nebraska, and the local banks could not handle the increased demand for loans. Farr recalled that the extension people advised that the feeders could not import corn and be competitive with the Midwest. But the extension workers apparently had overlooked the advantages of the Colorado climate plus the economies of scale from larger-sized operations. Each year, as Monfort, Farr, and others made profits, they increased their volume, and their advantages continued to grow.

In 1937 both Farr and Monfort were elected to the local bank board, giving them added insights into financing. This proved to be a valuable experience because it got them acquainted with bankers in Denver, and later in Chicago, and helped to secure credit overlines. Farr commented on those early years of rapid expansion: "We were in a fool's paradise and really did not realize what caused it."

As cattle feeding increased, sugar beets, vegetables, and other truck garden crops gave way to corn. Hybrids were developed for the area. Initially, corn was fed as grain in bunks in the center of the feedlot,

Fence-line feeding during the late 1930s. The team was trained to walk along the bunk while the man forked off the ration. This improved all-weather feeding by eliminating the need to enter the feedlot. (Courtesy of Richard Farr)

and hay was fed in fence-line bunks. After trucks began to be used for feeding about 1940, entering the lot was no longer practical, so fence-line bunks were adapted for feeding grain. Next, mechanical feed boxes were fitted to trucks, reducing the need for pitchforks and scoop shovels. Efforts were made to feed a total mixed ration, but the equipment was not yet available to make that possible.

After field choppers were developed, corn silage was produced and stored in large bunkers. Less hay was used, commercial protein was combined with silage and shell corn, and the mixture was fed in the bunks. With the local corn supply being consumed as silage, grain had to be shipped in from the Midwest. That, too, has changed, for in recent years Yuma County, in eastern Colorado, has become one of the largest corn producing counties in the nation. Some feedlots have moved there to take advantage of the local supply.

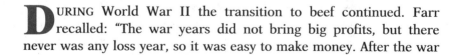

DURING World War II the transition to beef continued. Farr recalled: "The war years did not bring big profits, but there never was any loss year, so it was easy to make money. After the war the profits were high some years, but there were also some loss years. The government storage-loan programs helped Monfort pay for his elevators."

After World War II the portable electric welder became available. This improved the ability to alter feedlot equipment within the lot. Farr and Monfort were constantly innovating in an effort to make feeding easier, to increase cattle comfort, and to reduce labor requirements. Of the 150 people employed by Farr Feeders, many worked in the shop making new equipment or modifying equipment that was on the market but not built heavy enough for use in large lots.

As more knowledge was gained about nutrition, Farr and Monfort learned that they were feeding too much roughage and taking too long to finish the cattle.

The eye of the master determined when the cattle were finished. W. D. Farr in the early 1940s sorting cattle. He said it was one of his most enjoyable tasks. (Courtesy of Richard Farr)

Having accurate records helped them fine-tune their management. Commercial feeders led the transition from a heavy-silage ration to ground high-moisture corn blended with dry, chopped alfalfa. The commercial hay chopper and balers for half-ton and one-ton bales made that shift possible.

WHEN Farr was asked what he thought the future of commercial feeding looked like, he stepped back in time. He recalled that from 1945 to the 1960s, the supply of finished cattle was never enough to satisfy the demand, so it was a good period to be feeding. Then, in the 1960s, computers revolutionized cattle feeding. Farr immediately contracted with his bank to do Farr Feeders' computer work for several years before the company acquired its own system. Accurate, fast records gave those who used computers a real advantage over those who relied on assets rather than cash flow as a basis for securing financing.

He recalled innovations in veterinary science that made large concentrations of livestock possible and the changes this made in his career. Farr has no doubt that the industry will experience a sharp decline in the number of cattle feeders. He said, "The sheep business went by the wayside, and many cattle ranchers and small feeders will go the same way." Then, speaking of Weld County, he recounted that in 1957 agriculture used 95 percent of the water in the area and that in 1998 its share had fallen to 45 percent. In that time span the cost of water rights increased from $10 an acre-foot to $3,500. This, in his opinion, made startup costs for farming prohibitive.

Farr faced the inevitable family challenge. Son Bill wanted to develop the family's banking interest, and son W. Richard did not want to carry on the full responsibilities of commercial feeding along with other responsibilities. In 1995 Farr Feeders was sold to National Farms Company, and Richard continued to manage the business. Farr commented: "They [National] are on the cutting edge and will continue to succeed as long as they stick to the basics . . . and stay well diversified."

F ARR was riding the crest of success when he came to know Paul Engler. Both worked on implementing Cattle Fax, Engler as a newcomer to the industry and Farr as a veteran. Being involved in developing Cattle Fax enabled them to keep up with the rapid innovations and the rapid changes in marketing and demand in the industry. In 1970–71 Farr was president of the American National Cattlemen's Association, and he and Engler collaborated on industry issues. Farr liked Engler because he was so creative and willing to adapt. This was refreshing for Farr, who recalled his earlier days in the business: "I was constantly opposed by people who did not want to change." The bonds became sufficiently strong, and Farr hoped that Engler would succeed him as president of the Cattlemen, but that was not to be. Engler should probably feel fortunate that he was not given the challenge of leading the industry during some of its most tumultuous times.

As a parting thought, the 90-year-old Farr reminisced: "I often wonder what I could have done if I had had a better education. But I always surrounded myself with capable people."

Paul F. Engler

"It's Not Where We've Been, But Where We're Headed That's Exciting"

IF IT WERE POSSIBLE to credit one person with changing the pattern of the Texas beef industry, probably Paul Engler should be that person. Texas, by virtue of its enormous size, ranks first in farm real estate, with a 1998 value of $78 billion. It is followed by California, with $75 billion, and then by Illinois, Iowa, and Minnesota. The 226,000 farms in Texas represent more than 11 percent of all farms in the nation. The state's 147,000 cattle ranchers and/or feeders make it an easy leader ahead of number two Missouri, with 69,000 firms, and Oklahoma, with 62,000 firms. Its cattle and calf sales of $5.8 billion make up 44.2 percent of its agricultural income, well ahead of its cotton sales of $1.6 billion. In gross agricultural sales, Texas is surpassed only by California.

Paul Engler was born in Stuart, Nebraska, in August 1929 but grew up in nearby Bassett, where his father owned a gasoline filling station and had a bulk rural delivery route. At age nine Engler started working at the station

Paul Engler in 1992 at the Cactus Feedyard, Cactus, Texas.

while his father made rural deliveries. When Engler was in elementary school, his father purchased four acres on the edge of Bassett and moved an old shed onto the land to house one cow. His father taught his eager son how to milk the cow and then provided him with a coaster wagon so he could deliver milk to people in the village. The number of cows eventually increased to eight and produced more milk than the residents of Bassett needed. A $26 cream separator was purchased. The skim milk was fed to hogs, and the cream taken to the creamery. Engler was happy, for he hated the job of collecting. "Some of the customers were always short of money."

Engler started school at age five and then skipped a grade, so he was about two years younger than most of the freshmen when he started high school. His extracurricular activities in high school were limited because his father had 100 stocker cattle, whose feeding was his responsibility. (Stocker cattle are calves weaned but not ready for the feedlot.) That did not bother him, for he liked working with cattle, but he was unhappy that he was not paid a regular amount and had to ask his father for spending money.

His independent nature exhibited itself between his junior and senior years in high school when he purchased 100 head of 400-pound calves. He had worked at the sale barn and learned how to judge cattle. The sale barn owner knew the Engler family and lent him money for the cattle. Engler had them branded while still in the sale barn. When he told his mother what he had done, she said he would get a licking when his father got home. When Engler's father arrived, he immediately looked at the cattle and said that Engler had made a good buy, but the interest was too high. He agreed to cosign a note at the local bank and got a lower interest rate. Engler said: "I made good money on those cattle, which I bought for $16 per hundredweight and sold for $20. I had enough for one year of college."

Engler had planned to take the veterinary medicine course in college but was discouraged because so many returning veterans had priority. So he switched his major to animal husbandry. During his first year, he worked at an animal clinic, which required getting up at 5:00 A.M. to clean pens. After classes he washed dishes and waited on tables at a fraternity house for his meals. After his freshman year, he worked for a year on farms and ranches to make enough to continue his education. During his sophomore and junior years, he worked at the university hog barns, for which he received pay and had a "good room in one of the barns." Dr. Hanson, his mentor, often called him at night to help take blood samples and help with farrowing. He was married between his junior and senior years. Plans were for Engler's wife to work and support them both so he could take extra hours. This seemed necessary because in his junior year, he had taken law courses and fallen behind in his required courses. Those plans went awry when Mrs. Engler became pregnant, but everything worked out, and he graduated.

DURING the 1951–52 school year, Engler taught veterans' agriculture at Bassett. He had previously invested in an oil venture and used the money he made to purchase 150 cows from his wife's parents. He paid $240 for old cows and $300 for heifers. When his father came to see the herd, he said, "I thought I raised my son to be smarter." When the market dropped the following year, he sold the old cows as canners for $115 per head. He lost his oil profits and more. He liked teaching the veterans because his classes had such good discussions, but he could not tolerate the paperwork required by the government.

In the fall of 1952, Engler went to work for the Brickley Cattle Company at Valentine, Nebraska. Brickley was in partnership with Max Rosenstock, who had an order-buying firm at the Sioux City, Iowa, stockyards. Emmett Brickley "was the greatest country cattle trader—a real perfectionist and a super teacher. I was so fortunate to work with him. He could spot any defect on an animal." After a couple years, Engler realized that he did not want to be a cattle buyer for the rest of his life and let it be known that he was looking for another opportunity. Brickley recommended him to Rosenstock, but Engler felt the position offered him at the stockyards was a "dead end."

IN 1955 Engler learned that Lewis Dinklage, a renowned farmer-feeder at Wisner, Nebraska, was looking for someone with his qualifications. At that time, the Englers had two children and owned a home in Valentine, but Engler knew of Dinklage's great reputation in the industry and applied for the position. He was making $700 a month working for Brickley and planned to ask Dinklage for more. Dinklage was having some problems in one feedlot and did not come to his office until late in the day. After the interview, Dinklage offered to employ him for $300 a month. Engler was "thrown for a loop," but he wanted that job. Dinklage left the office. Engler spoke to the head cattle buyer, who advised him to take the job. He told Engler, "If you are worth more, you will get more very soon." Engler was not willing to risk moving his wife and son and daughter, so they remained in Valentine, and he rented a room in a funeral home at Wisner, in eastern Nebraska.

His first job for Dinklage was straightening out the books, "which were in horrible shape." The previous bookkeeper was also a cattle buyer who was supposed to take care of the records when he had the time, but he never had the time. After a couple weeks Engler asked if he could work on a farm after hours. He wanted to be outside, on a horse, and working with cattle. Once Dinklage recognized that Engler knew cattle, he suggested that he move his family to Wisner and live in one of his farmhouses. Soon Dinklage offered him a chance to feed cattle on one of his other farms. This was a new venture for Engler, but he was eager for the opportunity. The family moved to the Black Island farm, and Engler found himself in charge of farming a half-section of corn land and feeding about 1,000 cattle with a scoop shovel. At the same time, his bookkeeping job was still his chief responsibility.

The operation at Black Island went well, and soon Dinklage offered to help him finance a farm with a better house. He also offered him a partnership involving his elevators and feed mills at Beemer and Wisner. Engler rose to the challenge. He recalled: "Sometimes I went two or three days without taking my clothes off. Dinklage had no children, and he treated me like a son. I was doing great. We were really making good money." By then Engler was making about $1,000 a month but drew out only what he needed, leaving the profits in the partnership to build equity.

At the new location, which Engler now owned, the feedlot was increased to 8,000 head. Dinklage also had 13 other partners, with a combined total of 50,000 head on feed. His policy was to furnish the cattle and the feed and to have the partners provide the farms and the labor. If the partners were good managers, they did very well. Engler was among them. Then, Dinklage made a fatal mistake. In 1960 he sent Engler to Texas to buy cattle.

ENGLER was sent to Hereford, Texas, where Dinklage was well known, with instructions to purchase cattle at the stockyards there. He observed a train headed for California with half the cars filled with feeder cattle and the other half loaded with milo. He turned to Joe Reinauer, a local rancher, and asked why that was. Reinauer responded, "We are waiting for someone like you to come down here and show us how [to finish cattle]."

All the while he was driving back to Nebraska, Engler thought about what he had observed. He said, "I assumed that Dinklage would come down with me." Based on that assumption, he purchased land at Hereford for a feedyard. (For the remainder of this chapter, the term "feedyard" will be used instead of "feedlot." This is the term cattle feeders in Texas commonly use. Midwestern cattle feeders have traditionally used "feedlot." This apparently is a cultural difference, probably based on the fact that the commercial facilities in Texas are generally far larger than the traditional farmer feedlots of the Midwest.) But when Dinklage learned what Engler had done, he became very upset. Engler said, "I had done well with him and had about $200,000 equity in the partnership." When he and Dinklage settled their accounts, Dinklage was very disturbed because he did not like it that Engler was leaving. Engler recalled that after the family had left with the car and he and one son had the truck loaded with their belongings, they stopped where Dinklage was sorting cattle to say goodbye. Engler had to wait for over an hour before Dinklage approached, and then Dinklage would not shake hands. Engler commented: "I basically cried all the way to Texas because Dinklage had meant so much to me. But I could not back out at Hereford."

Engler was so impressed with the potential at Hereford that he put most of his money into the new venture. He was 55 percent owner. There were nine other investors, all of whom were local businesspeople. Several of them also had farming and ranching interests. A Small Business Administration loan was secured to complete the financing. When the yard was nearly finished, the only cattle committed were 120 head that Engler had on grass. He told his wife, "I believe I might have built my tombstone." His partners got cold feet when it came to putting cattle on feed. Then, Roger Brumley, a local rancher, stopped in and suggested that Engler should call Dinklage and ask him if he would buy Brumley's cattle and place them in the yard. When Engler called, "Dinklage said, 'Paul who?' But after a brief conversation, he drove all night to come to Hereford."

Dinklage purchased 700 head of two-year-old steers and asked to be shown around the pens. Then, he bought about another 3,300 head. When the news got around, Hereford Feedyard, Inc., was overbooked. Sometime later Dinklage turned to Engler and said, "You don't need me anymore." Soon the feedyard expanded, and the business was off and running.

Then, Engler "got backed into a deal" that involved a 140,000-acre ranch and 4,000 cows plus 10,000 stocker cattle. The ranch was on the Crow Reservation south of Hardin, Montana, along the Montana-Wyoming border. To put the business in order, Engler had to be on site, so the family lived there from 1966 through 1969. By 1969 the ranch operation was back in proper order, and a resident manager was in charge. In the meantime, Hereford Feedyard, Inc., merged with Producers Chemical Company on a 50-50 basis under the name ProChemco. In 1967 ProChemco acquired a 20,000-head feedlot at Pampa, Texas, and the 15,000-head Fartex Feedyard at Farwell. When Hereford Feedyard, Inc., went public with ProChemco, the public auditors said that it had to divest itself of Fartex. But the Pampa yard was owned outright by Hereford Feedyard, Inc., and remained in the company.

ENGLER'S success in the cattle business attracted the attention of Currier Holman, president, and others in Iowa Beef Processors (IBP), and they tried to recruit him. Engler liked his independence and was reluctant to give that up, but at the same time he was interested in doing what he could to make better connections with a packer. In 1972 he agreed to a 10-year contract with IBP on the condition that it would exchange his stock in ProChemco share per share for IBP stock. ProChemco was not doing well at the time, and its president was asked to resign. Engler was named president of that company and found himself "learning the oil business." As part of his employment with IBP, that firm requested that Engler remain on the ProChemco board, because it held a large block of that stock.

Because of his background in ranching, cattle buying, and cattle feeding, he was made Group Vice President of IBP's Carcass Division. This included procurement operations for the eight slaughter plants and the sale of beef carcasses, offal, and other by-products. Because of his sizeable holding in IBP, he was placed on its board, as was Lewis

Dinklage. The latter was quite uncomfortable seated with investment bankers but added a different dimension to the board.

Engler, who was used to running his "own show," soon found himself uneasy within the corporate structure. "It was just not my cup of tea. Had I started with Iowa Beef Producers out of college I probably would have made a career there." He asked to be relieved of his contract, and IBP agreed, providing he remain during the transition to his replacement. On August 16, 1975, he resigned but has continued excellent relations with the company. During his tenure with IBP, he initiated the establishment of the company plant at Amarillo, which, since the commencement of its operations, has purchased nearly all the production from six of the nine Engler yards.

AFTER Engler resigned from IBP but during the transition period, he purchased Cactus Feeders at Cactus, Texas. The $200,000 he had when he left Nebraska in 1960 had grown considerably by 1975. However, Cactus Feeders was much larger than any previous venture, so he needed a financial partner. Tom Dittmer, the owner of Refco, an agricultural commodities firm, was in a strong cash position and offered to come in on a 50-50 basis. As soon as Lewis Dinklage heard that Engler had purchased the Cactus yard, he called and offered to put in the first 3,500 cattle.

In 1975 the Cactus yard had a 40,000-head capacity. Later, that was increased to 75,000 head. The Cactus yard had been owned by the same people who also owned the yard at Stratford, Texas. Engler learned that they were in trouble because of their involvement with the "tax feeder deal" common during those years until the laws were changed. He sensed his opportunity and in 1977 purchased the 65,000-head Stratford yard and in 1979 the 40,000-head Frontier yard at Spearman. Stratford has since been expanded to a capacity of 80,000 head. In 1985 the 40,000-head Wrangler Feedyard at Tulia, Texas, was added to the growing empire and later expanded to a capacity of 50,000 head. Other yards were purchased during those years but were later sold, sometimes to their original owners.

From 1985 to 1990, Cactus Feeders was the largest cattle feeding company in the world. In 1990 Refco was expanding rapidly, and Dittmer needed cash. He sold his interest in the feedyards to Engler but retained his partnership in the ranches. This temporarily slowed the growth of Engler's operation.

By this time Cactus Feeders had exceeded the credit line of the Amarillo banks. When Engler purchased the Cactus yard, he had hoped John Hancock Insurance Company, which financed that yard, would continue to do so. However, Hancock decided that it no longer wanted to finance cattle feeders. Engler secured financing with First of Chicago, but when he wanted to buy the Frontier yard that bank, too, said that it wanted out of cattle financing. At that point Engler turned to Citibank of New York, with overlines with Rabo Bank, Credit Agricole, and Bank of America. This arrangement provided the funds to purchase Dittmer's interest. Between 1990 and 1993 Engler consolidated his position to prepare for his next expansion. He is a firm believer that one cannot stand still. In 1993 he was back in familiar territory when he purchased the 40,000-head Southwest Feedyard at Hereford. In 1996 he bought additional land at Perryton and constructed the completely new 60,000-head Wolf Creek facility.

In March 1999 an opportunity arose that he could not resist. Koch Industries had ambitions of establishing a totally integrated opera-

The four-story home office of Cactus Feeders in Amarillo, which houses 220 employees.

tion—ranches, feedyards, mills, and a beef processing facility. Unfortunately, its plan to do so came at the wrong time in the beef cycle, and Koch experienced huge losses. The company decided to divest itself of that venture. Beef was at the low point in the cycle, and Engler was ready for the opportunity. He purchased the 50,000-head Hale Center, Texas, feedyard, plus the 42,000-head yard at Syracuse and the 17,000-head yard at Ulysses, both in Kansas. For the first time he sold cattle to a packer other than IBP. The Kansas yards were closer to an Excel plant in Kansas, making it more economical to deliver there. With that purchase Engler surpassed Continental Grain and regained his position as the nation's largest cattle feeder.

On the average, every week more than 21,000 cattle must be brought into the 9 feedyards and more than 21,000 must be delivered to the packer.

The nine yards of Cactus Feeders had a one-time capacity of 460,000 head. The goal was to produce 2.4 turns of cattle annually. The three ranches containing 140,000 acres had 2,000 cows and at times up to 30,000 stocker calves for the feedyards. Nearly 500 employees were required to keep those facilities operating smoothly.

ENGLER never forgot the profit he made when he invested in oil stock in the 1950s. This gave him an affinity for oil and gas properties. But his involvement with ProChemco in the late 1960s tested his courage. About 1980 he and Dittmer bought a ranch at Roswell, New Mexico, within the Abao gas field, and the McFadden Ranch at Benjamin, Texas. Those properties were secured chiefly for their oil and gas potential, but they also served as a buffer to hold cattle on grass if the yards were full. However, they were not an essential part of the feeding business.

Shortly after those purchases, the Amistad Land and Cattle Company, a combination 10,000-acre ranch and 2,500-acre irrigated farm, was acquired at Amistad, New Mexico. Corn and alfalfa are grown there to feed cattle on the ranch. Because of its isolated location, the ranch is an ideal site to grow foundation milo seed for a major seed company. Amistad also serves as a place for stocker cattle until there is room in the feedyards. In 1996 Pablo Energy, Inc., another Engler venture, had six gas wells in production in the Texas Panhandle and drilled its first successful gas well in Hemphill County. In 1999 the Cactus board of directors voted to continue oil and gas exploration for another two years.

ENGLER has nerves of steel, but when cattle prices "were in the pits" for a prolonged period in the 1980s, even his desire to work independently was tested. From 1985 to 1987 he tried to organize the key feedyard operators who were members of the Texas Cattle Feeders Association into a marketing cooperative. About the time he thought they were ready to unite, prices improved. Everyone lost interest in doing something about pricing except Engler, who decided to approach IBP with his concepts. After he left IBP's employment in 1975, he remained on good terms with the company. Engler and IBP were comfortable in working out a mutually beneficial arrangement. Formula pricing was developed, and in August 1988 Cactus Feeders was the first major seller to market its cattle under the system.

The formula grid is designed to match quality with carcass weight. The ideal carcass is the most economical for the packer. The target is to have a choice or prime carcass, with a good yield grade, that weighs between 550 and 735 pounds. The top price is paid on carcasses that fall within those parameters. Outside that range, discounts prevail. The goal of the feedyard managers and the cattle marketers is to avoid the discounts. Cactus marketers are invited into the coolers to examine the carcasses on the rail. The history at Cactus is that most discounted carcasses result from scheduling problems. Sometimes they result from a genetic problem, a poor buying decision, or simply chronic never-do-well animals.

Cactus people want to avoid potential discounts, which is why every animal that enters a yard is personally observed by skilled individuals as it comes off the truck. The same holds true at selling time. The marketers are skilled enough to estimate where cattle will fall on the formula grid. The records can trace where the animals came from, the time on feed, and conditions during the feeding period. When a marketer decides that a pen of cattle is ready, the marketer gives the packer a two-week notice, along with the details of type of animals and number. During the first week, the Cactus marketer can amend the commitment but not during the second week. The IBP buyers do not go to the Cactus yards because they are so confident of the marketers' information.

The packer is as interested as the feeder in minimizing the highs and lows in the marketing system to keep the industry thriving. Many traditional producers are suspect of that notion and hold strongly to the mentality that cattle should be priced on a live-weight basis. Others argue that the seller is still a price taker, and some maintain that the cattle represent captive supplies. This is not true, for the animals are property of the seller until they are loaded onto the truck. Engler is a strong advocate of formula pricing, and his professionalism is carried to those who work with him to make it beneficial for everyone. However, he realizes that more must be done.

IN 1960, when Engler first arrived at Hereford, fewer than 200,000 beef animals were fed annually in the state. In 1997 more than 5 million cattle were fattened in Texas, and more than 700,000 of those came from Cactus Feeders. The state was surpassed only by Nebraska and Kansas in the number of cattle fed, thus ranking ahead of Iowa and Colorado. By 1995 large commercial feedyards had become so dominant that the state no longer counted yards that fed fewer than 1,000 head because they generated less than 0.3 percent of all production; however, they made up more than 80 percent of all yards. In

1998 Texas had 142 yards of 1,000 head or larger. Of that number, 47 yards with a capacity of 32,000 head or more fed 4.2 million cattle, 70 percent of the total. Texas was by far the largest producer of beef cattle, but a great percentage of them were exported to other states for finishing.

ENGLER'S experience as an employee made him aware of the concerns of the rank-and-file worker. As soon as he became the managing partner at Cactus, he instituted health insurance and profit sharing. While discussing the profit-sharing plan 25 years later, he happily commented that some of the original yard workers had more than $300,000 in their retirement accounts. In 1992 he took the next logical step and established an Employee Stock Ownership Plan (ESOP), which was reputed to be the first in the cattle industry. Upon origination of the ESOP, shares were given to all employees with six or more years of service. To make that possible, Cactus borrowed $1.43 million to establish the employees' trust account. After 1992 new employees became eligible after one year of employment, and in six years they were fully vested. In 1994, when the loan was paid off, the value per share had increased from $40.00 to $63.12. That year the profit-sharing plan and the ESOP contribution totaled 24.5 percent of payroll, only a half percent less than the maximum allowed by law.

Engler reminisced:

> I think the ESOP is one of the best things I have ever done, from both a personal and a company standpoint. During my experience of working for others and after getting the Hereford lot established, I did every job that needs to be done in this business, so I am not asking them [the employees] to do anything I haven't done. There is nothing I would like more than to have Cactus Feeders be known as a real people company. It [the ESOP] was designed to help employees of a private company gain shares [because] I don't believe there is any substitute for private ownership. When we put the plan into effect, I spoke to each employee individually and gave that person a stock certificate. I was in the Cactus yard, and a pen rider, about age 70, said, "I always wanted to own cattle, and now with this plan I do."

The ESOP has greatly reduced labor turnover and has helped [to improve] the work ethic and pride. The employees take better care of the equipment and facilities and keep things neater, all of which is very pleasing and rewarding to management.

One of the original company employees commented that although her work was stressful and she wanted to slow down, she did not want to quit. She added, "However, when I do, I have the satisfaction of knowing that I will have a good pension." That certainly is not true for the rank-and-file worker of most of the nation's smaller farms and feedyards.

Despite its good benefits program, probably one of the greatest challenges facing the company today is securing and retaining workers. Engler would like to be more selective in hiring and then improve the training program to avoid the loss of workers in the first 120 days. Because of the high percentage of Hispanic workers, the goal is to have an articulate bilingual worker at each yard devote a couple hours per day for two weeks to a month orienting every new employee. Cactus was one of the first cattle feeding companies to establish a position of employee development focusing on training and motivation.

Engler says that because of the risk involved, safety has to be a way of life. To achieve this, the company provides awards, ongoing seminars, and other incentives to accomplish its goal. Engler would like to have employees so aware of the need for safety that it becomes a "self policy" for everyone.

IN THE PAST, feeding cattle was not the exact science it is now. In the "good old days," once the cattle were on full feed, all the farm hand needed to know was how many scoop shovels of ground corn and how many buckets of protein to put into the bunk and how many pitchforks of hay to put into the fence-line manger. After checking that a block of salt was available and that the water tank was full, the job was done. The simplicity of feeding in those times carries its nostalgia, but the inefficiencies could not be tolerated today. Unfortunately, much of the perception about cattle feeding has not changed.

NOE TARANGO was born in Mexico in 1960. He attended school for six years and then spent most of his time working on his family's small farm. At age 18 he left for the United States and went directly to the Frontier Feedyard at Perryton, where he obtained his first job, washing the waterers and cleaning yards. After a few months Jack Rhoades, the yard manager, assigned Noe to clean-up work at the feed mill. Later, Noe was given the job of operating a loader to fill the hopper at the continuous-flow mill. This meant placing specific amounts of silage, chopped hay, and grain into the hopper. These ingredients were weighed, mixed, and transferred to the feed trucks. Next, Noe was promoted to operating the flaker rolls, which steam processed corn, milo, or wheat, whichever was the best buy at the time. He remained at that job for about seven years, after which he was assigned to feed preparation. This meant combining the grain, supplements, roughage, fat, hormones, salt, and protein into a total mixed ration. Every day the cattle were fed about 900,000 pounds in 3 feedings requiring 3 trucks that each hauled 10 loads.

By 1990 Noe had saved enough money to pursue the dream of running his own business. He returned to Mexico, where he purchased a limousine to "tour people around." He fought traffic in Mexico City for six years but did not do as well as he had at Cactus. Noe and his wife decided they wanted their two sons to learn English and eventually to get a college education. The best opportunity to accomplish that was for Noe to return to work at the Frontier yard. Rhoades had not forgotten him, and in rapid order Noe held positions as flaker operator, feed bunk inspector, and foreman of the feed delivery trucks. Then, Rhoades asked him to become the assistant mill manager at the Stratford Feedyard, where 11 people were employed in the mill. After he had performed well in that capacity, he was asked to interview for the mill manager's position at Frontier. The six-person crew at the Frontier mill is responsible for preparing feed for the 40,000 cattle in that yard. The promotion meant a good increase in pay for Noe, and he and his wife immediately purchased a home. His goal now is to become the best mill manager in the company. He wants to remain in Perryton because his family likes it there.

WHEN asked what he thought was the most practical sized feedyard, Engler replied that a yard had to have at least 30,000

head to justify a mill and auxiliary facilities. However, the experience has been that after 30,000 head, the economies of scale rise rapidly to about 60,000 head. After that, the economies continue to gain but less rapidly. The company has two yards of 80,000 head each but feels that beyond that, managerial ability and/or environmental problems, not economics, might be the determining factor. The company's computer network connects all nine feedyards to the central office. The centralized system relieves the feedyard managers of the burden of buying cattle, buying grain, and selling cattle, all of which are done at the corporate level. The Cactus philosophy is that the managers should be free to attend to onsite problems without having to be burdened with the above external duties.

Because the company is committed to maintaining full yards, the procurement of cattle is probably its greatest ongoing challenge. The one-time capacity of the nine yards is 460,000 head, and the goal is to have 2.4 turns of cattle annually. This means that more than 21,000 cattle must be bought and sold each week. Rhogene Easley, manager of cattle accounts, commented that the most cattle Cactus ever handled in one month was 104,000 head. This gives a good idea how smoothly the buying and selling operations take place throughout the year. One hundred commissioned agents report daily to get the market and instructions as to what type of cattle are needed and as to which of the feedyards they are to be delivered. The agents must check for clear title on cattle purchased and guarantee that the cattle are as represented. When the cattle arrive, company spotters observe each animal to verify that representation. If the cattle are not as represented, the agents are held responsible for the animals.

ENGLER, like Farr and Monfort, has spent much of his time working on equipment suitable for large-scale feedyards. In 1976, the second year of operation of the original Cactus Feeders, the company designed and manufactured feed-truck boxes capable of feeding 20,000 head of cattle a day. After Engler purchased yards in rapid succession in the early 1980s, the firm immediately started an in-house equipment fabrication program to build what best fit its needs and to have standardized equipment for all yards. A complete computer-controlled batch mill was introduced in 1983, and three years later the company designed and implemented a round steam cabinet to make flaking more efficient. The flaker was later improved and marketed by others.

In 1987 a completely engineered and computerized control design system was developed that provided for a continuous-flow system for feed preparation. Later, molded resin water tanks were designed to fit the company's needs. At the same time, manure handling technologies, automated cattle sorting facilities, and automated grain moisture control systems were perfected.

Each feedyard has a self-contained feed mill with a totally computerized processing, mixing, weighing, and loading-out system controlled by panel boards. Each truck also contains scales to weigh what is fed in respective pens.

A unique contribution from company shops is a feed-truck box with a kill switch that shuts down the operation of the feed box when the driver gets off the cab seat. This prevents the driver from becoming involved with any moving parts. To troubleshoot, a second person has to observe the working of the feeding mechanism while the driver remains on the truck seat. The company's innovations have been aimed at obtaining greater safety for the employees and the livestock as well as producing a more consistent ration to increase productivity and to lower costs.

Important as all of the above are, probably nothing has done more to enhance profitability than the in-house cattle health and feeding research program that started in 1985. By 1997 more than 60 separate trials had been conducted in what has become one of the premier research facilities in the industry. Research concerns the economic impact of poorly performing animals; animal behavior; beef tenderness; the effects of feeding various supplements; a vaccine to combat pasteurella; a total computer tracking system, from when the animal arrives until it is on the rail; and a quality-assurance program. In addition to its in-house programs, the company has been involved with ongoing research in international ventures in Argentina, Australia, Brazil, China, and South Africa.

WHEN Engler was asked to discuss the most difficult times he had experienced in his career, he answered without hesitation. It was apparent that he had learned a lesson from each occurrence.

He recalled that his financial loss on the 150 old cows and some heifers that he purchased from his wife's parents in the early 1950s completely wiped out what little assets he had accumulated to that date. "I had nothing when I sold most of those cows as canners."

His second low point came in 1960, after he had finished building the feedyard at Hereford and found the only cattle committed were 120 of his own. That was when he told his wife he thought he had built his tombstone. That time Lewis Dinklage turned the tide.

For about two decades, things went well. Engler experienced only a few skirmishes in buying feedyards that were having financial problems or in dealing with sellers who were less than honest. Occasionally he encountered minor problems related to his involvement with oil and gas drilling companies where he had to play a leadership role to rescue the ventures.

Then, in 1986 the dairy herd buyout set Cactus Feeders reeling, just as it did all beef feeders. "One day I was worth about $10 million less than I was the day before, not because of weather or the markets, but because of government action." Engler retold how the government had intimated that it would buy the cows and export the meat so the beef market would not be affected. But he soon learned otherwise, for he was acquainted with Dick Lyng, Secretary of Agriculture. Lyng informed him that "he [Lyng] was helpless." The American National Cattlemen's Association opposed the buyout and was told that the "buyout was a done deal." The cattle market dropped the limit in four out of the next five days because the market did not believe that "the government would export the meat." Like most entrepreneurs, Engler preferred to operate without government interference, and after the dairy herd buyout, he was more convinced than ever of his philosophy.

His most traumatic experience came in 1996, when he lost his wife to lung cancer. He said, "That was a day I will never forget." The cancer had been discovered only six months previously. Engler was 67 years old at the time. Those near to him related that up to that time he gave the impression that he felt invincible, because he had always overcome obstacles that were in his way. This time, however, he had no control.

After a prolonged silence, Engler commented on the cyclical nature of finance in agriculture, which he said is the same in any industry. Although financing has never been a major concern for him personally, he is aware that many smaller feeders have encountered problems when, for various reasons, their banks have decided to get out of agricultural lending. Engler's concern is for the long-run good of the industry, which is being impacted by the changes. Banks are consolidating, but as they increase in size, they are not always expanding their loan limits. Thus, in effect, they are reducing the available supply of funds.

ONE THING that was particularly noticeable was that Engler did not refer to mill fires, disease outbreaks, or accidents as some of his most difficult times. Those were events he expected as part of the routine in the cattle business and were happenings that his management people had to deal with. Handling them was their job, and the pressure was on them. They knew, however, that he was there to support them. He was not one to sweat the small stuff.

Those close to Engler describe him as an absolute workaholic "who gives 110 percent of himself" and expects others to do the same. But they were quick to add, "Once he hires you, he lets you alone. That's why you like working here and why those around you like it too. It's clear that he has a great compassion for people." Another colleague described him as a firm manager but not a micro-manager, who realizes that everyone, including himself, has physical limitations. "He wants to be thought of as a person who is a good people manager."

Another worker commented: "With him, service to others is a priority. We work in a comfortable office of intelligent people. . . . The industry is a casual one, but we approach things with real professionalism. With Paul Engler, integrity is the outstanding trait. . . . A handshake is still valid with Paul; it is as meaningful as a signed contract."

One person recalled that when he had applied at Cactus, he spoke to at least 10 people about working for Engler and that only one had given any encouragement. The negative arguments the others gave were exactly what were most intriguing. "I wanted to work for a tough businessperson who had a professional approach to the cattle business and for a company that had a future." He continued: "Engler's expectations were always high. He wanted you to be professional, but underneath you knew he was compassionate. . . . He was as good at communicating with the tank washers as he was with the investment bankers. He was a natural leader. . . . I doubt that there are many who could have done what he has done [in the industry]. He had the vision and the will."

WHEN asked what motivated him, Engler was stumped. After a long pause, he said, "I wish you hadn't asked that." He stalled while grasping for what he really wanted to say by recalling how his parents had reminded him that there was always a better way to do things. Then, after he had gathered his thoughts, he replied:

> I absolutely love cattle and love the challenge of the markets. That really pumps adrenalin. I wanted to see the cattle industry changed into a value-based marketing system, where the true value of the final product is recognized. Wanting to be the number one feeder is a big motivator, but it is more important to be number one in performance and profitability. When the markets are down and you are losing $100 a head, that is probably the time to buy more cattle. To be in the cattle business, you have to be a risk taker. Don't panic when you are losing money, and don't get too excited when you are making it. Money has never motivated me.

Then he recalled that he had purchased some of his feedyards because the previous owners had lost money or their courage. He would secure a yard when the market was down, and the first fill of cattle would pay for the facility. One of his colleagues remarked that Engler's cool attitude came with maturity. That is why he has taken the position of being a steady buyer and seller of cattle. "He has learned to manage risk."

WHEN questioned about what the future holds for him and the company, Engler made a statement from a prepared paper that indicated he was still thinking young: "It's not where we've been, but where we're headed that's exciting." Then he justified his reasons for expansion by stating that the company could not stand still, because if it did, it would lose many of its good people. There is no doubt that the managers of Cactus Feeders will continue to seek ways to expand. They will do so by acquisition or through alliances with entities either above or below them in the food chain. Internally, they probably will feed a greater percentage of the cattle in their yards than in the past, because Engler believes that is one of the easiest ways to expand. In Engler's mind, "custom feeding is a no brainer. . . . If you are a custom feeder, it's like running a motel with good occupancy."

He feels that the beef industry is overexpanded in the United States, so he has looked overseas for opportunities. Argentina appeared to offer the best choice for expansion based on its feed grain and cattle production. "We would like to think we have the potential to create another Cactus Feeders there like we have here." In October 1999, a feedyard was opened at Villa Mercedes, about 500 miles west of Buenos Aires, in a joint venture with Cresud. Miguel de Achaval, long-time employee of the company, manages that lot. Just as at Hereford in 1960, virtually no one is familiar with custom feeding, so the yard has difficulty securing cattle. Even though it appears that finishing cattle in Argentina should be cheaper than in the United States, cultural practices differ, and the lack of entrepreneurship appears to handicap progress.

ENGLER is a futurist and understands problems within the cattle industry that need to be overcome before a more consistent quality product will reach the market. Ranching is still the most fragmented sector of the industry, and that presents the biggest challenge to improving the product going into the feedlot and eventually to the consumer. It is expensive to buy 30 to 45 calves at a ranch and very costly from a trucking standpoint because at least 100 calves are needed to make a load. Many small-scale ranchers are not willing to upgrade their genetics, give the proper shots, or implant electronic ear tags that would enable tracing animals from the ranch to the carcass.

Engler would like to see a machine perfected that could measure tenderness in beef. This would provide one of the best ways to properly reward the original producer. If that happens and ranches are consolidated and become more business oriented, then it might be possible to have better working alliances with the primary producer and through the entire food chain. Progressive ranchers willing to form some sort of alliance with the feeder are the best hope of producing quality at less cost. Good genetic calves that are weaned at 650 pounds and that will finish at 1,150 pounds in 12 to 13 months will reduce the end cost of beef. When the producer, the feeder, and the packer can establish a good working relationship, the day may arrive when all can share the risk in a highly cyclical business. In the end, the consumer, as is usually the case, will benefit most.

IN DECEMBER 1999 Michael "Mike" Engler was named president of the company. His father continues as chairman of the board and chief operating officer. Mike, who earned his Ph.D. in biochemistry and did postdoctoral work in molecular biology, "was certain he would never come back [to the business]." He was being well paid for doing research with a large company and enjoyed what he was doing. Then, one day he was shifted to management and had to leave what he enjoyed. Sometime later his father called him to persuade him to come to work at Cactus Feeders. Mike decided that as long as he was immersed in management, he "might as well do it for Cactus." He went to work for the company in the fall of 1993. In April 2000 he reflected on that decision: "I know I made the right choice."

Cactus Feeders is poised to enter the new era in an industry that has long been known for its independence. Paul Engler liked his independence, but he realizes that a capital-intense business like cattle feeding needs to alter its traditional ways. He hopes that Mike will keep Cactus Feeders in the family.

Engler remembers the days when he was building the Hereford Feedyard. He said, "There was one six-week period when I never saw the children and we lived in the same house." But he can look back with great satisfaction to the abundance he has helped create for our society. He has not forgotten that he has been blessed by doing the work he enjoys. As in the past, he will share his talents in other ways. Lewis Dinklage would enjoy knowing what his $300-a-month former employee, who left Nebraska with $200,000 earned while working for him, had accomplished.

Bibliography

Interviews

Easley, Rhogene. Amarillo, TX, April 11, 2000.

Engler, Mark. Amarillo, TX, April 10, 2000.

Engler, Michael (Mike). Amarillo, TX, April 11–12, 2000.

Engler, Paul F. Amarillo, TX, April 10–12, June 6–8, 2000.

Farr, W. D. Greeley, CO, February 22–23, 1999.

Farr, W. Richard. Kohler, WI, June 10, 2000.

Knapp, Angela. Amarillo, TX, April 9, 2000.

Rhoades, Jack. Amarillo, TX, April 11, 2000.

Tarango, Noe. Perryton, TX, April 11, 2000.

Miscellaneous

AgWeek, April 5, 1999.

The Bulletin. An employee publication of Cactus Feeders. Various dates.

Engler résumé and application for the Cattle Business of the Century Award, 1997.

Miller, Bill. "Taking Aim at the Industry," *Beef Today*, June/July 1992, 9–11.

Smith, Linda H. "Cattle Feeding Visionary," *Beef Today*, March 1991, 14–15.

———. "Father of Formula Pricing," *Top Producer*, February 1997, 40–41.

Texas Agricultural Statistics 1998. USDA and Texas Department of Agriculture, Austin, 1999.

Jack and John Harris

Entrepreneurship and Managed Diversity

In 1992 my wife and I were taken to the Harris Ranch Inn & Restaurant near Coalinga, California. After lunch we made a brief tour of the farm. I knew at once that Harris Farms had to be included in my book on entrepreneurs of agriculture. Seven years later my interviews with the Harris people were very rewarding. The Harris Farms and personnel lived up to all the expectations I had about them.

Within minutes after I sat down for my first breakfast at the Harris Ranch Inn & Restaurant, I realized that my week with John Harris and his staff would be pleasant and informative. I told Ronnie, who served my food, that I would be there for the week working with Mr. Harris. With great enthusiasm Ronnie proceeded to tell me what wonderful people John and Carole Harris are. She said, "I have been a waitress for 35 years. The last 19 have been here, and I wouldn't want to change. Mr. Harris is just one of us. He makes us feel so at home, and we have such good benefit programs." Then she told about all the good things the company does for its employees and the community.

— Hiram M. Drache

HARRIS FARMS is a large diversified operation in Fresno County, the nation's leading agricultural county. In 1997 Fresno County produced $2.772 billion worth of agricultural products, of which $2.116 billion was crops. Livestock and poultry made up the other $656 million. Fresno County yielded 12.0 percent of the $23.032 billion worth of agricultural products raised in California that year and

111

1.4 percent of the $196.864 billion worth produced in the nation. The 1997 average income per acre of harvested land for Fresno County, most of which was irrigated, was $2,216.48.

During most of the 1990s, Fresno County was the nation's leading cotton producing county and had 7.7 percent of all orchard acreage. Its 73,000 acres of tomatoes made it the top tomato county, with 18 percent of the nation's total acreage, and its 3.6 million pounds of grapes made it the runaway grape county, with 33 percent of the total U.S. crops.

JAMES ALEXANDER HARRIS was born in Mississippi in 1872. In adult life he moved to Texas and became involved with a cotton gin at Stanton. In 1918 James and Kate Harris arrived in the Imperial Valley of California, where farmers were starting to produce cotton on irrigated land. Harris established one of the state's first cotton gins and became involved in other facets of the newly developing cotton business. During the next few years, he built several gins and helped finance cotton farmers. In 1925, after the Alamo River dried up and cotton production failed, James and Kate Harris, along with their son, Jack A., born in 1914, moved to the San Joaquin Valley and again started raising cotton and operating a gin.

By 1937 they were established in the thriving oil town of Coalinga, California. In the meantime, Jack A. and his wife, Teresa, had started a trucking business and were looking for opportunities in farming. They located 320 acres of virgin land at nearby Five Points that, up to then, had been used only for sheep grazing. Jack and Teresa leased the land from the Southern Pacific Railroad Company and secured financing from a pump company for a well and pump to irrigate the land. Soon they rented another 320 acres. James Harris was a very patient person who needed all his assets to operate his own land, but he realized that his son was a go-getter and very risk-oriented, so he farmed with Jack and gave him moral support.

The timing was right for Jack and Teresa Harris. Next, they were approached by another landowner who wanted to develop 320 acres for irrigation and crop production. They received some help from Teresa's parents and got financing from a cotton ginning company. Now they had a full section of land that was large enough to provide them with more than a subsistence living. The economy of the late 1930s was improving, and the 1940s proved to be a boom decade except for the fact that it was difficult to get labor. Fortunately, Jack Harris secured some relief from German prisoners of war, who were camped nearby.

In 1940 Jack Harris purchased his first half section of land for $100 an acre on a contract-for-deed, but he continued to rent as much land as he could profitably operate. His major crops were cotton and cantaloupe, with small grain, including flax, to complete the rotation. There were no restrictions on water use, but water was the limiting factor to production. His 12-inch well, powered by a 200-horsepower pump, provided only enough water for about one-third of the land at a time. This meant that on a section of land, about 200 acres of cotton plus some cantaloupe grown in the summer season and 25 acres of small grain grown in the winter season could be irrigated each year. The rest of the land was fallowed.

Originally, wells were drilled from 1,500 to 2,000 feet deep, but as more wells were dug and pumped, the water table fell. This necessitated drilling deeper wells, which added to the cost of both drilling and pumping. Other problems arose. Boron and salt in the water limited the crops that could be grown.

Even though cotton required a great deal of hard-to-secure labor, it was still the mainstay of local agriculture. The average cotton picker could pick only about 250 pounds a day. This meant that 12 pickers were needed daily to harvest an acre of cotton, which produced about 3,000 pounds of seed cotton necessary to secure about 1,000 pounds of lint. The going rate for harvesting seed cotton was 4¢ per pound, which was a fairly good wage at the time. This meant a harvest labor cost of $120 an acre to the producer. However, cotton was still the most profitable crop. Fortunately, the climate permitted the harvest season to last from September into March. Although the quality of the cotton deteriorated over the season, the lengthy harvest period made it possible to cope with the labor shortage by keeping a limited crew for a longer time.

During the late 1940s and early 1950s, Jack Harris gradually increased his operations to about 5,000 acres. He was becoming one of the major producers in the valley. Large farms dominated the area for two reasons: it was costly to secure water, and the Southern

Pacific, which still held much of the land, preferred dealing in larger parcels.

IN 1951 Jack Harris purchased several thousand acres near Chandler, Arizona, and another "couple thousand" farther south near Sahuarita. Both farms were irrigated, and Harris was interested in further developing them. In 1953 he sold his California holdings near Five Points to an eastern investor. Unfortunately for the buyer, the worst year in that era for California agriculture occurred. Ironically, this was a boon to Harris, for a few years later the land reverted to him and he retained the sizeable down payment.

In 1958 Jack Harris purchased 80,000 acres of irrigated land at Gila Bend, Arizona, which had been owned for several decades by the Gillespie family. He formed a corporation with the Giffens, who were neighboring farmers in California, and several nonagricultural investors. The operation did not work well, partially because it lacked a cotton allotment and there were no profitable alternative crops. To improve chances of making a profit, Harris moved a cotton gin from California to Gila Bend and grew cotton outside the government program. He received a considerable amount of publicity for that action.

In 1956 Jack Harris built a feedlot on his Sahuarita, Arizona, property and became "hooked" on feeding cattle. He partnered with a Mr. Spitalny, who knew cattle and did most of the marketing. Harris liked cattle feeding more than crop farming because of the greater risks involved. In the meantime, the operation at Gila Bend continued to be a problem. Cotton and wheat did not work, so a feedlot was added to utilize the grain and alfalfa produced there. When that did not work to everyone's satisfaction, the corporation broke up, and Jack Harris took over the feedlot.

By then he had become involved with Cliff Douglas, who became the key person in the Harris ventures in Arizona. Harris continued feeding cattle at Gila Bend and Sahuarita and kept his eyes and ears open for other ventures. He partnered with Douglas and purchased about four sections at Queen Creek, Arizona, east of Phoenix near Williams Air Force Base. The original intent was to produce cotton and nursery stock. The system did not work well, but Douglas turned a "lemon into lemonade." He developed Arid Zoned Trees, a nursery operation that specialized in desert horticulture, providing trees native to the area that were low water users. The business boomed as metropolitan Phoenix continued to grow. The enterprise also profited from the sale of some of the land for development.

JOHN C. HARRIS, who was born in 1943, recalled that his father, Jack, "was a good person for getting into partnerships. He was good at identifying people he could work with." That trait stood him well in the 1950s, when he was forced to divide his time between his Arizona and California interests. Harris Farms continued some of the Arizona operations and added other farms in California that were distant from the main operations at Five Points and Coalinga.

In 1964 Jack Harris quit the feeding operation at Gila Bend and shifted back to Coalinga. He chose California because he felt it made more sense to feed cattle there. The population base was considerably larger, and by that time he was farming far more land in California than in Arizona. He and a partner built a 15,000-head feedlot at Coalinga, but he soon purchased his partner's interest. In 1968, when John joined his father on the farm, Jack Harris turned the cropping operations over to him. At that point Jack rapidly expanded the feedlot to 60,000 head and set a goal of finishing 150,000 head per year.

To minimize the risk of feeding such a large number of cattle and to level out the swings in the cattle business, Jack decided to go one step further and become a processor. His search for a packing plant ended when he purchased the nearly bankrupt San Jose Meat Company. The long-time relationship of Harris Farms with Bank of America proved very beneficial, because the bank understood the concept of added value and supported the step into processing.

Par Kamangar, a native of Iran with a degree from Oregon State University, became the manager of San Jose Meat and did an excellent job. However, that was not enough to overcome the plant's limitations of an old, small facility and high labor costs. The plant had a capacity of only 250 head per day, which was less than half the cattle that the feedlot finished. But the experience convinced Jack Harris of the concept of being an added-value processor.

ARMERS in the San Joaquin Valley and others concerned with the long-range livelihood of the area realized that more water had to be made available for agriculture, industry, and the cities. In 1965 legislation was passed for an off-stream project (commonly called the San Luis project) that would hold 2 million acre feet of water to meet some of the water needs of the area. Jack Harris sensed what that project meant for him. He was one of the first individuals to sign a recordable contract under the provisions of the Westland Water District, by which he sold land to individuals and then leased it back so he could maintain an economically viable unit.

In 1969 Jack Harris again owned about 5,000 acres in the Coalinga area. Up to that time the Southern Pacific, which still had "a paranoia about selling land because it might be worth more some day," suddenly decided to sell. The 960-acre limitation on water, like earlier limitations, was necessary for political reasons, but that acreage was not large enough to satisfy the economies of scale of contemporary farming. To maintain and enlarge his farm, Jack Harris financed many of his employees to purchase half sections and then lease them back to him. Most of that land was purchased from the Southern Pacific at $800 to $900 an acre. The employees were assured of a return on their purchase as well as a good growth in equity.

In 1965, in anticipation of more water, Jack Harris became one of the first tomato growers in the area. He knew Paul Davis, chairman of F.M.C. Company, which was developing a mechanical tomato harvester. The machine removed a major impediment to large-scale tomato production and at the same time greatly reduced cost. This provided an opportunity for profitable production of that crop. Harris Farms started with a 200-acre contract.

In 1965 John Harris married Carole Glotz, who came from a family of farmers and cattle feeders. Carole understood and was supportive of farming. She preferred rural living, so in 1968 John and Carole moved to the main house on Harris Farms. With John on the farm and the knowledge that more water would be available in 1969 from the San Luis project, the Harrises decided to acquire land as quickly as it became available and they could finance it. The 1960s closed on a very upbeat note.

W HEN the 1970s arrived, Harris Farms was well positioned in its cropping, cattle feeding, beef processing, and other ventures to face the exciting decade ahead. It had 18,000 acres of irrigated land, much of it rented from the Southern Pacific on 5- to 10-year net leases. The tenant was obligated to put in the irrigation system, but Jack Harris was comfortable with that arrangement based on his long experience of leasing from the railroad. In the early 1970s, after the purchase of a few sections from the Giffens, the main operation of Harris Farms reached 20,000 acres.

In 1970 David E. Wood, a rancher's son who had just graduated from college with a major in animal science, was employed to work in the feedlot. He started as a cowboy but quickly worked up the ranks to a management position. In 1974 Don Devine, a CPA who had done independent audits for Harris Farms, was asked by Jack Harris if he would like to come to work for him. Devine had enjoyed doing audits for the company. He said:

> Jack Harris was an aggressive entrepreneur, and I saw a real future there. I was impressed with the company because it was so vibrant. Jack and John were a good team—Jack was aggressive, and John was innovative and he was very insightful and analytical.

Devine added that during the early 1970s the Harrises purchased "lots of operations." Jack was eager to buy at every chance and "would have owned the world if he could have gotten the financing. John looked at things more long term."

L ITTLE did anyone realize when the 1970s started that agriculture would experience one of its most profitable decades. The timing was right for the father-son team combining experience and youth and building on a solid foundation. Tomatoes proved to be very profitable, and the newly developed harvester worked well, so the Harrises rapidly expanded their contract. Their long-term average yield on tomatoes was 38 tons per acre, but they hit as high as 60 tons. Fortunately, they did not have to rely on hand-pickers to harvest a ton per person per day, for the mechanical harvester worked at least 150 times faster. The price swings varied from $30 to $60 a ton, but tomatoes nearly always returned a reasonable to excellent profit. By the 1990s Harris Farms was growing 4,500 acres of tomatoes, and tomatoes were one of its largest crop enterprises.

To enhance their profitability and ensure themselves of a market with tomatoes, the Harrises became a one-sixth partner in the Los Gatos Tomato Product Company. Los Gatos processed 850,000 tons of tomatoes annually and gave Harris Farms another means to diversify. Harris Farms also joined Tri Valley Growers, a large cooperative that processed tomatoes, peaches, and pears. The experience with Tri Valley was not as favorable as with Los Gatos and will be noted later.

Cotton was steadily losing favor with the Harrises. John Harris felt cotton's decline was caused by the development of synthetics and by "too much subsidization." To be profitable, about 40 percent of the national cotton crop must be exported, and California must rely on exporting 80 percent of its production. In any case, the return on investment in cotton production was steadily dropping. As the Harrises decreased their cotton acreage, it became less economical to operate their gin, which had long been a backbone of the company. They sold the gin to Huron Ginning Company, which handled about 40,000 bales annually.

In 1973, to offset the reduction of cotton acres, Harris Farms began growing almonds. It started with 300 acres, and before the decade was over, it had 1,000; by the 1990s it had 3,000 acres. Almonds have been a steady income producer, ranking with tomatoes as Harris Farms' best large-acreage ventures. Since the 1970s the price of almonds has varied from $0.60 to $2.00 a pound. In the early years the price was from $1.50 to $2.00 a pound, which encouraged rapid expansion in acres and caused prices to decline. Variability in yield and fluctuation in price have caused swings in gross income from $1,500 to $7,500 per acre.

By partnering with the Woolf family (neighboring farmers) in an almond processing plant, Harris Farms has again cushioned itself from the wild gyrations in the market. The plant processes 15 million pounds annually, and processing has become a major activity for the partners. When not busy with its own almonds, the plant custom processes almonds for other area growers. The crushed almond hulls are sold to local dairy farmers for roughage. As they are to producers of many farm products, foreign markets are important to almond growers, for historically a higher proportion of the almond crop is exported than any other farm commodity. Nationwide, exports take 60 to 80

A 1999 aerial view of part of the 3,000-acre almond grove, with the horse-farm buildings within the trees and the race track at the right.

percent of the total crop. In the case of the Harris-Woolf plant, about 90 percent is exported to Germany and other European countries.

Cotton, almonds, and tomatoes are the three major crops on Harris Farms, but the Harrises also grow lettuce, broccoli, onions, garlic, carrots, a host of other vegetable crops, and melons. Most of the lesser crops are grown under contract or are joint ventured with vegetable packers, who are in the market 12 months a year and provide the Harrises with better market access than they could gain on their own. In recent years Harris Farms has grown as many as 33 varieties of vegetables, fruit, and nut crops at one of its four locations.

ONCE Jack Harris was relieved of being in charge of cropping by his son, John, he was free to concentrate on his real love, "where the action was"—feeding cattle. Early in the 1970s he became involved in a feedlot and packing facility in Iran, but that venture did not sidetrack him from the feedlot at Coalinga. During those years he also purchased a feedlot at Blackfoot, Idaho. Like the Iran unit, this was a freestanding unit and had no bearing on the California operation.

By 1975 the feedlot near Coalinga had grown to a 100,000-head–one-time capacity, which meant that, depending on the "in" weight, the facility was capable of finishing up to 300,000 cattle annually. The Harris Feeding Company was the largest cattle feeder, not only in California, but on the West Coast, and the fourteenth largest in the nation. It fed about one-third of all cattle fed in California and was the sole source of animals for the Harris Ranch Beef Company.

Because of the inadequacy of the San Jose plant to handle the increased volume of cattle, Jack Harris considered the possibility of building a plant in Fresno County in the vicinity of the feedlot. Plans changed in 1975, when he learned that the Diamond Meat Company, near Selma, California, was for sale. The plant, which handled only carcass beef, had a daily capacity of 750 head and was near enough to the feedlot to be economical. Therefore, the Harrises purchased it. The San Jose plant remained in operation during the transition period.

John Harris commented on their outlook at the time:

> If you are going to be in the cattle business in California at our level, you have to be in the packing business. Probably the ROI [return on investment] in the [cattle] market during the last years has not been the best, but once you have the critical mass in farming, you are cushioned and have to keep going.

Since the early 1970s Jack and John Harris, later joined by Dave Wood, chairman of the Harris Farms Beef Division, have attempted to solve the problems of inconsistency and quality plaguing the beef industry. This writer believes that the extremely fragmented beef industry, with its many independent ranchers, has been slow in responding to changing demands of packers and consumers asking for a leaner and more consistent product. Jack Harris, Dave Wood, and particularly John Harris agreed that "quality should come ahead of volume," and they developed a branded beef program. Dave Wood, when asked to state his greatest challenges, replied, "Producing a consistent-quality beef product that is acceptable to our consumers and meeting all of today's food safety issues." In those respects the Harris operation is among the leaders in the industry. In 1975 the Harris Ranch Beef Company was established to produce beef under the Harris Ranch label. The beef produced was certified under the USDA Residue Avoidance Program.

IN 1972 Interstate 5 was built through the area adjacent to the Harris Farms property. The Harrises quickly realized the volume of traffic on that highway and in 1976 established a Texaco full-service station at the Coalinga exit. Business exceeded expectations, and in 1977, 63-year-old Jack Harris and his second wife, Ann, built an upscale restaurant/dining room and coffee-shop complex adjacent to the truck-stop service station. Financing for such a facility in the countryside was difficult to obtain at that time, but, fortunately, Harris Farms made good money during the 1970s and provided the millions required to construct the elegant structure.

Jack Harris (left) and John Harris (right) at the construction site of the Harris Ranch Inn & Restaurant along I-5 near Coalinga.

Jack and Ann Harris envisioned that the combination restaurant and coffee shop would have 700 to 800 customers per day. To everyone's surprise the restaurant business "took off." Why? Was it because of its location quite distant from any other stops, or was it because of its uniqueness of being within the boundaries of the farm and near the feedlot from which some of the fruit, nuts, vegetables, and beef came? The Harrises intended to prepare most of the food from scratch and to control quality and wanted to use as many of their own products as possible. The restaurant gave them an unexpected opportunity to sell their own branded beef, which was displayed in the "country store" housed in the same building as the restaurant. That, too, proved to be a success. Before the decade ended, a second station, which handled Shell products, and a convenience store were built adjacent to the Texaco station and the restaurant.

At the same time, John Harris searched for additional opportunities in farming. He succeeded in purchasing a 700-acre wine grape operation near Lodi that needed renovation. In succeeding years the vineyard was enlarged, and it has proved to be a worthwhile venture. Rather than battle the market in the wine industry, which is extremely difficult to penetrate, Harris entered into an agreement with Mondavi-Woodbridge for processing and bottling.

The final acquisition of the 1970s came with the purchase of 1,000 acres of rice land in Glenn and Butte counties in northern California. Although that property is leased out, it is part of the diversification program lending overall stability to the company.

WITH the advent of the 1980s, Harris Farms was on a roll, and it appeared bigger things were about to happen. However, suddenly in January 1981, Jack Harris, age 67, died of a heart attack. He was the spiritual leader of the company, but it was not long before everyone recognized that his 38-year-old son, John, had been well groomed for the responsibility that he had to assume. Father and son had worked together for 13 years, and John had a good understanding of the challenge before him. What he could not foresee were events beyond his control that would test his management ability. Fortunately, John Harris had two tried and tested associates in Dave Wood, chairman of the beef division, and Don Devine, chief financial officer, who served with him on the executive committee.

Wood was now purchasing about 200,000 cattle annually, in addition to custom feeding cattle for partnering ranchers. He also oversaw a crew of 225 to care for the cattle and the facilities. His duties included securing and processing 375,000 tons of grain and 50,000 tons of hay. More than 100 trucks were needed to transport the feed as well as to deliver 4,400 loads of cattle annually to the feedlot and later ship them to the plant. Because much of the grain was shipped in by unitrain to Hanford, 17,000 semi loads had to be trucked about 40 miles to the feedlot. Nearly 200,000 tons of manure was scraped by 10 loaders and mounded in the yard or hauled to compost piles or fields.

John Harris immediately set to work to fine-tune the company. He sold the feedlot at Blackfoot, Idaho, closed the San Jose meat plant, consolidated sales forces, and slowly reduced the main farm at Five Points to 15,000 acres, at the same time further diversifying the cropping program there. Harris was more interested in quality than quantity. He wanted to get into the branded beef program and realized that Diamond Meat was not large enough to compete with the major packers. Therefore, he decided to establish a premier branded product and find a niche market. In 1982 the name Diamond Meat was dropped from the packing facility and products and was replaced with Harris Ranch Beef Company.

The early 1980s were a difficult time in the beef business, and the situation was especially serious in 1986, when the federal dairy buyout program caused the beef market "to drop like a bomb." It was the only year in the history of the company that it lost money. Problems were compounded from 1986 to 1992, when the state of California experienced a widespread drought that forced authorities to reduce the amount of water available to agriculture. This was part of the reason for the reduction in the number of acres operated at Five Points.

Dave Wood spearheaded the Harris Farms' effort on behalf of the beef checkoff program, which cost the company $1 per head but appeared to be one of the best ways to improve business and to avoid getting government involvement in the industry. At the same time Harris Farms started a Partnership for Quality Program, which involved a group of progressive ranchers who were interested in providing tested calves for the Harris premier beef program. Such a program was seriously needed in the beef industry to improve uniformity and quality of product.

But the 1980s were not all gloom and doom. George Hunt, a brother-in-law of John Harris, became a partner in the closed-down San Jose meat plant and developed it into a self-storage facility. It became a good money maker, providing another buffer to the risks of farming. At the same time, the search for land continued, and in 1984 Harris Farms purchased 5,000 acres of range land from long-time neighbors the Giffens. Located about 15 miles east of Fresno, the land was used as a holding pasture for range cattle to be placed in the feedlot and to carry the 500-head cow herd owned by the farm. The purchase was part of the Kings River unit and was the key to securing additional land from the Giffens at a later date.

John Harris's stepmother, Ann, operated the Harris Ranch Inn & Restaurant for a few years after Jack Harris died, but John had the ultimate responsibility. The business continued to thrive, so it was decided to enter the hotel business to make the total complex more profitable. Several million additional dollars were invested to build a 123-room inn, to increase the size of the three restaurants so they could handle conferences and banquets, to add a bar and a country store, and to create a private landing strip for fly-in customers. The facility expanded until 1986-87, when the restaurants served about 500,000 patrons annually. Business continued to grow, and by the late 1990s more than 2,000 patrons were served daily. In 1999 a $10 million expansion program was announced that included a middle-price-range motel, a fast-food restaurant, and facilities for RVs. John Harris commented that the company is approached frequently by parties that would like to have the Harris Ranch Inn & Restaurant duplicated elsewhere. He is hesitant to do so because of the large amount of capital involved and the intensive management necessary.

In 1984 Erick Johnson met John Harris while working for a U.S. senator. Johnson soon joined Harris Farms as a computer operator to help the controller make analyses of various projects. Later, he worked at Harris Ranch Inn & Restaurant as a floor manager trainee. After nine months he was put to work on computer projects that involved special programs and budgets. One of the projects concerned partnering with Getty Agri-Business and involved wine grapes, walnuts, and other crops not grown by Harris. The total project was for about $120 million, but the lender backed out, which proved fortunate for all parties.

Guest swimming pool in front of the 123-room upscale motel located adjacent to the restaurant and conference center complex at the Harris Ranch Inn & Restaurant.

In 1986 Johnson asked Harris about returning to college for a masters in business administration (MBA). Harris encouraged him, with the understanding he had a position with the company when he finished. When Johnson returned after earning his MBA, he was assigned to the packing plant, where he worked at several different jobs, including director of sales. He later became the first dedicated purchasing manager of the plant.

WHEN John Harris was asked what the biggest challenge of his farming career was, he did not refer to any of the previously mentioned difficulties that probably would have crushed an individual with a weaker constitution. Instead, he referred to his battle with the Internal Revenue Service (IRS) while settling his father's estate. Jack Harris had started some estate planning by turning over much of the common stock in Harris Farms to his son, but he retained the preferred (voting) stock. The IRS maintained that the transfer of stock had not taken place long enough before Jack's death to satisfy its guidelines. The IRS wanted to tax the estate at a 70 percent rate, but Harris contested the IRS on the appraised value. Once the proper valuation was established, he was given 15 years under the special family farmer transfer legislation to pay the inheritance tax, with interest.

Harris closed his comments about his toughest battle by contrasting it with his father's biggest challenge of trying to raise money to operate. He said, "Dad was really leveraged, but that bothered him little. As it turned out, except for the first year, we had a series of good years that enabled us to pay off the IRS." Harris Farms, like all family farms, had to face the inevitable of coping with the IRS when generational transfer occurs.

JOHN HARRIS entered the 1990s bent on diversification and on continuing his program of acquiring more land. He partnered with Donald Valpredo, and each purchased 2,000 acres in Kern County, near Bakersfield. Vegetables are the major crop, particularly carrots, lettuce, and tomatoes. Much of that land is double-cropped, giving it a larger gross than would otherwise be possible from the number of acres involved.

In 1992, when the Giffen estate was being settled, Harris Farms purchased another 1,700 acres of citrus orchard, 300 acres of irrigated pasture, and the magnificent Giffen home. The acreage is the remaining parcel of the Giffen land that adjoined the 5,000 acres purchased in 1984, called the Kings River unit. The orchard has three varieties of oranges, two varieties of grapes, and grapefruit. The irrigated pasture is used chiefly for horses from the horse division but at times may be used for the herd of 500 beef brood cows. Rod Radtke, manager of the Kings River unit, oversees a year-round staff of 28, which expands to 300 in peak season. Carole and John Harris have made the Kings River unit their home since 1992. Harris has a secretary and a fully equipped office at Kings River, but by using his plane he has an easy commute to the main office at the Five Points farm.

HARRIS FARMS has its own research and development (R&D) department and, like most food producers and processors, is trying to keep pace with the changing desires of consumers. Harris commented that people want three things when they consider food—nutrition, taste, and quick preparation. Consistency is a fourth trait. The R&D department considers those factors when determining the best way for the packing facility to provide grocery stores with products that can compete with those available at fast-food restaurants. The first Harris-branded beef product was released in 1982. Finally, in 1993, the fully cooked tri-tip roast (part of the sirloin) entered the market and became an immediate hit. Since that breakthrough, fully cooked pot roast, beef stew, Swiss steak, and barbecued short ribs have been developed.

Most of the cuts involved, such as chucks, rounds, and briskets, were undervalued in the market and took considerable time to prepare. By further processing those cuts, the packing plant improved its profitability. The key was to pre-cook them, make them tasty, and make them quick and easy for the consumer. A standard 2-pound roast could be on the table in 10 minutes. The company did most of the preparation for the consumer and, by packaging the cuts, made it easy for stores to display and sell them. Harris customized pre-cooked meat products were more costly to produce, but once customers experienced superior products and service, they willingly paid the difference.

One of the most intensive and prolonged R&D projects involved the use of vitamin E in the cattle ration. Although vitamin E added cost to the ration, it not only helped produce a better-appearing product, but it also made the meat stay red longer. The company feed mill, which prepares 75 tons of total mixed ration per hour for the feedlot, is the final step in the complete integration enabling the company to make this important product differential. Under the leadership of Dave Wood, the beef division has been very innovative. By January 1990 the company had about 100 ranchers "who want to do things right" signed up to produce 32,000 cattle annually for the feedlot.

A genetic consultant encouraged the use of Angus breeding stock because that breed has the largest data base available. The American Angus Association's record program provided the best chance to offer predictability of uniform quality for the end product. The ranchers are paid a premium to come into the program using the proper genetics. Additional incentives are offered for following the vaccination program, for delivering cattle in the off-season (when other cattle are not available), and for following a post-weaning program. There are also grade and yield premiums. The ranchers have four years to phase into the program. This allows them a normal turnover of bulls. Ranchers may retain ownership through the feedlot to the packer if they desire. The beef industry is in a period of consolidation, and the partnership of rancher, feeder, and packer is part of the necessary trend.

Because the feedlot needs at least 200,000 cattle annually, it will be some time before all the cattle it requires will come from ranchers willing to join the genetics program and partner with the company. California range cattle are available from April through June. Harris Feeding Company buys about one-third of all the cattle grown in the state; the remainder come chiefly from Arizona, Nevada, and Oregon. The Harris Ranch Beef Company has a daily capacity of 750 cattle, which are delivered in 17 loads throughout the day as needed by the plant. The excess cattle from the feedlot are sold to two other packers. When the demand for Harris Ranch–labeled beef increases, the packing plant will be expanded to handle 1,000 animals per day. Sales in 1999 reached $265 million, and company leaders expect sales to increase another $65 million by 2004.

Harris Ranch–labeled products are delivered by the company's fleet of 45 trucks, mostly to small chains of 10 to 30 stores. The largest customer is Nob Hill Market, a chain of 50 stores. This fits the packing facility's program of store-door delivery, because the operation is not large enough to fill the warehouses of the big chains. To date, Harris has sold to 1,200 retail outlets in Alaska, California, Nevada, Oregon, Washington, and the Pacific Rim.

Harris Feeding Company has had a longstanding business arrangement with Japanese customers. The company produces an American version of Kobe beef considerably cheaper than the Japanese can produce their own product. The biggest challenge has been overcoming Japan's tariff, which, after recent reductions, is still 40 percent. The Japanese have consistently purchased about 8 percent of the plant's output and did so even through the economic downturn of the 1990s. The Japanese have been reliable and good customers.

Harris Ranch Inn & Restaurant takes about 1 percent of the daily beef output and is therefore able to keep a close tab on the product coming from the plant. Harris Ranch Online Country Store markets a considerable volume of products directly to the consumer. Online buyers, like all other Harris customers, buy with the assurance that if any product is not satisfactory, it will be replaced.

In 1998 Harris Feeding Company and Harris Ranch Beef Company were recognized by the National Cattlemen's Beef Association (NCBA) as leaders in new product development with their fully cooked, ready-to-heat-and-eat beef. David Wood and John Harris shared the honors. The Harris Ranch Beef Company received a $250,000 award for its fully cooked pot roast in natural gravy. Harris money was used to split the cost of a national beef promotion campaign with the NCBA. In 1999 sales of the added-value products increased 100 percent more than 1998.

IN 1994 Larry Chrisco, who holds a degree in animal science and agronomy and who had spent 19 years as farm manager on a 23,000-acre unit of intensive cropping for the J. G. Boswell Company, was employed as farm manager by Harris Farms. He is responsible for 225 permanent employees, including 25 in the office, and for up to 1,500 seasonal workers to plant and harvest the 15,000 acres at Five Points. The farm uses 75 wheel tractors ranging from 85 to 200 horsepower and 9 track-type tractors of about 275 horsepower.

Because all the land is irrigated and most of the crops require handwork, the direct labor cost is about 25 percent of the total input cost. Harris Farms uses three types of irrigation—flood, sprinkler, and drip. The overall average cost of water and equipment ranges from $300 to $380 per acre. Labor cost alone per irrigated acre is $88. When water is available, up to 3,000 acres will be double-cropped.

Minority workers make up more than 90 percent of the farm's labor force. Most of them speak only Spanish, which creates a problem in communicating. However, in recent years some workers have become bilingual and have been promoted to supervisory positions, easing the communication problem considerably. Chrisco commented that generally it takes about three generations for the minorities to learn English. Because of the difficulty in screening for illegal immigrants, the company minimizes the problem by using contractors to furnish most of the short-term workers. Seasonal work starts in March, with more workers being added each month to the peak demand in June. Generally the seasonal work lasts through August, but in some years it continues until October. The company still maintains trailer parks and 85 homes for workers. Some of the homes have had the same families living in them for three or four generations.

Harris Farms, like all of agriculture except for small, one-person operations, is plagued by the shortage of labor, particularly seasonal labor. Chrisco commented that because of the prospect for the continued shortage of labor, the best solution is for more labor-saving technology. With 225 people on just the cropping unit and 1,500 seasonal workers, management's greatest concern is the age-old problem of trying to motivate people to do their best. This is particularly critical when timing is so important, as in irrigation and harvesting of perishable crops. Harris management has a reputation for being very conscious of the needs of labor, but it finds that labor's changing expectations cause a real challenge. That problem is more difficult on the farm than in the feedlot, packing plant, restaurant, and inn because those divisions have a larger percentage of permanent positions.

Securing married college graduates for managerial positions on the farm and in the feedlot is difficult because their spouses often cannot find positions within easy commuting distance. "The new generation of graduates have different expectations than those who came into the labor market when jobs were not so plentiful. They expect a $5^1/_2$-day work week, while the older people have to be encouraged to take time off." Because of the difficulty in securing college graduates, some of the submanagerial positions are filled with individuals with less formal education but with previous management experience. John Harris personally has a positive feeling about finding good management people. He said, "We have some very sophisticated people looking for ag jobs."

WHEN John Harris was asked to enumerate some of the challenges he had encountered during his farming career, he started with his struggle with the IRS in settling his father's estate, as detailed above. His second greatest challenge involved the farm's membership in Tri Valley Growers, one of the largest farmer cooperative vegetable and fruit processors in the nation. Tri Valley operates nine processing plants with 1,500 annual and 9,500 seasonal employees. Harris Farms is one of the biggest producers of the 500-member cooperative, which, in the early 1990s, was in financial trouble and held up payments to Harris Farms on 100,000 tons of tomatoes. This represented a potential loss of several million dollars. Erick Johnson, chief operating officer for Harris Farms, serves on the board of Tri Valley, which is faced with the challenge of recovery. John Harris commented, "We are hopeful that it [Tri Valley] can make a comeback."

Harris is very understanding about the daily challenges of running a business, but he added, "My [greatest] challenge is with the reclamation law on how to comply and keep the farm a viable entity. We . . . are one of the few farms that pay full price for water." Although Harris thinks the law will probably stay, most farmers would like to privatize the water districts so they could control their own water. To do this, the districts would have to buy out the portion now owned by the federal government.

Harris likes the day-to-day hands-on problems but makes a conscious effort not to micro-manage. He encourages the managers with the most knowledge of a problem to come up with its solution. Harris obviously prefers the free-enterprise system and would not like to see government play too active a part in the pricing structure. He feels that government should help in technology and research and in leveling the trade playing field. John Harris is a free trader and favors trade agreements if they are fair. His company has a big market for vegetable products in Canada and has great potential for beef sales in Mexico.

His chief concern about low prices, as experienced in the summer of 1999, is that they prevent funds from coming into the business for capital improvement. He realizes that agriculture is in a real crisis because of the low prices. Harris Farms is well diversified, so it can cope with the problem better than most farmers, particularly those with mid-sized or smaller operations. Harris hopes that government "does not get too involved because that would just prolong the problem." If the government does take steps, he hopes it will use a market-oriented approach.

Chrisco and Devine were asked how large a farm had to be for it to be economical. Their answers were prophetic. They felt that a husband and wife could survive on a modest-sized farm, but they would likely have a substandard living. Chrisco and Devine were quick to add that if most small farmers went beyond what they could do by themselves, they would probably be in trouble unless they were such exceptional managers of people and finance that they could really grow.

It was the consensus of the Harris management that the financial community is directing change so that agriculture will be more in keeping with the pattern of other industries. Out of necessity, agriculture is consolidating, which will make it like other sectors of the economy.

Basically, Harris Farms has not experienced any trouble in financing. Bank of America has long been its lead bank, closely followed by Wells Fargo. Today the company is enjoying "lots of competition" for its business. Harris stresses that good accounting provides a basis of credibility. He added, "Credibility is sometimes more important than the numbers with finance people. . . . It is critical that you keep them informed. We have had basically the same loan officer for 20 years." That is an advantage.

JOHN and Carole Harris have no children, so, unlike many farmers, they do not have the problem of deciding which offspring will most likely succeed in continuing the business. John Harris says that his personal challenge is to enjoy what he is doing. "I do what I want to do, and I like what I am doing, but I am not driven to do. [However,] even when I am gone from the farm, I keep in contact."

John Harris obviously likes farming, but he readily admits that he has a favorite enterprise. In the mid-1950s Jack Harris started a thoroughbred horse business on a small scale. By 1966 that business had grown sufficiently that the Harris Farms Horse Division was established. Both father and son liked raising thoroughbred race horses and decided to devote some of their energies to that enterprise. By the late 1980s the horse division had more than 400 horses, including some of the top stallions in California. In a normal year more horses are raised than the company cares to have, so Harris has sold some top-ranking horses. In 1998 Harris Farms Horse Division horses won more than 20 percent of all races entered. The horse division was named the Outstanding Breeder of 1998 by the Thoroughbred Owners and Breeders Association, a significant recognition considering that California ranks third in production of foals. It is a top-ranking state in the number of race tracks and is one of the highest in purses.

John Harris with one of his favorite horses. Some horses are kept at the Kings River Ranch, where John and Carole Harris reside and enjoy the benefits of a hobby that became a profitable enterprise.

Probably one of the truest measures of a person is found in what he or she does to help others. This chapter opened with comments from Ronnie, who served breakfast my first morning at

Harris Ranch Inn & Restaurant and who spoke in such glowing terms about what caring people John and Carole Harris are. During the interviewing process I met with Diane Johnson, who started her career with the company in 1991 working primarily on public relations for the restaurant. Next, she found herself doing fund raising at rallies for water rights, a major issue for California farmers.

BEFORE joining the company, Diane Johnson had worked as a fund raiser for United Way. Apparently that experience caught the attention of John Harris, and soon she was involved in one of his two most important charities—the Jack A. Harris Memorial Scholarship. Because many of the company's employees had little education beyond the eighth grade, Jack Harris made provision in his will to help employees, their children and grandchildren, and needy local people further their education. Since its establishment in 1981, the fund has grown through earnings, memorial gifts, and contributions from John Harris, who also manages the fund.

In the 17 years up to 1998, about 340 people received scholarships. In 1999, 32 applicants from a pool of 35 received $27,000 in aid. One of those was the child of a non-employee with a disability. As long as their grades are satisfactory, persons may receive aid until they achieve their end goals. In addition, the company gives employment to the scholars during their breaks from college.

In the history of the program, about 35 percent of the recipients have graduated from a two-year college or technical school, about 25 percent have graduated from a four-year college, and 5 percent have earned graduate degrees. The remaining 35 percent have taken courses of specific interest. One person was 50 years old when she received a four-year degree while working for Harris Farms. Johnson ended her comments on the scholarship program by stating that some former employees and many of their children have gone on to very successful careers.

Harris Farms also has a very strong United Way campaign, John Harris's second big charitable interest. In the first year the company became involved with United Way, it received the award for the newcomer with the greatest dollar volume contributed in Fresno County. It has continued to be the largest company contributor in dollars raised in the county, which has a population of about 600,000. John Harris gives a partially matching gift for every employee. Meetings are called for every division to encourage giving. The beef plant shuts down the line for its meeting. People from United Way are invited to come for presentations. Many employees give personal testimony about how United Way has helped their families.

John Harris spends a considerable amount of his time working on civic and political issues, dealing with local boards to presidential races. His particular interest obviously centers on agriculture and water rights. Johnson noted that Harris was a great citizen lobbyist, especially for the Pacific Legal Foundation, California Women for Agriculture, and Agriculture in the Classroom.

My first impression was a good one, and everything that the late Dr. M. E. Ensminger, world-renowned animal scientist, had told me was "right on." Ensminger said that Harris Farms was one of the best diversified and best managed farms that he had ever encountered. I respected his opinion because, in his long and active career, Ensminger had become acquainted with many farms and farmers throughout the world.

Initially, John Harris's demeanor gives the impression that he is very quiet and laid back. But it takes only a few minutes of observation to realize that he is a serious and thoughtful person. He cares about his business and, equally as important, he cares about the people he works with.

Bibliography

Interviews

Chrisco, Larry. Fresno, CA, June 23, 1999.

Devine, Don. Fresno, CA, June 23, 1999.

Harris, John C. Fresno, CA, June 20–23, 1999.

Johnson, Diane. Fresno, CA, June 23, 1999.

Johnson, Erick H. Fresno, CA, June 22, 1999.

Radtke, Rod. Fresno, CA, June 22, 1999.

Wood, David E. Fresno, CA, June 22 and October 5, 1999.

Miscellaneous

Agricultural Statistics for Fresno County, CA, 1997.

Crow, Pete. "Genetic Focus Guides Harris Ranch Alliance," *Western Livestock Journal's North America Bull Guide*, January 11, 1999, 42–46.

Fee, Rod. "Beefing Up Demand," *Successful Farming* 97, No. 12, November 1999, 36, 38.

Harris Farms web site.

Lamp, Greg. "Consumer Conscious," *Beef* 35, No. 6, November 1998, 20–21, 24.

Nunes, Keith. "Fully Cooked Value," *Meat & Poultry*, February 1999, 32, 34–35.

Sisk, B. F., and A. I. Dickman. *A Congressional Record: The Memoir of Bernie Sisk*. Davis, CA: Regents of the University of California, 1980.

The Kreiders

IN COLONIAL TIMES Pennsylvania was called the breadbasket of the colonies. Lancaster County, in the heart of the state's rich agricultural region, has maintained its position of leadership. Today it produces the largest dollar volume of agricultural commodities of any nonirrigated county in the nation. It is outranked in dollar volume by 12 counties in California, Colorado, Florida, and Washington, all of which rely heavily on irrigation. Lancaster County ranks sixth in dairy, fourth in hogs, and first in eggs of all counties nationwide and, as recently as the 1950s, was the leading tobacco producing county.

The county's 937 square miles are shared by 454,063 residents and by 4,930 farms occupying 654 square miles. These farms have 52 million broilers, 10 million laying hens, 100,000 milk cows, and 584,000 head of other livestock. The farms contain 418,500 acres, of which 320,310 are planted to crops. The average-sized farm in the county is 85 acres versus the state average of 154 and the national average of 466. The intense concentration of poultry and livestock produced $2,931 of income per harvested acre in 1997. The average price of Lancaster County farm land is about $7,000 per acre (varying from $5,000 to $11,000) versus $890 nationally.

In 1990, when Noah Kreider, Jr., was asked if he would consent to be interviewed for a book on entrepreneurs of agriculture, he replied, "Twenty years ago I would have said no, but a great deal has changed in that time. Today very few farms in this county exist without the income from at least one person working off the farm. Yes, I will agree." From 1 farm of 102 acres in 1934, the Kreider family farm had grown to 27 farms and 3,233 acres by 1999. A labor force of 159 individuals raises alfalfa, corn, and potatoes; cares for 1,600 cows, 1,400 young stock, 85,000 pullets, and 2,125,000 laying hens; and maintains the machinery and truck fleet needed for the farm and subsidiary enterprises. Another 300 work in the processing plants, stores, and restaurants. The Kreiders have an integrated, industrialized farm situated in Lancaster County, a stronghold of small Amish farms that still use horses for their field work and transportation.

I**N THE FIRST DECADES** of the 1900s, Tobias Kreider farmed 57 acres at Bird-in-Hand, Lancaster County. In 1921 Tobias died, leaving his widow, Emma; a daughter, Minerva, age 18; a son, Noah, age 16; and another son, Allen, age 9. At his father's death, Noah left his job at a silk mill in nearby Lancaster to help his mother run the farm. The farm had six to eight milk cows, a team of horses, a couple hundred chickens, and a few pigs. Five acres of tobacco provided the main cash crop. A few acres were used to produce garden crops for home use and for sale. The remaining acres provided hay and grain for livestock and poultry feed plus some pasture. In 1927 Noah married Mary Hershey (a distant relative of the candy family), and the young couple moved in with Noah's mother and siblings.

In 1934 Mary's grandfather put his 102-acre farm near Manheim up for auction. Mary was sentimental about the farm, which had been deeded to her ancestors in 1739 by the sons of William Penn. She urged Noah to bid on it. Noah was reluctant because the couple had no money. However, Mary convinced her husband that they had nothing to lose. Noah was the successful bidder for $13,965, or $136.91 per acre. The farm had a 2½-story stone house built in 1792 that measured 34 by 72 feet, two hand-dug wells, a two-story summer home, a partially submerged barn with a driveway to the hay loft above the cow and horse stable, a 20-by-60-foot tobacco shed, a hog house, a garage, a corn crib, and smaller buildings.

Noah's 22-year-old brother, Allen, took over the Kreider homestead, and during the spring of 1935 Noah and Mary moved to their newly purchased farm. Noah immediately remodeled the barn so that it had room for 10 cows and younger stock. Next, he converted the hog house to a chicken house for 1,000 laying hens. He used the area above the garage to start 1,000 chicks. He immediately purchased an engine-powered Surge milking machine so that Mary never again milked cows by hand. Noah contracted with the Lancaster creamery to deliver whole milk from his farm and five other farms. After a few months Noah enlarged the cow barn for 10 more cows and built a stable to feed some Hereford steers. He started farming with four horses

A 1986 view of the farm that Mary and Noah Kreider, Sr., purchased for $136.91 an acre in 1934. Note the density of farms and rural homes. The original stone house dates to 1792.

and four mules, but in 1935 he traded some of the animals for a new Allis-Chalmers tractor and a two-bottom plow. His neighbors told him that tractors would never replace horses.

Noah and Mary were good managers, and although they lost much of the second year's six-acre tobacco crop to rust, they managed to pay for the farm in two years. The 20 milk cows and the 1,000 chickens under Mary's diligent care were very productive, and the milk route income was "especially helpful" in accomplishing a goal that most farmers needed a lifetime to achieve. The Kreiders had two year-round men, and in the busy season three, who lived with the family.

Richard Kreider, born in 1929, and Noah, Jr., in 1932, had regular chores as soon as they were able to work. Mary managed the chicken flock and all the housework, which, prior to 1937, was done without the aid of electricity. In 1937 an 11-by-50-foot cement stave silo was erected. This necessitated purchasing a silo filler. However, standing corn was still cut by hand. The silo and the silo filler were major investments that were considered risky by many, because silos were still a novelty in the area. But silage proved to be a superior feed. Two

Noah Kreider in September 2000 at the farm shop where as many as 14 employees work to keep equipment in condition when not busy on the farm.

years later a larger 12-by-50-foot silo was built. In 1938 the barn was enlarged again to provide for two rows of nine cows each, increasing the milking herd to 38 cows, twice the number of cows most area farms had. That year Noah contracted with the New Holland creamery to deliver milk from 15 farms daily. He purchased a larger truck that could haul one hundred 10-gallon cans.

Each morning, before going to school, Richard and Noah, Jr., went to the arch (root) cellar to clean eggs. The cellar, with its hard-packed dirt floor, was used to store eggs and vegetables and to preserve them through the winter. The Kreiders were in the dairy business, but they seldom had butter on the table, for with three hired men and six family members, it was good economics to use margarine. In 1939 Noah purchased a second tractor and in 1940 rented a neighboring 121-acre farm for $1,500 annually. The rented farm had a complete set of buildings, so a married couple was employed to live and work there. The Kreider farm was now 223 acres, very large for the area at that time and still 2½ times the average-sized Lancaster County farm in 1999.

In the 1930s the tobacco acreage was increased because it was the best means of providing work for the farm hands during the slack season. Noah also increased his beef fattening herd. He purchased a Bear Cat hammermill, and the corn that was not needed for silage for the milk cows was shocked and ground into fodder for the steers. Richard and Noah, Jr., both helped with shocking corn, which was "real work" for children their ages. In 1941 Noah wanted to purchase a new square twine-tie baler, but before he could get the machine he had to agree to do custom work with it. Farmers were recovering from several years of reduced income, and new machinery was in short supply. After World War II started, the government limited the amount of steel that farm equipment manufacturers could secure, thus reducing the availability of machinery even more. Noah, Jr., was nine when the Kreiders got the baler. He spent most of the summer either riding it to watch for and clean dirt from the knotter or cleaning the knife.

In the fall of 1941 Noah purchased a 6-foot Allis-Chalmers combine, which eliminated the costly practice of shock threshing. The neighbors anxiously hired the Kreiders to harvest for them. Noah made no changes during the war because little machinery and little land were available. In 1946 a neighboring 137-acre farm was purchased at public auction for $233.58 an acre. Two years later the 121 acres Noah had rented since 1940 was for sale, but because of his debt load and rising prices, he was reluctant to bid. The two sons both urged their parents to buy, so they yielded and paid a new high of $307.79 an acre for that farm.

Unlike most farmers, both Mary and Noah were good record keepers. The boys recalled viewing the large journal sheets used to tabulate expenses and income and commented that they were let in on money matters from an early date. Probably it was the parents' way of keeping the boys interested. Richard and Noah, Jr., described their parents as cautious risk takers who were not afraid to borrow money if they could see their way ahead clearly.

Kreider's final purchase for 1948 was a 10-foot Massey-Harris self-propelled combine. The neighbors lamented that it was too big for the fields, but they were happy to hire Kreider's to do their harvesting. Tobacco had become very profitable, and because it still fit in with other farm work, planting was increased to 20 acres, a sizeable acreage for tobacco at that time. The other cash crops were 85 acres of potatoes and a few acres each of lima beans and tomatoes. The remaining 245 acres were used for alfalfa, corn, and oats necessary for rotation with the cash crops and for livestock and poultry feed and bedding. The feedlot was enlarged to handle more steers for fattening.

The truck milk route continued, as did custom baling, combining, and field chopping. The two teenage sons carried the brunt of the custom work when they were not in school. The potatoes were hand-harvested by workers who followed the harvest season north from Florida.

WITH their agricultural course in high school, Richard and Noah, Jr., both enrolled in the Future Farmers of America (FFA). Noah, Jr.'s project was 1,000 laying hens and 2,000 broilers. He did very well on that project, which, ironically, led to his winning a dairy heifer from the county Holstein association. As a result, he started a dairy project. The success of his two projects helped him earn the Future Farmer Award, which he credits as being a real confidence builder. Richard's projects were more along mechanical lines. He was not as outgoing as his younger brother and, even though he was three years older, preferred to let Noah, Jr., be the leader. Richard, who was quite reluctant about being interviewed, stressed that they were opposites. He said, "That is why we got along so well together. We had special talents, and each of us went his way and did his thing. We never had formal planned meetings—we were always very informal about doing things."

Richard continued:

> Usually Dad and the two of us met on the run, and things seemed to evolve. Frankly, that [planning] business was not important to me. I loved farming because there was always something different to do. My favorite task was fixing things. I liked to keep things going. I did not like working with employees in later days because I did not like the leadership [role]. In the early days the work was all side by side, so it did not appear like someone was a boss.

The good working relationship of Richard and Noah, Jr., is clearly the reason the Kreider farm prospered during the next four decades. Anyone familiar with family businesses, and farming in particular, appreciates why most family farms have so much difficulty with generational transfers. Noah, Jr., was clearly the idea person who always had a list of projects on his mind waiting for the proper opportunity or the money to make them happen.

Richard graduated from high school in 1947, and Noah, Jr., in 1950. Both recalled that during their high school years and in the following decade they took many trips with their father to learn about improved ways of farming. They traveled to Illinois and Iowa when they wanted to learn about new ways of feeding cattle, to the Red River Valley of the North to gather information on the latest in mechanization in the potato industry, and elsewhere to learn about the latest in dairying and poultry.

THE KREIDERS traditionally sold their potatoes on the cash market each fall as soon as they finished harvest. They soon realized that once the harvest glut was sold, the price started a gradual rise through the winter. In 1946 they decided to take advantage of the winter price rise. They used the basement of the tobacco barn to store potatoes for winter sales. This proved very profitable, especially in 1949, so in 1950 they decided to build permanent storage for winter sales. For reasons of economy, they thought a combination potato and chicken building would work. The potato cellar was 16 feet into the ground. On top were two floors that rose 35 feet above ground for 16,000 chickens. Just after they had finished pouring the basement walls, a 7-inch rain caused the walls to collapse. Neither the contractor nor the Kreiders had insurance, and construction was delayed.

The problem was compounded when the Kreiders were unable to get their usual harvest crew from Florida and had to rely on local labor. This delayed harvest, so some of the potatoes were frozen and others were infected with blight. The inexperienced workers inadvertently harvested some frozen and/or diseased potatoes, which started to rot as soon as they were placed in storage. The heat from the chicken barn above compounded the potato storage problem, while the chickens got sick from the wet litter caused by the moisture given off by the potatoes below. Noah, Jr., said of that experience, which came in his first year out of high school, "This was one of my greatest challenges. We hauled the potatoes back to the field in a manure spreader. It came early [in my career] and was a great learning experience. It put us in a financial bind, and [it] took us a few years to work out of that." Richard and Noah, Jr., realized their mistake in construction. As soon as they could afford it, they insulated the ceiling of the potato cellar, which corrected the problem, and their combination structure worked well.

IN 1949 Richard and Pauline Miller were married, and in 1954 Noah, Jr., and Marian Landis were married. Both young women were very involved in farming—driving tractors in potato harvest, doing handwork in the tobacco fields, and performing virtually every task expected of farm hands in the days of labor-intensive farming. Marian preferred farm work to house work, so as a young woman she had always helped her father on the farm, even working with him when he did custom combining and field chopping. Tom Stoppard, who started with the Kreiders in 1950 as a general farm worker, recalled that with tobacco, lima beans, tomatoes, chickens, and 45 milk cows, there was ample manual labor for everyone.

The 1950s was the only decade from the 1930s through the 1990s in which the Kreiders did not purchase land. This was partly because of debt from the collapse of the combined chicken and potato structure; partly because of debt from extensive upgrading of their machinery line; partly because of expenses from expansion of the dairy and poultry facilities; and partly because of their ability to rent 150 acres, which gave them a total of 510 acres of crop land. In any case, they were not standing still. They continued to do extensive custom work as a means of generating extra cash to pay off debt. Larger combines and field choppers greatly reduced their workload and gave them the advantage in getting custom jobs as well as freeing men from field work to spend more time on the dairy and poultry enterprises.

It was dry in early 1954, so Richard and his father searched their property for water and found an underground water source where three steams crossed. They dug a well that provided 1,000 gallons a minute, sufficient for irrigation, and purchased a hand-movable irrigation system to use on their potatoes. Moving irrigation pipes was a very demanding job.

After a trip to the Red River Valley of the North, the Kreiders became convinced of the virtues of a potato harvester, a machine that was beyond the imagination of most farmers in Lancaster County. On their many trips together, Noah and his sons also visited with other farmers about their arrangements with family members who wanted to work into the business. In 1956 Noah decided that Richard, 27, and

Noah, Jr., 24, were ready to become partners in the business, and a three-way partnership was formed. In 1958 the brothers learned another lesson that prepared them for the "real world," when they speculated on potato futures and lost!

BY the 1960s the Kreiders' debt was down sufficiently that they felt comfortable about making some major advancements in the business. They converted a cattle barn to use for 10,000 laying hens, which expanded their chicken operation significantly. Next, they enlarged another barn and increased the milking herd to 60 cows, and they entered the retail milk business. After a trip to the Midwest Corn Belt, they decided to expand their beef operation to 600 head. On that trip they learned how to harvest and store high-moisture ear corn, which became the mainstay of the cattle ration.

At the start of the 1960s, the Kreider farm was well over 530 acres, about seven times the county average. Five full-time men were needed to care for the 60-cow milking herd, which was three times the size of the average herd in the county, plus the extensive beef cattle, poultry, and potato enterprises. The sizeable operation attracted considerable attention. The Kreiders were interviewed by *Hoard's Dairyman*, a leading dairy magazine, for an article entitled "How to Keep a Good Man." Among the criteria they cited was: "We do not care what the rate of pay is as long as he is worth it. We want . . . [him] to be satisfied and happy with the bargain." The Kreiders continued that fringe benefits were a house, milk, eggs, potatoes, a percentage of the tobacco income, and a bonus paid at the end of the year after an individual evaluation. They stressed that loyalty was developed by giving responsibility and by encouraging the workers to think, come up with new ideas, and then use them whenever possible. The Kreiders cited their success with assigning jobs without giving details about how to accomplish them. Instead, they let the workers decide how to handle the tasks. They noted that employees most often gave one of two reasons for leaving—either they wanted to get away from farm work or they wanted to start farming on their own. Some employees, like Tom Stoppard, spent their entire working careers with the Kreiders.

The 1960 improvement in the dairy operation proved so successful that in 1964 the Kreiders decided to upgrade to free stalls and a milking parlor with all new equipment for 100 cows. In 1967, with the purchase of the fifth farm of the decade, they acquired a good dairy facility and increased the milking herd to 150 head. The five farms purchased during the 1960s varied from 56 to 111 acres, for a total of 431 acres. The lowest price paid was $720.72 an acre, and the highest was $1,447.62. At this point they were farming 791 acres, milking 150 cows, and had 10,000 laying hens, a large flock of broilers, 600 feeder steers, and 22 acres of tobacco. The basic reason for staying with tobacco, a very labor-intensive crop, was that whenever employees were not needed on their regular jobs, they could strip tobacco from the stalk. Employees were paid on the piecework basis, which proved to be a bonus for them.

By 1965 the Kreiders had expanded to 300 acres of potatoes, a huge acreage for their area. Noah liked producing potatoes, and as more farms were purchased, he increased the potato business to 440 acres. In 1982 the Kreiders discontinued the wheeled-pipe irrigation and replaced it with three center-pivot systems. To achieve a three-year rotation for the large acreage of potatoes, they eventually purchased nine center pivots, which provided drought insurance in the rotation of alfalfa, potatoes, and corn for silage to sustain the larger milking herd.

Richard and Noah, Jr., worked well together. Noah, Jr., had the ideas and liked managing people, whereas Richard put things in motion. Richard commented, "My job was to make it happen and keep things working." Noah, Jr., related that when the two of them were buying farms in the 1960s, their mother, who was then in her sixties, often asked, "Why are you doing this?" Noah said his reply was always the same: "I don't want to move irrigation pipes and milk cows all my life."

The two sons were seasoned businesspeople by the 1950s, which pleased Noah, Sr., who understood the direction agriculture was going. He reduced his role in the day-to-day participation in the business. Starting about 1960, Marian (Mrs. Noah, Jr.) began working in the office paying invoices, filing, and writing out the payroll checks. Next, she did accounts payable and receivable, then balanced the books. As Noah, Sr., became more comfortable with Marian's work, he turned all the farm office transactions over to her, and in 1965 the office was moved into Marian and Noah, Jr.'s home. Noah, Sr., devoted his efforts to church and civic work, which he continued through the final decade of his life, and let the sons make their own decisions. In 1971 he was one of six chosen as a Pennsylvania Master Farmer.

THE 1970s opened with Noah, Jr., and Richard "on a roll." Instead of relying on their father, they turned to working closely with their banker, who, according to Noah, Jr., was extremely helpful in providing advice, support, and financing. As the 1970s evolved, Richard managed the day-to-day cattle feeding, poultry operations, and field work. Noah, Jr., was in charge of the milking herd, milk processing and retailing, and the potato business. The last two enterprises involved dealing with the public, which he enjoyed and Richard shunned. All the time Noah (from this point on, "Jr." will not be used because Noah Kreider, Sr., was no longer actively involved in the business) looked for ways to expand.

Kreider Farms added four more farms during the 1970s, totaling 306 acres. This gave the Kreiders a grand total of 1,097 acres, nearly 13 times the county average. The price per acre ranged from $699.22 to $3,775.00, for an average of $2,039.00. The national average price per acre at the start of the decade was $196.00. Richard commented about their experience when the neighbors heard they planned to buy another farm: "They said we would never make it work. But we just set our minds to it and went ahead. We had to learn to ignore the negativists."

Noah, Sr., was delighted to watch Kreider Farms grow. He had a favorite phrase in that respect: "The business cannot stand still." But he realized that the three-way partnership formed in 1956 was no longer fair to his sons. They were building his estate without any direct input from him. A cap was placed on the Noah, Sr., and Mary Kreider estate, from which they received their share of the income for the remainder of their lives. After their demise, proper settlement was made so the farming operation remained intact, avoiding the pitfall that so often happens with the death of the founder.

Since 1961 Noah had wanted to start a retail dairy jug operation in which the customers provided their own containers. In 1972 the operation was finally begun. Richard Shellenberger was employed to take charge of the new $125,000 milk processing center. By Pennsylvania law, producer-processors could sell their own product for 25¢ a gallon less than the established retail price. By January 1973 the jug operation had increased to where it took the total production from 100 cows. In 1974 the Kreiders purchased a farm with facilities for 45 cows to keep up with the growing demand. In January 1975 they opened an entirely new 450-head dairy and then were milking 550 cows. The new barn was the key to expanding the dairy business.

Noah had thought that a full-service dairy outlet might some day be a profitable venture, but he was not sure how to take the next step. In 1979 Shellenberger wanted more challenge and started taking courses in ice cream making. Next, the Kreiders hired a person to teach ice cream making. Soon the retail store was selling ice cream and milk. Noah commented on what happened next: "Probably the biggest unexpected opportunity was evolving into the restaurant business because it was the best way to keep a good man [Shellenberger]." The 45-cow dairy was dismantled on the farm purchased in 1974, and a combination dairy store and restaurant featuring dairy products was opened on that farm site in 1979.

In 1975 the Kreiders reduced their tobacco acreage, and in 1977 they quit tobacco entirely because it no longer was needed as a schedule filler and they "did not like what tobacco did to people." They learned that the dairy operation was a way to obtain and keep good year-round farm help. At that time they had 14 farm employees, who took care of a 500-head feedlot, 550 milk cows, more than 500 young stock, 10,000 laying hens, and 20,000 broilers and farmed 1,097 acres.

The poultry business increased sharply in 1977, when two caged layer houses for 82,000 birds each were built at Manheim. The original 10,000-layer unit was discontinued because it became outmoded. Office space was provided in one of the laying houses. Marian recalled that it was a "happy day when we no longer had the office at home." At that time she spent up to 60 hours weekly doing all the office work for the farm, the milk processing unit, and the restaurant. She said of the move, "Now I could at least get away from the job."

In 1979 two more houses for 82,000 layers each and a large new egg-packing-machine facility were constructed on the same farm. To spread the cost of the packing facility, the Kreiders contracted to case eggs, which they purchased from other farms, for Agway. Another 16 employees were added to the poultry house staff, plus a full-time bookkeeper and a part-time CPA who did all the government regulations work, tax work, and "other technical jobs."

Ronald Kreider (Ron), son of Noah and Marian, like most farm children, had worked at about every job he could manage during his early years. Ron was 14 when the dairy store opened in 1972 and "was very involved in its opening." After high school he spent one year in Bible college and then returned to the farm. Ron had spent most of his time doing field work and working with the dairy, but on his return to the farm full time, his father asked him to help run the poultry operation, which was being expanded from 10,000 layers to 170,000. Ron was sure he was back on the farm to stay, but he was not 100 percent sure he wanted to be involved with chickens. Within a year another 170,000 layers were added. The advent of Richard, Jr., the second person of the third generation in the business, sparked the start of a prolonged transitional struggle so typical of most family businesses.

DURING the 1980s the Kreiders continued their practice of buying land whenever it became available by purchasing nine farms totaling 1,375 acres at a cost of $5,788,023. The farms ranged in size from 22 to 495 acres, with a per acre price range from $1,895.43 to $6,740.71. The national average price per acre for land at the beginning of the decade was $725.00. The Kreiders paid nearly six times the national price—a clear reflection of the strong agricultural economy of Lancaster County. They now farmed 2,472 acres, 29 times the average-sized farm for the county, but acreage comparison does not give the total picture of the difference in the scope of operations.

They expanded in the dairy and poultry businesses at the same time they purchased those nine farms. This made the bankers nervous, but Noah saw the opportunities and was determined to continue moving ahead. In 1982, when the banks got "too difficult" to deal with, they turned to E. F. Hutton and Company to lease buildings and equipment for five years. These were placed on the newly purchased Donegal farm. Noah commented about that arrangement: "It was easier than going to the banks. We have never had a problem getting money. When we decide to do something, we do it while other people are thinking about it. Bankers are taught to think negative."

The Kreiders purchased those farms based on their financial strength and not from farm profits during the decade, for the 1980s were not kind to them financially. When asked how they fared in the depressed farm economy during the years 1981 through 1986, the answer given was quite different from the experience of many farmers, especially those of the Corn and Wheat Belts. During 1984 they suffered from the avian flu epidemic and had to depopulate their poultry houses for about four months, the length of time it took to get rid of the birds and thoroughly clean and disinfect the facilities. All income was lost for those months, but all the workers remained on duty to prepare the houses for new disease-free chickens.

The second major blow of the decade came during 1987 and 1988. They had just purchased the 495-acre farm at Middletown and finished construction of a 750,000-layer unit when the egg market collapsed. This was caused, in part, by negative publicity about eggs. For a prolonged period, they lost $20,000 a day, chiefly from the poultry business. They were financing with a bank out of Philadelphia at the time, and some of the officers put on pressure to sell land and empty the laying houses. There was family discussion about downsizing. Noah resisted because selling chickens and not being able to provide customers with eggs meant running the risk of losing their place in the market. Noah understood the risk involved, but downsizing would have been disastrous. He dropped the broiler enterprise because that industry was too integrated. He preferred the freedom of the laying hen business, even though the risk was greater. He said that made it more of a challenge. The bankers gave no consideration to the Kreiders' equity because they were so disturbed by the negative cash flow. After the market turned around, the president of the bank learned of the problem and asked Noah, "Why didn't you come to me?"

The Kreiders prided themselves on being diversified, and it is no secret that diversity helped them on their roller-coaster ride through the 1980s. In 1980 they started a new 250-cow dairy, and in 1982 they increased that herd to 450 head. Their dairy herd now contained 1,000 milking cows. Production from the expanded herd was greater than what their two dairy stores and two restaurants sold as milk or ice cream. Just when they were looking for an additional market, the Baltimore Jewish community approached them about supplying kosher milk. A rabbi was moved to a site near the dairy barns so he was available to do the blessing at the appropriate times. The Kreiders milked 24 hours a day, which gave the rabbi a busy schedule. Later, New Square, the New York Jewish community, started buying private-labeled kosher products. This gave the Kreiders a market when they needed one, and it has lasted to the present.

Noah continued his goal of expanding the farm at every possible opportunity. By 1987 the laying flock had grown to 1.75 million hens.

The Kreiders' $20 million poultry enterprise was ranked thirty-fifth among the layer operations of the nation. But 30-year-old Ron felt he had a monster by the tail and was not sure where it was headed. The problem was compounded because Noah's younger son, Jim, now joined his brother, Ron, and his cousin, Richard, Jr., making three of the third generation. The farm had grown to where it needed professional nonfamily management using business principles to bring the farm into the future. As is the case in most family businesses, the pending internal crisis was far more difficult to resolve than the external challenges.

S*UCCESSFUL FARMING*, in its January 1992 issue, ranked Kreider Farms as 172nd on its list of the nation's top 400 farms and credited the business with $30 million in annual sales. The Kreiders continued growing by enlarging their poultry operation, expanding their dairy herd, and adding stores, restaurants, and a motel. Also, they bought more land and made the transition into the third generation of management.

In 1991 they opened a company pullet house. Heated trucks deliver 85,000 day-old chicks, which are immunized and debeaked before they are placed in a house heated to 87 degrees. The pullets remain there for 20 weeks, at which time they are moved to laying houses. Then the pullet house is cleaned and disinfected in preparation for the next shipment of day-old birds. Because of improved genetics, layers are kept for 15 months instead of 12 months as formerly, but the single pullet house provides less than 10 percent of the replacement hens needed each year. The other 90 percent come from seven contract pullet growers. This enables the Kreiders to concentrate their management on layers, partly as a safety precaution against disease and partly as a way of using capital where they most need it.

At the same time they built the pullet house, they laid plans to construct six layer houses, each 600 feet long and each with a capacity of 83,000 hens. The hens have the potential for producing 65,000 eggs per house per day. In 1997 the Kreiders added three more houses, which increased the laying flock by 375,000 to a total of 2,125,000 layers. Plans are under way for construction of a 1-million-hen laying unit with a projected cost of up to $10 million. Even though egg sales

One of several complexes operated by the Kreiders. A half million laying hens are cared for, and eggs are placed in cartons by a crew of 12. Note two semitrailers at the loading dock and two on standby. Every day each of five trailers are loaded with 30,000 dozen eggs to be delivered directly to the supermarket.

Twelve hours each day 32,500 eggs flow from the adjacent laying houses into the grading, sorting, and packing room. The 3-foot-wide belt moves continuously and usually is as crowded as shown.

are increasing faster than production, that unit will be brought "on line" slowly to minimize the risk of start-up problems.

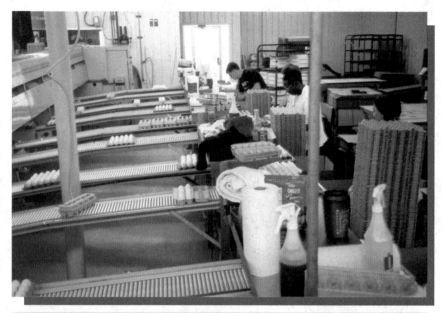

After the eggs have passed through a sensing device where dirty or cracked eggs are removed, they are sorted into five sizes and then carried to the proper spot on this conveyor where they are placed in cartons. Except for operation of the sensor, the first time a person is involved is when the cartons are put into shipping cases.

Noah commented that the cholesterol issue, which hurt the egg producers for "a while" in the late 1980s, made things especially difficult for the small producers, but that scare passed. He added that during the downturn in the egg market, the larger farms, especially those like Kreider's, had an advantage because they did their own marketing. Marketing was becoming more difficult because the number of packers was declining. Starting in 1977 Agway marketed most of Kreider Farms' eggs, but in 1982, when the Kreiders opened a new 500,000-layer unit, they sold those eggs to Kellers Creamery. This was a good move because it kept both buyers "on their toes." As early as 1982 Kreider's could process, pack, and deliver semi-load lots as eggs were ordered. In 1984 Agway sold its egg marketing business in the area to Kellers, and Kreider Farms stayed with Kellers until 1997, when it purchased its market from Kellers. Then, Kreider Farms employed a salesperson to represent it directly with the chains. In

1999 Kreider's was marketing 35 semi loads of eggs each week, 11 to one chain. Each semi held 850 to 960 cases of 30 dozen eggs each. The semi loads represented about 11 million eggs weekly, in addition to the cracked eggs sold to specialized buyers, who pasteurize them for use by food processors.

THE DAIRY ranked far behind poultry in gross sales, but for the Kreiders it was the best way to maximize the return from their large land base. After the 1982 expansion, the milking herd reached 1,000 cows, and all the Kreiders' production was sold through their stores and restaurants or was koshered for sales to the Jewish communities. By 1992 the demand for koshered dairy products increased, so another farm was purchased and 200 cows were added to the herd. In 1998 construction began on a completely new dairy facility to include a 100,000-square-foot dairy barn and a milking parlor with a 54-cow milking carousel. The site, finished in early 1999, houses 1,600 cows. By using the carousel, two employees can milk 324 to 375 cows per hour. Only 17 employees are needed in the dairy. Every cow is tagged with a computer chip, which greatly enhances record keeping, includ-

The 54-cow milking carousel rotates every eight minutes, enabling two people to milk about 350 cows per hour. The cows all carry computer chips that provide daily production and health data. They seem to enjoy the ride.

ing heat detection for breeding purposes. The next step will be shifting to three-times-per-day milking, and if production increases sufficiently, the carousel is adequate to handle the herd four times every 24 hours.

To ease environmental concerns, the manure is put through a process in which the solids are separated, dried, and stockpiled to be used for bedding for the milking herd or to be composted for the retail market. The urine is held in a large cement reservoir until it is spread on crop land via irrigators. The immediate goal is to process and market all the increased milk production either through the owned outlets or to chains under the Kreider brand or private label. Recently Kreider's contracted with a chain to market three semi loads weekly in privately labeled cartons. That equals the production from about 400 cows. Milk products, like poultry products, are contracted to deliver a fixed amount on a regular basis with the price fluctuating with the market. Chains prefer working directly with the producer as one means of preserving identity. The dairy will continue to expand as new markets are found.

BESIDES the milk and ice cream processing facility that retails their products, the Kreiders have four combination stores and restaurants located in Manheim, Hershey, Lebanon, and Lancaster, Pennsylvania. These outlets feature poultry and dairy products, including 28 flavors of ice cream. The demand from tour groups to visit and observe the egg packing facilities and the milking operations has become so great that Kreider Farms has established fees and regular times for tours. A 60-unit motel is located adjacent to the Manheim restaurant to service tour groups to the dairy and poultry facilities. The Kreiders believe that the motel is a "relatively risk-free investment" and will be an asset to the restaurant. The processing plant, the stores, and the restaurant employ about 300 of the more than 450 who are involved with Kreider Farms. Although not as profitable as the poultry and dairy operations, the retail outlets bring value-added prices for about 5 percent of the eggs and 20 percent of the milk produced on the farm. The Kreiders feel that the outlets are good advertising and enhance their image.

DURING the 1990s the Kreiders continued to purchase land whenever it became available and fit into their existing operation. They purchased 6 farms totaling 668 acres to increase the farm to 3,233 acres, 38 times the average-sized farm for the county. Even more dramatic is that the gross dollar volume of the Kreider farm is more than 190 times the average income per farm in the county. This is accomplished with the labor of 159 people employed in the cropping, dairy, and poultry enterprises.

Land prices continued to rise. The lowest price per acre paid during the decade was $4,033.33, and the highest $11,914.89. The 27 farms have 25 houses used by farm employees and their families. Richard recalled that over the years many people commented that his father had foresight to buy land when he did. Richard replied, "We did not think it was foresight, but when a neighboring farm came for sale, it was an opportunity not to be overlooked. Dad always said it was the Lord's doings that made it happen."

The 17 employees who do the farm work fill in at the shop when not needed on the land. The farm is planted to about 2,000 acres of corn, which is harvested primarily for silage but also for grain. Corn silage is harvested with a machine that had a list price of $280,000. It is capable of chopping 11 acres an hour, requiring 9 trucks to haul silage to the storage pits. Another thousand acres are used for alfalfa. Until 1999 the remainder of the land was planted to potatoes. Over the years the potato acreage had been reduced as the demand for feed products increased. Potato production no longer benefited from economies of scale and was not as profitable as formerly, so it was discontinued.

In 1994 a 21,000-square-foot shop with nine bays, each capable of holding the very largest piece of equipment, was built. Everything contained in the 2,000-square-foot two-story parts room, including 200 sets of tires, is inventoried on a computer. The farm shop crew is capable of repairing 90 percent of all mechanical problems that occur on the 14 over-the-road semi-rigs, 10 triple-axle trucks with heavy-duty forage and manure boxes, 4 trucks with feed boxes, the field repair trucks, and a large assortment of pickups.

Because of its heavy concentration of livestock and poultry, Lancaster County has an abundance of animal fertilizer and poultry litter. The laying hens produce more than 18,000 tons of litter annually (about 18 pounds per hen per year), so to maximize its fertilizer value and to minimize environmental problems, the Kreiders sell and deliver their excess poultry litter as far as 50 miles away.

As integrated as the Kreiders are, they have never processed their own dairy or poultry feeds. The dairy requires processed feeds with protein to supplement the alfalfa hay, shell corn, and corn silage produced on the farm. All the poultry feed is purchased. The total volume of purchased feed required weekly exceeds 2,000 tons—about 100 semi loads. The Kreiders realize that a mill could quickly pay for itself, but it would also increase management problems. They feel that they can better spend their time and money managing what they do best and rely on local firms for their feeds. Because of the heavy poultry and livestock population, Lancaster County is a grain- and soybean-deficient area, which means that processors must purchase and store those commodities as well as process them.

THE OFFICE STAFF at the poultry house had increased to four by 1983 to keep up with the growing business. Marian commented that when she took a vacation she had records with her to work on and called her data in to the office. In 1987 the office was moved into a stately early-1900s red brick house on the farm by Manheim. In 1998 a staff of 10 handled the business of Kreider Farms. The beautifully remodeled farm house also serves as base for Marian, Noah, and Ron Kreider and seven management people. The fully computerized office is far different from the shoe box of records, small pocket-sized notebook, and bullet pencil so reminiscent of farm record keeping of the past.

Going from a loosely operated farm to a more structured one was a necessary but slow process. Noah related, "It was important to us to have key people in each enterprise, but the enterprise had to have adequate volume to justify the manager. We can shift people to different positions, and we offer opportunities for internal promotion. This is how you keep good employees."

The Kreiders have a strong religious heritage and have a great concern for their community and their employees. That led to the adoption of a program called Marketplace Ministries, which offers assistance to employees on the job. Starting in 1997 the Kreiders engaged six part-time chaplains to work with their employees at their stores and restaurants. The program cost per employee is nominal, and the results are very encouraging. In addition, they have a program of recognition, awards, and bonuses for length of service, accident-free driving, and/or service in the workplace.

KREIDER FARMS, like most family-run businesses, experienced several years during which the transition to the third generation evolved. When asked what was the greatest challenge he faced, Noah answered, "Probably the family situation in recent years. Turning it [the business] over to my son Ron." As stated previously, Noah and Richard both acknowledged that basically they did not have a problem. Noah liked to work with people and lead, whereas Richard was content to implement and then keep things in motion. They managed to settle any differences they had amicably.

Then Ron and his brother, Jim, and their cousin, Richard, Jr., entered the business. Ron was concerned that the business was drifting and not especially profitable, and in 1988 he convinced Jim that the problem was serious. About the same time, the banker became concerned and, during the next eight years, frequently called meetings aimed at settling the transition problem. The banker realized that the transition issue had to be resolved or it could result in a financial problem. Ron stated, "Things had to change, or I would have left." He was caught between his uncle, his father, his brother, and his cousin. Probably each of the others felt the same way. The resolution gradually evolved. Richard, Jr., was killed in a car accident, Jim decided that he wanted to pursue a career in gospel music, and Richard sold his interest to Ron, leaving Ron and his father to continue the transition.

Like his father, Noah has been active in church, community, civic, professional, and business organizations, which will ease the transition. Marian said of her husband: "He is a visionary, always looking ahead for new opportunities. . . . He never hesitated to buy land or otherwise expand. He always says you have to expand so you can get away from milking cows and pulling irrigation pipes. He will continue to brainstorm because that is his nature."

When asked what impact the government has had on the Kreiders' business, Noah replied: "The government basically has not had an impact on our operation. The dairy buyout and subsidy programs were never a concern. . . . Our philosophy has been to work on volume and efficiency and forget about the programs. We have found our markets and produce for them."

Ron used his experience from the transition years as the basis for a speech entitled "Building a Team Beyond the Family," presented to the National Agricultural Bankers Conference in November 1997. His basic theme was changing the farm organization from family centered to team centered, which meant that everyone had to participate at every level and that everyone had to work with a common set of principles. He stressed: "We must change to grow, and change is constant. Tomorrow's problems cannot be solved with current thinking. We must build a framework for management growth, but our beliefs must remain as core principles because they develop a set of common ideas for communication." He continued the theme of the two previous generations that people are the key to success: "There must be a common framework built on trust and shared principles; each employee has freedom to act; we grow as our people grow. That philosophy will provide the framework for Kreider Farms as it enters the twenty-first century."

Bibliography

Interviews

Kreider, J. Richard. Manheim, PA, January 26, 1999.
Kreider, Marian L. Manheim, PA, January 27, 1999.
Kreider, Noah W., Jr. Manheim, PA, January 25–27, 1999; telephone conversations, April 21, May 3, 1999.
Kreider, Ronald. Manheim, PA, January 27, 1999.

Miscellaneous

Agricultural Statistics for Lancaster County, PA, 1998.
Chronicle. A publication of Kreider Farms. Various dates.
Freiberg, Karen. "They Cut Out the Middleman by Running Their Own Restaurants, Dairy Bars, and Convenience Stores," *Farm Journal*, May 1986.
"How to Keep a Good Man," *Hoard's Dairyman*, December 25, 1963.
"How Marketplace Ministries Brings Assistance to Employees on the Job," *Lancaster Intelligencer Business Monday*, April 6, 1998.
Kreider, Ronald. "Building a Team Beyond the Family." Address at National Agricultural Bankers Conference, November 10, 1997.
Manheim Sentinel, February 25, 1975.
Lancaster Farming, January 1, 1994.
Statistical Summary and Pennsylvania Department of Agriculture Annual Report 1997–1998. PASS 123.

Louis E. "Red" Larson

"I Really Loved Those Cows"

IN the 1500s, when Spanish explorers came to what is now the state of Florida, they brought horses, hogs, and cattle with them—the horses for transportation and the other animals for food. What is unknown is whether the cattle were for milk or purely for meat. They also brought seeds of various crops, including oranges and grapefruit. They had no real desire to establish agriculture in the area, because they intended to have food brought in from the nearby islands and from Europe. Little was done to propagate livestock or crops. The first known cultivated citrus grove was started between 1803 and 1820, and as late as the early 1900s there were virtually no dairy-type cattle in the state.

In 1954 the then 79-year-old John G. DuPuis, a medical doctor, wrote *History of Early Medicine, History of Early Public Schools and History of Early Agricultural Relations in Dade County: Some of the Experiences and Activities of the Author.* DuPuis reported that many babies died because their mothers did not have enough milk to keep them alive and no cow's milk was available. Until about 1914 the ticks that carried Texas fever were so prevalent in Dade County that it was commonly believed that cows could not be raised there. A combined state and federal eradication program proved so effective that by January 1916 the entire area of Florida south of the Hillsborough Canal was declared tick free.

DuPuis was such a believer in the need for milk that he raised a herd of Dutch Belted milking cattle. In a paper he read before the annual meeting of the Florida Medical Association, May 18, 1917, he strongly urged doctors to "prescribe, endorse, and encourage the production and use of clean fresh cow's milk." He also urged the medical people to demand better conditions on farms and reminded them that milk was the least expensive source of protein.

Even with the ticks eradicated, Florida's climate made the state far from an ideal place for dairying. The intensely hot, humid season that starts in July and lasts through October decreases production 15 to 25 percent per cow. In 1999 *Dairy Today* polled leading dairy farmers about states that were the best locations for dairying. Florida was not included in the results. That should surprise no one familiar with the climate and the crops necessary for dairying. What is surprising is that Louis Larson, a pioneer large-scale commercial dairyman, who in a 1997 issue of *Successful Farming* was listed as operating the nation's largest dairy, has spent his entire career in the state.

Florida ranks thirteenth nationally in cash receipts from milk, fourteenth in the number of dairy cows, and sixteenth in milk production. Even though Okeechobee County is on the edge of the Everglades, it is Florida's leading dairy county. The county has only 418 of Florida's 35,204 farms, but it has 22 large dairies with 36,000 cows, which makes it the state's leader in livestock receipts.

Citrus pulp, a by-product of orange and grapefruit processing, was originally sent to Europe, where it commanded a higher price than Florida farmers were willing to pay for it as cattle feed. However, as the population of Florida increased with the influx of service personnel during World War II and the demand for milk grew, dairy farmers began to feed citrus pulp. During the 1950s frozen concentrate juices became very popular and made citrus pulp even more abundant.

After the Cuban Crisis in the early 1960s, many wealthy Cubans came to southern Florida. They plowed under land that grew saw grass, which had supported a limited number of deer, and established the sugar cane industry. Molasses, a by-product of sugar cane processing, became available and was blended with citrus pulp to improve its palatability and nutritional value. Pulp became a valuable locally produced dairy feed. But despite the availability of molasses and pulp, Larson maintains that Florida "is one of the poorest milk producing areas of the nation but one of the best markets."

L OUIS EDWIN LARSON, SR., lived with his parents on a farm near Brandon, South Dakota, where his father supplemented his farm income by lathing (plastering laths on walls of houses). About 1919 Gudrin Henjum came from Minnesota to teach rural school near the

Larson farm. Gudrin roomed and boarded with Larson's parents and met Louis. The couple married and started farming on their own. On March 9, 1924, Louis E. "Red" Larson, Jr., was born to the couple. The 1920s and 1930s were not good to many farmers, and in 1932 the bank foreclosed on the Larsons. Young Louis, Jr., recalled that the full impact of what foreclosure meant did not hit him until the day the trucks came and took the 25 milk cows. That was a sad day, he recalled, because "I really loved those cows."

Gudrin's parents left Wells, Minnesota, and went to Hollywood, Florida, where Mr. Henjum established the Hollywood Foundry and Machine Company. Louis Larson, Sr., made a trailer out of an old farm truck and hitched it to the family's Graham Paige car. With all their belongings packed into the trailer and the car, the Larsons left to join the Henjums. Louis went to work for his father-in-law but soon found a job in a dairy, which was more to his liking. In 1935 he started his own cement making business. His son, Louis, Jr. (hereafter referred to as Larson to distinguish him from his father) helped him when not in school or working on his paper route, which he assumed in 1933 at the age of nine.

The July 10, 1940, *Miami Daily News* featured the 16-year-old Larson, who had just been elected secretary of the Florida Future Farmers of America (FFA) and was looked upon as a local entrepreneur. His newspaper route had grown from 18 customers to 196. His earnings from the paper route enabled him to purchase a pony for $35, which he rented out for rides. He soon sold the pony for $40 and, with another $20 of earnings from the paper route, purchased a cow for $60. The cow soon had a calf. He purchased other calves from townspeople who had family milk cows and did not want to keep their calves. His father cosigned his son's note at the bank for the calves, 300 chicks, and 100 plants. Larson's little enterprise experienced hard luck when a pack of wild dogs killed about half his calves and a cold snap destroyed his chicks and plants, but that did not stop him.

His father discontinued his cement business because of poor health and rented a five-acre farm from a local dairyman who had gone out of business. The Larsons had six milk cows of their own and rented eight more from townspeople. In addition, Larson had recouped and owned 21 calves. He helped his father with the herd mornings and evenings during the school year. In the summer months his heifers were on pasture, so he took a job at MacArthur Farms, which had a very progressive dairy. Larson recalled, "Mr. Mac was a real friend and mentor to me."

BECAUSE Larson worked so much while going to high school, his graduation was delayed one year. When World War II started, he enlisted in the United States Army Air Force (USAAF). He sold his milk cows before going into the service but kept his heifers and paid someone to care for them. When he was discharged in September 1945, he had 20 mature cows, but he decided to take advantage of the G.I. Bill and enrolled at the University of Miami, where he majored in business. He quit college in his second year and decided to go into dairying. On October 5, 1947, he married Reda Buchanan, a native of Tennessee, who was employed by a local bank.

In December 1947, he entered into a partnership with Wilbur Grant, who had an 80-cow dairy and shipped all his milk in 10-gallon cans to Foremost Dairy in Miami. Larson had 20 cows, but because Grant wanted an equal partnership, Larson had to purchase another 60. A friend financed him for the purchase of 60 Holsteins and Guernseys, all "heavy springers" (i.e., about to calve) from Wisconsin. Soon after the cows arrived, they were diagnosed as having acetonemia, a sugar deficiency. The veterinarian reported that no medicine was available, but Larson gave them whiskey from a soft-drink bottle and then Karo syrup. He virtually lived with the cattle until they recovered and lost only one. Larson and Grant never learned whether it was the climatic change or the ration change that caused the cows to get sick. Larson had borrowed $15,000 to buy those cattle and pay the freight on three carloads from Wisconsin. This was big money to him at that time.

The partners each worked in 12-hour shifts. The cows were fed outdoors by a helper, while Larson or Grant used the little barn with six stalls for milking with the standard pot-type Surge machine. The milk was cooled and placed in 10-gallon cans, which were picked up daily. Reda's income from her bank job provided for all the newlyweds' living expenses because the income from the dairy went to pay off the note on the cows. About the time that Larson had paid for the cows, his partner, Grant, decided that he wanted to quit.

When Foremost Dairy learned about Grant's decision, it approached Larson and asked if he would be willing to buy the 123-head Weatherlee Dairy in Dade County. Larson's herd had increased to 100 head while working with Grant, and he agreed to buy the Weatherlee business, which he financed on the strength of the Foremost contract. He and Reda made their first move.

Larson secured permission and support from the State Highway Patrol to drive the 100 cows five miles down Highway 441 to their new location. Suddenly he was the owner of 223 cows. This was a philosophical turning point in his life. When he first entered the dairy business, his aim was to build a 100-cow herd. He had been in business just over a year and had already surpassed his goal. Larson said: "I learned that you could add 100 cows to a herd of 123 cows and the only real increase in cost was feed and similar ingredients. It worked just as well when doubling again."

The 223 head were milked twice daily in a 23-head cow barn, with one person bringing the cows in, cleaning the udders, and feeding them while another ran the milker. The herd averaged 6,000 pounds per year on a native pasture of wire grass, Bermuda grass, and palmettos (a native palm brush with seeds), plus a 20 percent dairy ration supplement. This was considered good production for that time and in that area. Soon Larson increased the herd to 250 head.

In November 1949, Jack Wahtasack, whom Larson had worked for while in high school, became ill and wanted Larson to partner with him on Fareway Dairy Farm, Inc., in Broward County. Wahtasack had 200 cows and had a contract with Adams Dairy in Key West. Larson owned 250 cows, had lots of energy, and knew how to manage a dairy, while Wahtasack had money and experience. Larson and Wahtasack pooled all their assets and liabilities. The herd was rapidly expanded to 600 head.

In 1951 Larson was one of the first to sign up with the extension service to do a cost analysis of his operation. He learned that his records were tops and realized that he had a net worth of $50,000 after four years of dairying. In 1952 he started using artificial insemination (AI), and in a couple years, when the first heifers freshened and started to lactate, he learned the true value of that program.

But life was not without challenges, for on Christmas Day 1952, only two of a crew of five came to work. Larson was forced to call on neighbors for help. He decided that he had to hire more dependable people—that he had to get good employees and keep them. In retrospect he said, "They [good employees] not only stay, but they also bring family or friends to work with them." It was the first of many learning experiences.

From 1947 through 1954 Larson had made about 10¢ per gallon profit. He especially enjoyed the last three of those years, when he was milking more than 600 cows. Unfortunately, Wahtasack's health deteriorated further, and he sold out. Larson commented that his experience with Wahtasack had been like gaining "a Ph.D. in dairying in three years." The business venture had been very successful, and when the partners sold out, Larson had cash "for the first time."

IN 1955, Larson started over for the third time by using the cash to make a down payment on 124 cows and a 140-acre rundown farm near Del Ray. "That year we actually had an operating loss." But that experience served only to motivate Larson to do better, and in 1956, despite the added cost of increasing his herd to 800 head, he made a profit. The biggest improvement that year was switching to a pipeline system, which enabled him to do away with the pot-type milkers and greatly improved the efficiency of the milking crew. Larson learned well, and every year from 1956 to 1982, he made a profit.

Larson entered into a contract with a processor in Palm Beach, where the market was growing rapidly. This encouraged him to produce more milk, but soon he was producing more than the processor would take, so Larson located another outlet in Miami. Based on the strength of the two contracts, he was able to secure money to buy more cows and improve his setup. He remodeled the barn, installed a 1,000-gallon bulk tank, improved the cooling system, and built five houses for workers. He purchased a tank truck and delivered two or three loads to the processors daily. This was less labor intensive and more sanitary than the old method of using 10-gallon cans. Much-improved varieties of grass made it profitable to convert all the tillable land to pasture for year-round grazing, which reduced feed cost.

LARSON expanded his land base to accommodate the ever-growing milking herd. By the 1960s he realized that his operation was going to be crowded out by people, so to protect himself he purchased land in Highlands County. He paid $78 an acre for 1,870 acres that belonged to the Burroughs Adding Machine Company. He was not ready to move there, so he rented that land to a cattle rancher. Little did he realize at the time what a stroke of luck that purchase would prove to be. In 1963 he bought the 350 cows and equipment of a neighboring farm in Palm Beach County and moved a fourth time. He had a 15-month lease on the land. When the lease expired, the landlord tripled the rent. Within 120 days, Larson had a modern 72-stall milking barn built on the Highlands County farm and, in July 1964, moved those 350 cows to that farm. This became Farm No. 2, the start of his multi-unit operation. He immediately built eight houses for workers because Larson had learned that by having houses on the farm, "people will be there to milk the cows."

An added benefit of moving to Highlands County was that he was now on some of the most productive land he had ever farmed. He immediately set to work to upgrade the soil and to improve the environmental waste system. The transition to AI was nearly complete by this time. Average annual production had increased from about 5,200 pounds per cow to about 10,000 pounds, and butter fat content of the Holsteins was also improving. Larson attributes both to AI and to much improved forage.

Because he expanded through the purchase of other herds, Larson sometimes secured cattle that were not healthy. In 1962 he acquired a herd from Canada that had been infected with tuberculosis but had been cleared for five years. A young veterinarian misread a test and pulled 60 head out of 700 because he thought they were reactors. Larson recalled that he did not sleep for a few nights. The Florida state veterinary specialist reexamined those 60 cattle and detected 7 that had signs of the disease, so they were slaughtered, but not one had tuberculosis. Nine years later he had another outbreak, and this time the veterinarians determined that the best thing to do was to dispose of the total herd of 2,400 cows. The entire operation was shut down for 18 months, the crops were plowed down, and all the facilities were idled. In the 1960s, tuberculosis was widespread, and science could not cope with the disease as well as it can today. However, assistance was available to make such a drastic step financially endurable. Larson still relies on outside cattle to keep his herd at full size because the hot weather makes breeding difficult. His operation has been unable to reduce the breeding cycle to less than 14 months. About 12 months would be normal.

IN 1968 Farm No. 3 was added to the growing dairy. This time a family with only one child, who had no interest in farming, put its farm on the market. The 1,100-acre farm, with a total herd of 600 good cattle and with good buildings, was located in northern Okeechobee County. The operation also had a fine work force and "proved to be a winner." With the purchase of Farm No. 3, the dairy had grown to 1,700 milk cows and had 35 employees. Annual milk production was in excess of 15 million pounds. At that time Larson reported that the record system consisted of a daily log on each cow's milking, on other animals, and on labor and feed cost. Since 1965 total cost of feed, labor, and production was known for each animal at every location.

When asked how he was able to cope with the rapidly rising cost of production in the 1960s, Larson explained that he became involved with price discovery to learn what the cost of imported milk from other states would be. This information was shared in the cooperative in which he played a leading role and was then presented to the processors. From his early years and into the 1960s, a net profit of 10¢ a gallon was almost assured, but then costs rose rapidly. Interest rates increased from 6.5 to 9 percent; supplies and services jumped about 20 percent; feed cost rose 16 percent; and labor cost increased 25 percent, due chiefly to a lowering of productivity. Larson was forced to mechanize and to automate wherever he could, a trend that has continued. Always a very precise record keeper, he became an advocate of wise and disciplined use of money. He tightened his management methods like never before.

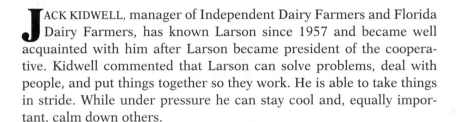

JACK KIDWELL, manager of Independent Dairy Farmers and Florida Dairy Farmers, has known Larson since 1957 and became well acquainted with him after Larson became president of the cooperative. Kidwell commented that Larson can solve problems, deal with people, and put things together so they work. He is able to take things in stride. While under pressure he can stay cool and, equally important, calm down others.

During Kidwell's 39 years as a leader in the cooperative movement in Florida, he has experienced the decline of milk processors in the state from 29 to 5, which he says is nice in one way but dangerous in another: it has increased efficiency but has reduced competition. During the same period, the number of dairies in southern Florida has declined from 160 to 20 due to expanding population overtaking the

land. Environmental regulations removed another 35 percent of the dairies because they were not able to cope with the cost of conforming to the requirements. Others left because they did not have the management abilities needed to compete on a larger scale, even though the average-sized herd was nearly 200 head in the early 1950s, far above the national average. The biggest exodus came when farmers left the urbanized East Coast areas to move to Okeechobee County, utilizing the tax-free land exchange. Fortunately for the remaining dairies, they worked together well enough that their cooperative was able to control production and, hence, the price of milk.

IN SEVERAL RESPECTS 1971 was an eventful year for Larson. He secured the services of Bill Bradley, who possessed a master's degree in agriculture and had taught vocational agriculture, been a Dairy Herd Improvement Association (DHIA) supervisor, and been a herd manager. Bradley was employed as a production record keeper and breeder and quickly became involved in herd health work. Bradley was amazed at the excellent records and cost accounting system Larson had. In all his experience, he had seen none that could compare.

In 1971 Farm No. 4 was purchased from a bankrupt estate. It included 975 acres of fine land and 300 head of good milk cows. On the other hand, the buildings were run down, and all five of the employees had a "side deal that siphoned from the farm, so they had to go." Fortunately, by then Larson had enough reliable employees that he quickly replaced those five.

The Larsons moved (their fifth time) from the Del Ray Beach Farm No. 1 to the newly acquired farm, which was renamed Farm No. 1. It was a good year to move as far as the family was concerned. Mrs. Larson liked the house purchased in Okeechobee and said that her anchor was cast for good. The farm office was moved to Okeechobee, which remains company headquarters. Larson commented that when he improved the pastures on the Del Ray Beach farm, he planted fast-growing Ficus trees along the fence line to provide shade for the cattle while on pasture. Today those trees give shade on the fairways of the golf course that has replaced the pasture.

W HEN the dairies in southern Florida were forced out of urban areas, some of them moved to Okeechobee County, which was sparsely populated and had a sizeable land base. There they occupied the highest land, and soon the livestock waste found its way into the waterways leading to Lake Okeechobee. In 1970, in response to the problem, the University of Florida started a waste disposal research project. That resulted in the determination that each farm should use a two-stage lagoon for animal waste and then drain the urine into ditches from which it could be used for flood irrigation. Tests indicated that the liquid was saturated with nutrients, especially phosphorous, which was the probable cause of algae in Lake Okeechobee.

For many dairy farmers, that project led to the most far-reaching event of 1971 and, for Larson, at least a temporary setback. The Department of Environmental Regulation (DER) called a meeting of all Okeechobee County dairy farmers. They were informed that each had to establish a two-stage lagoon, one an anaerobic and the other an aerobic, at the point of pollution—i.e., the barn and the cattle yard. The Soil Conservation Service, the DER, and other government agencies—federal and state—toured the area dairies and determined how to cope with the regulations. The farmers' choices were adapt, move, or quit. The government agreed to cost sharing of the improvements that had to be made, but the cost was still more than 19 of the 42 dairies involved felt they could afford to make, and they discontinued farming in Okeechobee County.

Any dairy impacted by the decree was paid $602 for each cow that had to be removed from production, and the farmers were free to reestablish in another area, sell the animals and crop farm, or discontinue farming. Larson was forced to reduce his milking herd from 12,000 to about 9,500 cows and close two of his eight barns, which significantly decreased his economy of scale and cost him several million dollars over and above the payment received for his cattle.

Over the years the phosphorus content of Larson's land has been reduced from 12 parts per million to less than 0.5 parts per million. The cost of maintenance of the pollution control system is about 50¢ per hundredweight of milk.

D ESPITE the environmental problems, the 1970s were very good to Larson Dairy, Inc. Larson's had achieved 90 percent AI breeding efficiency, and it was able to raise enough heifers from its herd to reduce its purchases of replacements. The only purchased cows and heifers needed were those for expansion herds. Larson's has altered its cropping programs in recent years because it learned that more chopped grass improved herd health over the heavier concentrate ration formerly fed. It also discovered that grass consumes more phosphorus than row crops, which enables Larson's to keep that nutrient more in line with environmental regulations. It can chop grass for silage at least 10 months during most years. This not only increases production per acre but also reduces cropping cost, because annual tillage and planting are minimized. It is hoped that using more chopped grass silage will increase the number of lactations per cow, thereby reducing replacement cost. Larson's has one dairy on a heavy grazing program. This reduces milk production, but the lower purchased feed requirements make that farm a good net profit center. However, environmental regulations would make it virtually impossible for Larson's to switch its entire herd of 9,000 cows to grazing.

By 1973, just as more room was needed for heifers, the nearby Dixie Ranch, with 3,366 acres, a working canal, and two large houses, came up for sale. It had been owned only three years by a nonfarmer investor who was concerned about the economy and wanted to sell. It was ideal for raising dairy calves, with enough acreage for some beef cattle. This was a well-laid-out farm, with fair land and good grass, so the $1,000-per-acre price was reasonable. It is on that farm where the grazing herd is located.

ONE OF THE INTERVIEWEES commented that the strongest point of the Larson Dairy was the business leadership. He added, "It is the best run business I have ever known." But Larson's active mind needed additional challenge. By the 1970s, when his business had acquired sufficient size and his family was of an appropriate age, he felt that he could repay society for the good that had come to him. He was active in his church from the earliest times. He served on several boards, was a deacon, and served a prolonged stint as a Sunday School teacher. Larson also was active in civic organizations in the communities where the family lived.

From the start, he participated in industry organizations and in 1965 was elected to his first significant leadership position, the board of directors of the National Milk Producers Federation. In 1978–79 he served as president of that organization. In 1970 he became a director of the Federal Land Bank and a director of the Production Credit Association of Miami. He remained in those positions until 1976, when he was named to the Federal Land Bank District Advisory Board of Columbia, South Carolina, which headed the Southeast District of the Land Bank. In 1977 the Larsons were named Farm Family of the Year by the Farm Bureau Federation. Larson served a term as president of the Florida Agricultural Council, was elected to the Dairy Hall of Fame, and served on the boards of the Future Farmers of America Foundation, the Florida Agricultural Museum, and the Sun Bank.

Larson served two terms as president of Dairy Farmers, Inc., an organization of dairymen active in governmental public relations, advertising, and nutrition education. This led to his being selected to serve on the National Dairy Advisory Committee of the United States Department of Agriculture (USDA). It was on that committee that he was able to impact high-level government individuals on behalf of the industry.

Larson credits the cooperatives for much of his success, especially in marketing. As president of the National Milk Producers Federation he stressed, "Cooperatives keep milk utilization at its highest, and they add stability to the market by putting milk where it is needed when it is needed." Cooperatives protected producers from being cut off by processors as they had in earlier times.

WHEN Larson was first rapidly expanding his milking herd, he had to concentrate his financial resources on that enterprise to get the quickest return on his investment. He had to purchase all his feed and supplements from local feed companies. By 1973 the business was large enough to justify Larson's investing in his own feed processing facilities. He felt it was necessary to get into feed preparation to better control the quality of the ration and "save some money." It did not take long to discover that by doing his own direct feed purchasing and owning a processing facility, he could save enough money to equal the income of another dairy. After becoming involved in large-scale direct buying, he protected himself by taking positions on the futures market to cover about 50 percent of his annual needs.

In 1979, Larson Dairy, Inc., and McArthur Farms, Inc., both of which were about the same size, formed Dairy Feeds, Inc., and erected a complete mill at Okeechobee with a capacity of 40 tons of mixed feed per hour. The business made the Okeechobee freight receiving station the largest in southern Florida. The partners were unhappy with the railroad service, particularly the rate charged on their products, and could not get any concession for volume. In 1990, to reduce the cost of incoming feeds, they erected a feed receiving facility at Fort Pierce, about 30 miles to the east along the Florida East Coast Railroad. They quickly recovered the cost of that terminal because they gained a $10-per-ton rate reduction. The mill prepares 140,000 tons of feed annually, and Dairy Feeds, Inc., does about $20 million volume per year.

Other than chopped grass and corn silage, the only locally produced feed the dairy uses is citrus pulp, which initially made up about 40 percent of the ration because it was so cheap. Today citrus pulp makes up about 25 percent of the ration, but cottonseed hulls are fed to neutralize the acid in the pulp. Soybeans come from the southeastern states, and corn comes from the Midwest. Alfalfa is shipped in from Idaho at a freight cost of $85 per ton. Because the northern-grown alfalfa receives optimum sunlight, which makes it very digestible, it is well worth the added cost.

L IKE most entrepreneurs, Larson is a real people person. In 1965 a group health insurance plan was added, and later it was expanded to include accidental death and life insurance. All the premiums for the employees, plus a major portion for the dependents, were paid by the dairy. Larson said: "We really like the program because it takes care of the employees. They look for that coverage. It is expensive, but it is a must." In 1973 profit sharing was inaugurated, and a pension program was added soon after. The two benefits combined can add up to as much as 25 percent of an employee's annual pay.

The company also provides more than 120 two- or three-bedroom houses for full-time employees. A three-person maintenance crew works full time keeping the houses in good condition. Larson feels that good housing is the key to keeping qualified help. Housing is in addition to wages or salary, and the employees do not pay income tax on the rental portion of their pay. One farm has 49 houses plus a church that the farm provides. Besides religious services, the building is used twice weekly for English and Spanish classes and other community events.

Each April, under the direction of Reda Larson, the farm sponsors a picnic at the Okeechobee Civic Center, which is generally attended by about 400 people. It is especially appreciated by retired employees and employees with families. Awards are given for safety and longevity. A band provides entertainment. In addition, an annual safety barbeque is held, at which a speaker addresses some key issue on safety. A $10 safety bonus is given at the end of each month if no employee is injured. A Christmas bonus and a "good year" bonus top off the benefit program, which, in total, adds about 34 percent to the payroll cost. In 1999 the health insurance program alone cost more than $1,000 a day.

The company has 145 employees, of whom about 80 milk cows; 28 are in herdsmanship, management, and administration; and the balance are in calf raising, maintenance, and crop production. The Larson philosophy is "Treat people right, give them good pay, good benefits, and housing, and they will stay." Living proof is the many employees who have spent their careers at the Larson Dairy and today are able to retire in relative comfort.

In addition to the above programs, Larson's has ongoing training courses for every level of job. These have proven very beneficial for the company and the employees and, without being facetious, also for the animals. Larson's entrepreneurial traits are clearly exhibited by the fact that he engages professionally trained individuals in every phase of his business. He is clearly in charge, but unlike many farmers who profess to be jacks-of-all-trades, he understands the importance of relying on professionals. He readily admits that the biggest administrative headache is securing a good computer system capable of coordinating all phases of the business for everyone involved.

L ARSON took the lead in developing a program to create a fund for financing dairy science research projects at the University of Florida. He also headed the Florida Agricultural Council at the time it developed SHARE (Special Help for Agricultural Research and Education).

He obviously is looking into the future and wants to see dairying, as well as other sectors of Florida agriculture, survive. He is well aware of the challenges the industry faces in that rapidly urbanizing state. When Larson was asked how he benefited from being involved on so many boards, he responded:

> You stay tuned to what is taking place in the dairy industry; the exposure to other industry leaders enables you to learn from them; it helps the image of your business. On the regional Federal Land Bank board, I learned about government intrusion into business. The biggest mistake we made there was when we allowed the making of loans up to 85 percent of the appraised value, which was an injustice to the borrower.

THE FIRST ADDITION in the 1980s came in 1981 when Larson purchased a 1,102-acre farm with 700 young cows (no young stock), poor buildings, including the milking barn, and no housing. "It was a homemade setup." The farmer was elderly and had no one in the family who was interested in continuing in farming. Soon after the original purchase, Larson secured another 366 acres from the same family. He immediately built a state-of-the-art double-24 herringbone milking parlor and six good tenant houses.

In early 1982 he doubled the operation with the purchase of the bankrupt McArthur Farms, Inc., from Bank of America for $4.9 million. J. N. McArthur was a cattle dealer and got stuck with some cattle, so he started a dairy. He hired "a gem of a manager from Wisconsin and did very well, and if Mr. McArthur had lived, they would have passed us." Later his son Charles took over, but, unfortunately, he was killed in an accident. In 1978 "an entrepreneur" from Iowa purchased the farm, and by 1982, when it was foreclosed on, losses were about $40,000 a month. Larson, along with five other dairymen, had originally planned to buy the 9,000 cattle, but four of the others backed out in the process. He bid $5.5 million for the cattle and planned to lease the land. The big attraction was that McArthur Farms, Inc., had a 28,818-gallon daily milk base, plus a land lease on 8,474 acres and the equipment. Monthly payments were $72,000, of which up to 40 percent could be diverted to make repairs.

———◆———

LOUIS E. "WOODY" LARSON, JR., majored in dairy science in college because he knew he wanted to be in the dairy business. After he graduated in 1973, he applied at dairies other than Larson's for a job. When Larson found out what Woody was doing, he said he wanted Woody to come for an interview. During the next nine years, Woody learned his lessons well and developed a more conservative management style than his entrepreneurial father. In any case, when McArthur Farms, Inc., was purchased, Larson tossed the task of getting it under control to Woody. When asked what was the biggest challenge he had encountered to date, Woody immediately retorted, "Trying to straighten out the McArthur fiasco." Some of the 102 McArthur employees were not exactly on the level, and they attempted to run some of the cattle through twice when Larson's tried to get an accurate count.

Left to right: John Larson; Louis E. Larson, pioneer large-scale commercial dairy farmer; and Louis E. "Woody" Larson, Jr. Taken at the Louis Larson home.

Fortunately, Larson's was able to employ McArthur's former controller, who knew the business, and he proved to be a real help. Everything was very rundown, which handicapped the task. Woody was 31 at the time he was assigned to get the place into shape and stop the losses. For the next three years he virtually did not see his young sons because he left before they were awake and got home after they were asleep. Larson suffered losses in 1982 and 1983, his first since 1955, but Woody met the challenge and made the place profitable. Fortunately, the other Larson dairies were profitable and subsidized the McArthur deal. Larson proudly stated that every monthly payment was made on time. In 1984 Larson purchased another 3,000 acres of the McArthur farm from Bank of America. In retrospect, Larson said of the initial McArthur purchase: "The timing was poor because interest was high and all other costs were skyrocketing. You cannot have losses go on too long on such a big operation." The bright side of the McArthur purchase was that Woody proved himself as a manager and learned that not every year in the dairy business is profitable. That experience made him a solid, cautious manager.

In 1984 Larson proved that he had not lost his perspective on values when he purchased a herd of 70 Jersey cows from the widow of a dairyman who was too old to manage the business. The truckers had loaded the cows onto two semis, and six Jersey calves were left. The widow and her daughter and grandson sadly watched the action. Larson observed them and thought of the day in 1932 when trucks hauled the cattle away from his father's farm in South Dakota and how sad he felt. He looked at the little grandson and asked him if he would like to have the calves. Sixteen years later, that "little boy" is running his own dairy.

BY 1984 the Larson Dairy consisted of 6 milking farms and 3 heifer farms encompassing 16,000 acres of land with 13,000 milking cows and 8,000 replacement heifers in Highlands and Okeechobee counties. A writer for *Top Producer* proclaimed that Larson's probably was the biggest dairy in the free world, with 220 employees and a gross of more than $30 million annually.

In the early 1980s the USDA had a diversion plan in which Larson Dairy felt it was obligated to participate, so Larson's reduced its herd by 15 percent. In 1986 the USDA had the dairy buyout plan, but

Larson's chose not to participate. This plan created a good market for milk and virtually doubled the price of whole milk, but it severely impacted the price of beef, causing major damage to that industry.

Larson's was "on a roll" when suddenly the governor of Florida decided to make environmental concerns a campaign issue involving the counties where Larson's was operating. Because Larson's shipped more than 350,000 pounds of milk a day, its herd was closely watched. The phosphorus content of its ration was considered too high by about 1 percent. But phosphorus is a key to reproduction and must be at a 2-to-1 ratio with calcium. Through reworking the ration, Larson's has been able to lower phosphorus to 0.55 percent. The discharge from the heifer operation was too high in phosphorus. The heifers had to be moved, and their former facility was converted to a beef grazing operation.

ED SMITH grew up on a small dairy farm in Indiana and, after receiving a degree in animal science, took a position with a 150-head dairy in the Midwest. In 1985 he started with Larson's, working on the night shift milking cows in barn 6. He soon became a supervisor and then a barn manager. In 1989 he was moved into administration. One of his assignments was the ever-growing task of coping with environmental regulations. Smith has a very philosophical view of his challenge: "You have to accept the fact that the bureaucrats do not understand the problems of producing milk. Part of the problem as a society is that we do not totally understand how we impact nature. From an economic view, society has to understand what it wants from the environment."

In 1984 the chief concern of the environmentalists was that phosphorus was affecting Lake Okeechobee. It had to be determined how to stop the seepage. Fortunately, much of the area has a hardpan at about 6 feet, so there was little chance of subsurface drainage into the lower elevations. The challenge was to stop the buildup of phosphorus in the soil and surface drainage. The state and federal governments cooperated to cost share with the affected farmers. Much of the early delay came from the fact that the soil conservation specialists were so overwhelmed with the task as it related to the dairies that they did not know where to start. Three major things had to be considered—land-use history, soil types, and annual rainfall. At this point Larson's went to a three-stage lagoon, from which it could dispose of the liquids through irrigation. At the same time many area dairies discontinued operation.

The cost on the Larson operation was about $1,000 per cow, of which Larson's had to bear nearly $4 million. That made 1990 the fourth year Larson's had an operating loss. It could have been made even more severe if an Internal Revenue Service (IRS) ruling had been sustained. The IRS wanted to tax the cost-sharing portion provided by the governments, but, fortunately, a ruling from higher authority determined otherwise.

Larson ended his comments on the environmental problems: "It seems like the environmental people would like to get us out of business in the Everglades. So lots of milk will have to be hauled in [and the cost will be greater]." Smith said that he was baffled when a college group toured the farm and one young woman broke down and cried when she saw the cows in the para-bone milking parlor. "It was her impression that animals should be left free to roam in nature." A former extension employee commented: "In my opinion, until some of the population gets short of food, it will never back off agriculture. The activists will do everything they can to prevent the industry from producing food at the lowest possible cost."

The Larson farms are all inspected twice monthly to determine if the waterways are fenced off and the manure handling practices are being followed. Woody added that the inspectors are very familiar with the operations and everything Larson's does. Larson's has managed to cope with the change. In 1989 the county's remaining 22 dairies had 36,000 cows, which outnumbered people four to one and produced 65 million gallons of milk. That year dairying provided 80 percent of farm income and provided 1,548 full-time equivalent jobs generating $29.8 million in earnings. In 1995 payroll income had fallen to about $10 million, 512 full-time equivalent jobs had been lost, and the county's unemployment rate hit 15.7 percent. The county was looking for a miracle to solve its problems.

STEPHANIE GARLOUGH'S parents had worked at Larson's, and because they liked it so much, Stephanie wanted to work there after finishing school. Her first job, starting in 1986, was recording the production and history of each cow. This entailed following the

DHIA method of recording each cow's production, with the barn manager evaluating the cow's record. This task kept her busy in the milking barns 16 days each month; the rest of the time she updated her card file. In 1988 Larson's acquired hand-held computers for use in the barns, but the number of cattle exceeded the capacity of the computers. However, it was a step in the right direction. In 1991 Larson's upgraded its computer system and was able to place everything on the software. This provided a complete history of each cow, including its dam and sire. Stephanie proudly pointed out that cow No. 12670 had produced 129,941 pounds in four lactations. In her third lactation she produced 38,807 pounds, proof that good genetics and good management can help overcome the climatic disadvantages of Florida.

The Larson Dairy is well managed, with an excellent record-keeping system. The six dairies, the heifer farm, the feed processing facilities, and the forage production unit are all separately cost accounted. Feed production is charged to each dairy, and no crop is carried on inventory after the fiscal year. Larson commented that when he was totally responsible for the operation, he got up at 3:00 A.M. each day to do the record keeping before he went into the dairy. After gross income exceeded $25 million, the IRS forced Larson Dairy to go to the accrual method. Larson said: "It is cruel, but it gives a truer picture of your finances. I know that if the IRS came tomorrow, we would have our records in the best possible shape." A complete financial statement is prepared monthly.

IT WAS PREVIOUSLY MENTIONED that one of the greatest cost-reduction innovations was constructing the feed mill at Okeechobee and the grain terminal at Fort Pierce. In 1998 Dairy Feeds, Inc., shipped in 127,392 tons of grain, soybeans, and citrus pulp, in addition to trace elements and 10,000 tons of alfalfa. That is part of the price of being located in a grain-deficient area. Five people are needed to prepare and process the feed, and the manager is responsible for all the buying. Dairy Feeds, Inc., contracts the trucking from the mill to the individual dairies because it prefers not to own the trucks or be responsible for the liability connected with that business. Larson's and McArthur's prefer to concentrate their efforts and capital on what they know best—producing forage and milk.

The added cost of importing much of the feed is offset by the good local market for milk. On the average, Florida has to import about 15 percent of its milk. That milk has a transportation cost of about 12½¢ per quart, and it takes from five to seven days for a delivery truck to make a round trip. This gives the local producer a competitive edge. In most years Larson's also has the advantage of being able to chop grass silage 10 out of 12 months. Its feed cost at 43 percent of total production expenses is commendable, as is its 10.5 percent labor cost. The greatest disadvantage Larson's has is the long period between calving because of the inability to breed back during the intense summer heat.

The Florida dairy producers are probably the most supportive of their cooperative—Dairy Farmers of America—which enables them to maintain and effectively control their supply. With fewer processors and food chains, it is imperative that the farmers stay united. The supermarkets, which, next to the final consumer, have the most influence on how the product is received, have three concerns: steady supply, top quality, and competitive price, in that order. Larson's is satisfied that it has the most to gain by staying with its cooperative rather than establishing its own processing plant and attempting to independently market its product. This is true even though it produces nearly nine semitrailer loads of milk every day.

The individual Larson Dairy farms are large enough that each of them can operate in a different manner. The average annual herd production varies from 13,800 to 19,000 pounds, which is good for

One of the nine semitrailers with a capacity of 6,000 gallons taking on a load at Farm No. 3.

Front view showing one-half of a para-bone 68-cow milker that enables three people to milk 250 cows per hour. This is the newest Larson dairy, with 2,000 cows in production. The cows are kept in a free-stall barn, where they are fed a total mixed ration.

Rear view of the cows in the preceding photo, showing how the milk flows directly into stainless-steel pipes leading to a pre-cooler, which quickly reduces the temperature of the milk before it flows to large storage tanks.

Florida. The lower-volume herd relies on grazing for 65 percent of its intake, has a lower turnover in cows, and has a very satisfactory profit margin. The key to that herd is keeping other expenses down. On its largest and newest dairy, about 40 percent of the herd is milked three times daily; the other 60 percent is given BST but is milked only twice daily. In layperson's language, BST is an expansion of a natural hormone already in a cow's body. The booster causes the cows to consume more feed so they are able to give more milk. The production of those two herds is very similar. Experimenting has taught Larson's that beyond genetics there is an upper limit to which a cow can be pushed.

Environmental regulations have forced Larson's to make some major changes in its operations. Besides reducing its milking herd by nearly 3,000 cows, Larson's has contracted with a farmer in Georgia to feed its heifer calves. The Georgia farmer is paid a set fee per pound to raise the heifers from shortly after birth to 300 to 350 pounds. The calves are placed on pasture, which is supplemented with by-products of a peanut butter processor, such as peanut hulls and peanut skins, plus imperfect cottonseed, sometimes candy that is not up to specification, and products from a distillery.

Since 1999 the 300- to 350-pound heifers have been moved to North Carolina, where a major hog producer, Murphy Family Farms, has an abundance of grass land fertilized with manure from its facilities. For the next 18 months, the heifers graze the pasture land and are bred. About 60 days before calving, they are returned to Larson's. In this manner, the Georgia farmer gains a needed supplemental income, and the Murphys gain income from their pasture. But, more important, the Murphys are able to comply with environmental regulations for their hog operations.

Because of the less stressful climate in Georgia and North Carolina, the growth and genetic potential of the heifers is more fully realized. Larson's feels that heifers produced in this manner are better than most replacement cattle it could buy. The cost of moving the heifers to a cooler and less humid climate is more than offset by the superior animals returning to the milking barns. Using AI and the above heifer boarding program serves to reduce cost and to improve the genetics, just two of the management practices necessary to increase overall milk production. In spite of its heifer raising program, Larson's still needs to purchase about 10 percent of its replacements.

If the contract heifer raising program continues to be profitable, it is possible that the former heifer raising farms can be converted into two more dairies. This would enable Larson's to regain the economies of scale it had when it was milking more cows. Larson's is confident that it could double its current operation and have everything under control. This would enable it to produce more than a million pounds of milk every day.

LARSON'S 18-hour work days did not prevent him from looking beyond the immediate tasks for ways to improve his business and the dairy industry. The Larson herd was offered as a living laboratory to study a virus that resembles HIV. Because of Larson's success in dairying and his reputation as a good businessperson, he was called upon to serve on many boards of organizations. From those experiences he learned how the "system worked when it came to implementing national policies." He recalled that sometimes the decisions made on the national level did not work for the best interest of those they were intended to help—specifically, decisions involving legislation that ended in a surplus of dairy products.

Larson was forced to move his farming operation several times because of urbanization, so he was asked to comment on how the appreciation of land impacted his business. He stated that the higher cost of labor in areas of higher land values had offset some of the gains. In areas where land was less costly, labor was more abundant; hence, the cost was less. The gain from appreciated land was used each time to construct new facilities on the next farm, and in almost all cases the replacement cost was higher because of inflation and the fact that the technology was more advanced.

Money for its own sake was not a concern for Larson, but rather he enjoyed overcoming the challenges that his visionary and restless mind conceived. When Reda was asked what her husband's strongest traits were, she replied, "Motivation and timing. He took advantage of every opportunity, but timing is everything." She said her husband never bothered her about the day-to-day problems, but she was his "sounding board" for major decisions.

When asked what the future holds for commercial dairying, Larson recalled that milk was $0.50 a gallon in the 1950s, $0.60 in the 1960s, $0.86 in the 1970s, and $1.93 in 1999. The adjusted rate for 2000 was equal to what he had received in 1978, and he anticipated that it would be lower in the immediate future. The industry was experiencing a constant reduction in the number of producers, but the best cattle from the retirees generally were absorbed by expanding dairies, so production did not decline. A major price increase was not likely unless production was reduced, and he did not foresee that happening. In light of that statement, he was asked what, in his opinion, was the smallest herd on which a family could survive. He ventured that on good land in the Midwest, using family labor, a producer with a herd of 60 good cows might survive and that in Florida survival would require 400 cows. But he volunteered that "neither would have more than a life of hard work, with few amenities, and the children would not stay."

When Woody Larson, the eldest son, was asked what his greatest challenge was, he replied that it was working in the family business. He continued: "I have two sons in college, and both are interested in coming into the dairy. Business survival is the biggest challenge ahead for this operation." This is proof positive that the Larsons are no different from most families in business. Once the entrepreneur has made his or her mark and is ready to step aside, a far more challenging problem arises.

Bibliography

Interviews

Bradley, Bill. Okeechobee, FL, January 25, 2000.

Garlough, Stephanie. Okeechobee, FL, January 25, 2000.

Kidwell, Jack. Okeechobee, FL, January 25, 2000.

Larson, John. Okeechobee, FL, January 26, 2000.

Larson, Louis E. ("Red"). Okeechobee, FL, January 24–27, 2000.

Larson, Louis E., Jr. ("Woody"). Okeechobee, FL, January 25, 2000.

Larson, Reda. Okeechobee, FL, January 24, 2000.

Price, Kent. Okeechobee, FL, January 25, 2000.

Smith, Ed. Okeechobee, FL, January 25, 2000.

Miscellaneous

Daily Okeechobee News, June 4, 1999.

DuPuis, John G. *History of Early Medicine, History of Early Public Schools and History of Early Agricultural Relations in Dade County: Some of the Experiences and Activities of the Author*. Privately published. Miami, 1954.

Florida Agricultural Statistics. USDA and Florida Department of Agriculture, 1999.

Florida Grower and Rancher, April 1981.

Freeman, Louis. "Larson Dairy One of the World's Largest," *The Sunbelt Dairyman,* January 1981, 34, 35, 38.

Johnson, Charles. "Bigger Than Huge," *Top Producer* 7, No. 1, January 1990, 18–20.

Larson, Louis E. Speech dated 1970.

McCary, Miles R. "Business Almost as Usual at Larson Dairy," *Hoard's Dairyman,* April 25, 1992, 354.

Mclachlin, Mary. "Our Century," *Palm Beach Post,* December 19, 2000.

Miami Daily News, July 10, 1940.

Miami Herald, January 9, 1940.

"National Milk Producers Federation President, Red Larson, Speaks Out for Cooperatives," *Hoard's Dairyman,* November 10, 1979, 1411, 1425.

News Tribune (Fort Pierce), June 27, 1982.

News Tribune (Tampa), July 5, 1981.

Ward, Mary. "Okeechobee County Looking for a Miracle," *Florida Agriculture* 54, No. 6, June 1995, 1, 12, 13.

Wendell H. Murphy

"I Was Biding My Time, Looking for Opportunities"

AGRICULTURE has always been relatively important in North Carolina's economy. Initially, the forests provided timber and hides, but by the late 1700s tobacco and cotton had become the major crops. In post–World War II economy, first poultry and then, in the 1980s, hogs became the dominant commodities. In 1997 hogs led agricultural output with $2.017 billion in gross sales, followed by poultry with $1.372 billion and tobacco with $1.193 billion. Cotton, with a total sales volume of $329 million, was in sixth place. The state's $7.676 billion gross agricultural sales made it the eighth-ranking agricultural state, producing 3.9 percent of our nation's total output. It ranked second to Iowa as the national leader in hog production. The hog industry increased so rapidly that the state's total dollar sales in 1997 represented a 59 percent increase over those in 1992. In 1997 poultry and livestock made up 66 percent of the state's total agricultural income. The state saw an 11 percent increase in the number of full-time farms because of the rapid growth in the number of contract hog producers. This was a sharp contradiction to the nationwide trend of declining numbers of full-time farms. A 1998 survey indicated that the average income of contract producers was $110,000.

Walt Cherry, Executive Director of North Carolina Pork Producers, stated: "In North Carolina, if you want to stay on the farm, you have to contract in chickens, pork, or turkey. If you live east of I-95 in the state, how else are you going to make a living? People want big industrial complexes, but the skilled labor force is not there." Iowa counties that were leaders in hog production have been replaced by North Carolina's Duplin County, which has become the number one pork producing county in the United States, and by Sampson County, which has become number two. Duplin also holds the distinction of being the number one turkey producer, followed by Kandiyohi County, Minnesota. Before the growth of hog and poultry production,

Sampson County ranked seventy-fifth in per capita income in North Carolina, but in recent years it has climbed to tenth.

Six firms—Carroll Foods, Goldsboro Milling Company, Lundy Packing Company, Murphy Family Farms, Nash Johnson & Sons Farms, and Prestage Farms—have emerged as the dominant players in North Carolina and in the industry. They have made the 25-county area of eastern North Carolina a mecca for chicken-, pork-, and turkey-oriented agriculture. There are several reasons for this rapid rise. In the 1950s the Lundy Packing Company, of Pennsylvania, provided breeding animals to farmers in the area because it needed a larger supply of hogs. Most of these smaller operators were looking for an alternative to raising tobacco, and feeding hogs on the range appealed to them because it was an easy and rather inexpensive business to enter. Goldsboro Milling Company, once the largest turkey producer in the southeast, had moved its turkeys to confinement. This freed up the former range land for confined hog production.

The Carroll, Johnson, Prestage, and Murphy families were native farmers looking for new opportunities and realized that they were in a pork-deficit region. They also understood that the area has a moderate climate, which makes year-round pasture production possible. This was before the days when confinement facilities had proven their superiority in both hog and poultry production. Probably as important as any other factor, these families were new to large-scale hog production. "They looked upon themselves as food producers and were willing to provide what the consumers demanded. This was in contrast to the traditional hog raisers, who looked upon themselves as pig farmers raising pigs to sell wherever they could." That clique of progressive families was not held back by traditional methods of raising hogs, which handicapped growers in the Midwest. These families understood that consumer demands were changing the food industry and that greater integration was inevitable. Today, those leading producers in Duplin and Sampson counties generate more than 750 semi loads of hogs for market each week.

NORMAN HOLMES MURPHY was born in 1919 near Rose Hill, North Carolina. He attended school until the eleventh grade, which was the top grade offered by the district. He married and immediately started farming on a small acreage. Whenever he accumulated the cash, he purchased another plot, until the farm eventually grew to 47 acres. His eldest son, Wendell, said: "He never borrowed to buy. If there was no cash, we did without. Each fall when we went to buy school clothes, he always tried to get the price down. He had to deal." A very sentimental Wendell, with moisture in his eyes, continued: "Daddy never looked for the easy way to do things. He really resisted change, but I have never met a harder worker or one who enjoyed working more. He was totally in control."

Wendell's younger brother, Harry, nicknamed Pete (hereafter Pete), commented: "Daddy never slowed down. His greatest pleasure was getting up and going to work. He had a heart condition, but he was getting up to go to work when he died." Wendell's and Pete's sister, Joyce, added: "Our parents were both hard-working people of high principles and were devout. They were excellent managers who lived within their means. Our parents never had high school diplomas, but they were insistent that the three of us were going to college—no ifs, ands, or buts."

The 47-acre farm was powered by two mules. Wendell fondly recalled the hours he followed his father in the furrow as Mr. Murphy operated the walking plow. Wendell, Pete, and Joyce all had jobs assigned to them—"real jobs, not just 'make work'"—which did not allow them much time to play with neighborhood children. The three children and their parents spent a great deal of time weeding the seven acres of tobacco, which was their chief cash crop, and the large garden, which provided much of their food supply. They also raised corn for the hogs and hay for the mules. Joyce recalled that when Wendell was a junior in high school, he asked his father if he would buy a cow because he wanted to milk it. Wendell milked the cow morning and evening even after he became the school bus driver during his senior year.

In addition to farming, the Murphys owned a country store about one-fourth mile from their home. Mrs. Murphy ran it when her husband was busy with farm work or when he was gone buying tobacco or collecting tobacco from neighboring farms with his small truck and delivering it to the company for which he worked. When the farm work was over, he ran the store each evening until 10 o'clock. Everyone walked from the farm to the store because the family did not have a car and the truck was only for business. That included carrying lunch to whoever was working there. The store was the neighborhood social center, with "lots of card playing and visiting on weekends."

◆

WENDELL MURPHY came home each weekend to work on the farm while he attended North Carolina State. In 1960, after earning a degree in agricultural education, he interviewed with a tobacco company. Even though their father was a heavy smoker, the children knew that both parents were very much opposed to their using tobacco or otherwise being involved in that business. Wendell wanted to stay in his home area, and teaching was his best opportunity to do so. "I was biding my time, looking for opportunities. Teaching was not my goal. I thought I might get an ag job that would allow me to farm evenings and weekends." But he took a teaching position where he earned $4,080 his first year, and his wife received $3,600 working as a secretary. He was not turned on by teaching and kept a constant lookout for something more to his liking. That opportunity presented itself during the 1961 Thanksgiving break his second year of teaching. He and a lifelong friend, Billy Register, were driving through nearby Warsaw when they spotted a grind-and-mix feed mill. "Billy said, 'I think that would work in our community.' The idea hit me."

Wendell inquired about the Register Mill (no connection to Billy) that was for sale near Rose Hill and learned that the price was $13,000. He and his wife had $3,000 in savings from their first year's earnings, but where could he get the rest of the money? He had never borrowed money. "I talked to Daddy about the mill. He was petrified by the

Wendell Murphy in his mid-50s, shortly after moving into new company offices at Rose Hill, North Carolina.

[thought of the] debt, but he finally agreed to cosign to get me off his back." His father agreed on the condition that Wendell would continue to teach until the loan was repaid. The Rose Hill bank initially agreed to make a loan under those conditions, but when the Murphys arrived at the bank to get the money, the banker informed them that the board had decided that they wanted a mortgage on the Murphy farm. Wendell realized that that was too traumatic an undertaking for his parents. He was upset because the banker had not kept his word, so he refused to agree to the bank's request and searched for another source of funds. The contractor who was to install the feed mill equipment suggested that he could get a loan for him from General Electric Finance Corporation. The interest rate was higher than at the bank, but his father was not asked to cosign, which made them both feel better about the deal.

Wendell worked on the mill project whenever he had free time throughout the rest of the school year, and on Labor Day 1962, the Murphy Milling Company, at Register, was opened for business. Wendell taught at the nearby East Duplin School and worked at the mill nights until about 9 or 10 and on weekends. He dreaded those last three years of teaching, but he commented that after the mill started, he was so "turned on by what I was doing there that I just couldn't have been happier. I was excited about going there each day." It was not long before the mill was making more money than he earned teaching. The mill was totally owned by Wendell, but his father was interested in its success and was there to help out. Wendell said that his father's presence was important because his name was like gold in the community and it helped attract customers.

One of the main services at the mill was shelling ear corn, which was sold by the local farmers. The farmers were paid 10¢ a bushel less than the market price for shell corn. The mill's sheller was capable of shelling 100 bushels an hour and was kept very busy from about mid-September until after Christmas each year and after that whenever farmers needed cash and delivered more corn. The Murphys quickly realized that they could profit by grinding the husks and the cobs and selling them at $20 a ton for cattle feed. Cobs and husks weighed from 10 to 12 pounds per bushel. This meant that each hour the sheller operated, between 1,000 and 1,200 pounds of cobs were produced, yielding $10 to $12 an hour additional income. In a 15-hour day of shelling and grinding, the daily income was increased by as much as $180. Competing area mills let their cobs and husks accumulate to be given away or burned, so the Murphys collected them to grind for resale.

The Murphy farm was adjacent to the mill, and Norman Murphy decided to finish feeder pigs on his own. It was easier and more profitable than raising tobacco, so he soon leased his quota to a neighboring farmer. Next, Wendell started his pig raising venture and found that he enjoyed the hog business even more than he did the mill. Neither man took any income from the mill for his labor, and both paid for the feed they purchased.

The mill business increased as the hog business expanded. Father and son discovered that the sideline hog business was a profitable one, and on Labor Day 1964, they formed a partnership, Murphy Farms, for the purpose of feeding hogs. The 1964–65 school year was Wendell's last, so starting in the summer of 1965, he was able to devote full time to the "opportunity" he was looking for.

Wendell was well aware of how successful contracting had been in the broiler business and could see no reason it would not work in the hog business. This prompted him to contact area farmers, most of whom were in need of supplemental income, about contract feeding for him. Murphy Farms used its generated profits from its feeder pig enterprise to purchase pigs, feed, wire and posts for fencing, and other necessary equipment and provided the veterinary services. The contract farmers provided land, water, and labor and were paid a dollar for each hog fed. Generally, they fed about 1,000 hogs per batch and finished three batches a year. This meant they could make as much as most of them earned with their tobacco contracts. Murphy stood all the risk. The North Carolina farmers liked getting into a business where they did not have to assume any risk, and contracting grew rapidly.

THE YEAR 1968 provided several related turning points for the fledgling business. In 1964 the first corn combine entered the area, and Wendell sensed that as more were introduced, his company's shelling and lucrative cob grinding business would soon be at an end. So it was decided that the mill output should be entirely for Murphy hogs, and the business was closed to the public. Starting in early 1968, Murphy Milling Company provided feed and financing only for farmers who contracted with it to feed hogs.

Pete Murphy decided that he had spent enough time in college and was eager to join his father and his brother in the business. Wendell admired and respected his father, but the two had constant disagreements about how to run the business. His father always resisted expansion and never looked for the easy way to do things, whereas Wendell always looked for ways to become more efficient and charge ahead. Wendell decided he did not want a partner in Murphy Milling nor did he want a third person in Murphy Farms. Instead, he paid Pete $100 a week to work at the mill, grinding corn and delivering feed to the hog farms. By then, the Murphys owned their own stock semitrailer rig, and when Pete was not busy at the mill, he delivered semi loads of hogs to the packer at Smithfield, Virginia.

Pete recalled a slogan on the wall of his economics room at college: "It is easier to become successful in an industry that has as much inefficiency as agriculture." He said that stuck with him and he decided he wanted to be an entrepreneur too. He was not satisfied being an employee of his brother and started feeding hogs on his own. Pete had beginner's luck, for he netted $7,500 on his first 500 hogs and another $7,500 on the next 1,000 head. In six months' time he made $15,000. In 1969 he married and, with his wife, Lynn, formed Quarter M Farms, which was their separate enterprise. However, he still worked for Murphy Farms buying and selling hogs and driving truck for the company.

By Labor Day 1968, both the milling and the hog enterprises had grown sufficiently that it was deemed necessary to establish a more formal structure for both. Wendell, as sole owner of Murphy Milling Company, incorporated that business. He and his father formed a class "C" corporation for Murphy Farms.

BY the late 1960s, buying feeder pigs to supply their own lots as well as for contract growers became a major task, and all three Murphys bought feeders when not otherwise occupied. About 1970 the telemarketing auction came into use and made buying easier. The state provided a grader for the auction, which gave the remote buyer some feeling of security about what was being bought. Then came the Quality Feeder Pig sale, which furnished a person who graded the pigs and proved to be another step in the right direction. Pete recalled that the best auction days were when the weather was bad and most farmers failed to show, so some "real buys" were available.

Buying feeder pigs from every available source exposed the buyer to a great deal of risk. In 1968, after four years of indiscriminate buying, the Murphys experienced their first severe setback as a result of that practice. They lost 3,000 pigs to hog cholera, which shut them down for a few months and tested their finances. But they could not turn back. The demand for feeder pigs outpaced the local supply, so they started buying in Tennessee, the number two feeder pig state in the nation. In 1972 that state was quarantined because of hog cholera, and the Murphys faced another challenge. Pete and his wife took their camper to one of the farms they leased in Tennessee and improvised some feeding facilities. Pete's father and some employees also moved there for the "duration." They purchased feeders and moved them to the leased farms, where the pigs were quarantined and fed for 30 days. If the veterinarian certified the pigs to be in good health after that period, they could be shipped to Rose Hill. During the epidemic, the Murphys had as many as 10,000 feeders on hand at any one time. In all, about 30,000 feeder pigs were procured in that manner.

Wendell was not aware that any other operator made the same effort to secure feeders. In any case, the hog market went into the $60 range, and the Murphys made more money in 1973 than in all previous years combined. But in 1974 things quickly changed when they were caught with high-priced feeders on hand and the market for finished hogs dropped sharply. Wendell said of that experience:

> I thought we were going broke, but the prices turned in time to give us a breakeven year. I quickly saw that if we stayed in all the time, the cycles would even out and we could float. From 1962 through 1997, we never had a losing year.

THE MURPHYS purchased most of their soybean meal from the Central Soya Company branch in South Carolina and by 1967 were one of Central Soya's biggest customers. Although they were making good profits, they could not generate capital as rapidly as they needed. By 1967 they had outgrown the Rose Hill bank, so they did short-term borrowing from Central Soya. In the process they became well acquainted with Jim Stocker, finance manager for Central Soya.

Wendell wanted to employ a professional finance person. He was not sure, however, if he had attained an adequate volume to justify such a move. Wendell was impressed with Stocker's financial expertise and approached him about coming to work for Murphy Farms. Stocker liked the challenge of being in on the ground floor with a business that was leading the way in changing the pork production industry and was pleased to be a participant in a stock incentive program. In July 1973 he became Murphy Farms' twenty-fourth employee.

Shortly after employing Stocker, the company started converting to confinement finishing, with Marion Sasser the first contractor to do so. In 1977 the company built its first confinement structure. The Murphys' experience caused them to encourage their contractors with farrowing and nursery facilities to switch to confinement. Stocker commented that going to confinement was "a no brainer, because the animals were so much healthier, they grew much faster, the death loss declined sharply, it was environmentally friendly, and it was far more economical." In 1985 the last pigs were finished outside, and in 1993 the last sows farrowed outdoors. By then the entire industry was rapidly shifting to confinement.

Constructing a new pig feeding facility in the 1960s. This is a typical early outdoor feeder pig lot used by the Murphys. The upright feed storage was used to hold a total mixed ration. Feed was taken from the storage bin and placed in individual feeders stationed near each shed. Fences were not yet erected, nor were waterers or feeders in place. A great many acres were required to feed a limited number of pigs using this type of facility.

Local farmers soon realized how well contracting worked. Wendell shielded them from the market risk by giving them a guaranteed return, and initially he advanced money for confinement facilities when the local banks shied away from doing so. Farmers flocked to secure contracts because contracting was the best way to supplement their marginal farm incomes.

The demand for feed increased with every newly signed contract, and because the Murphys wanted to assure themselves of the best possible ration, they had to control its manufacture. In 1974 they purchased elevators at Mount Olive and La Grange and shortly after leased the elevator at Chadbourn, all in North Carolina. These facilities served as assembly points as well as places for additional feed manufacture. In the meantime, Pete, with backing from his father, incorporated Quarter M Farms and expanded his contracting ventures rapidly. He continued to buy feeders for all three Murphys, which was becoming an increasingly more difficult task because the available supply was not expanding as rapidly as the growing demand.

In November 1979 Murphy Farms took the inevitable step and entered the farrowing business. It contracted with Sands, of Nebraska, to build it a complete complex, starting with a 650-sow farrowing facility, then the needed nursery barns, and finally the finishing units. Sands was engaged because at that time the builder was the most experienced in the country. Sands was accustomed to the more extreme Midwest climate and built to suit those conditions. After using the facilities through a full weather cycle, the Murphys learned that they could get the same performance with less costly structures because of the milder conditions in North Carolina. They formed a construction department for building future facilities both for themselves and for contract producers.

For several years the Murphys had worked with Smithfield Foods searching for premium hogs that were more profitable for both the producer and the processor and more satisfying to the consumer. Now, with their own facilities, they could breed the hogs that not only performed best in production but were most suitable to the packer. With complete confinement facilities, they were able to control totally what took place from the time breeding stock was selected until the carcasses were on the packer's rail. They could pinpoint exact cost through the entire life cycle of their animals. This provided them with guidelines for their contract producers in all three phases—farrowing, nursery, and finishing. The breeding and farrowing facilities required topnotch managers, so the company took great care in selecting those contractors and seeing that they were the best paid. The sow units needed and received longer-term contracts with better built-in incentives for performance.

Wendell commented that a sow farm was very management intensive but that he did not know of any investment a farmer could make that would pay out as fast. All contractors were paid on pigs run through their facilities. He proudly stated that the Murphys have not had a single contract grower fail because of their contract operation. At the same time, contrary to all the negative publicity about contract farming, he was well aware of many who became millionaires in the process. He also stressed that regardless of the production phase, the company facilities have always produced for less than the contractors, so the Murphys knew they were doing something correct. Initially, they had hoped to produce 50 percent of each phase—farrowing, nursery, and finishing—but because so much less management is needed in finishing, most contractors preferred doing that. This led the company into operating a greater portion of the sow units.

Until they constructed the combined facility in 1979 and had to stock it with sows, the Murphys always operated on their own money or what they borrowed from the Rose Hill bank or Central Soya. They had not partnered with outside investors, and they never applied for or received a government loan. However, after they got into the farrow-to-finish business, the turnover of capital was much slower. This was because they had to wait through the full breeding and finishing cycle of at least nine months instead of the feeder pig finishing cycle of three months plus. They made a profit every year and felt confident about asking local investors to do contract feeding as a sideline activity.

Because of all the construction that the Murphys and the independent contractors were doing, very large sums of money were needed, and neither the Farm Credit System nor area banks wanted to finance fixed assets for contract producers. It was not until about 1986 that those agencies realized that contract producers "were making good money" and became receptive to helping.

In the meantime, the International Paper Credit Corporation, a leasing company, was approached to help individuals with limited resources get started, provided they had contracts. The contractors who received financing did very well for themselves and at the same time helped Murphy Farms become one of the nation's top hog producers. Confinement took much of the volatility out of the hog industry, because the more business-oriented operators did not go in and out of production with each swing of the rapidly changing cycles, as did those who used open range and minimum facilities. The old-style hog farmers went in and out of business so frequently that financiers could not keep accurate track of what they were doing.

Soon after the Murphys got into contracting sow units, they experienced a surprise about economy of scale in that sector of the hog business. Until the 1970s, most breeding and farrowing had taken place on open range, and the Murphys had no idea what problems they would encounter by going to confinement. Initially, they contracted sow operations of 200 to 250 head. This was an easy size unit for a husband and wife to handle, but the couple had to be present seven days a week. That size unit worked so smoothly that the Murphys increased to 500 sows. They discovered that three people could handle twice as many sows as two and could operate the larger unit more smoothly than the smaller one. More important, this arrangement provided a break from the seven-day schedule. Then, the Murphys went to 1,000 sows, which required a crew of five. This allowed for another per unit reduction in labor cost. They increased to 1,200 sows with a work force of five. That seemed to be a very good ratio. They continued expanding to units of 2,400 sows and then 3,600.

Every expansion brought greater economies. By the mid-1990s the larger sow units had grown to 11,000 head requiring only 35 workers. Accurate records, combined with a good training program and improved handling facilities, worked together to achieve new levels of economy.

HARRY PARKER is one of the Murphys' very successful contract farrowers. After Parker had completed 3¹/₂ years of college, he said he was ready to spread his wings. He worked at different jobs until he found one to his liking as a fire fighter for the government. That job gave him time to farm on the side. He grew up on a farm that had hogs, and he liked working with them. He realized that even with all the benefits his job offered, he did not want to continue working for the government. Parker was 14 years younger than Wendell Murphy but had known him since he was a boy and had always liked him and the rest of the family. When he decided to change careers, he went directly to Wendell, who convinced him that his 80-acre farm would be adequate for a sow operation large enough to meet his financial goals.

In 1989 Parker started contracting with a 525-sow unit. His wife, a registered nurse, worked with him for four years "to learn the ropes" and is satisfied that she could run the business if something happened to him. At one time he was incapacitated because of an encounter with a boar and was thankful that she could be there to oversee the work.

Parker worked closely with his employees, so everyone knew all positions. After six years he had virtually eliminated his debt, so he went to four barns and 1,425 sows. He has four men and two women working with him. Although he stressed that he had good employees, his biggest task has been keeping them motivated. He commented that he has a friend in Chile who owns a very large hog operation. This friend has little trouble getting people to do exactly as they are told and suggested that the high unemployment rate in Chile was probably the reason for the difference.

Parker is a very precise manager and pointed out how important it is to do each procedure properly. "Hogs are very intelligent animals and respond to proper treatment." He gave an example about how his crew watches young gilts when they first arrive. If the gilts are nervous, various workers spend 15 minutes several times a day just walking among them and being friendly to them. In about four days, their nervousness is gone. His facilities are spotless: "There is no room for unsanitary conditions in this business." To get people to do exactly as told is so necessary because "the hardest job to do is a job that wasn't done correctly the first time."

Parker has an incentive program that is worth up to $8,000 per employee per year. All the employees have to work together. If things are not done correctly, all share the responsibility, but if all is well, they all share the rewards. His workers are on a 42-hour week and have a three-day weekend every third week, with one day off the other two weeks. One of his female employees is the assistant manager and has total responsibility when he is not there. He has never advertised for help because his workers are very good about referring their friends if there is a vacancy.

Parker enjoys the freedom the Murphys give him but is quick to add that if he has a problem, they are excellent in providing all the support needed. He is totally pleased with his decision 11 years earlier to enter the hog business. When asked how he was affected during the downturn in the hog business in 1998 and 1999, he replied that the contract called for a guaranteed income, which was not changed in those years. He added that that was not true for independent producers.

BY the mid-1980s the company had grown large enough that the Murphys considered diversifying. Not all hogs gain weight at the same rate. After the first thousand hogs were removed from a building because they were market ready, the Murphys were obligated to pay the full cost of the finishing barn and labor for the remaining 200 head. If they had their own processing plant, they could sell the stragglers without discounting them. They felt the pork industry was solidly based and looked at the possibilities of constructing a meat packing plant. So many synergies are involved by controlling every link in the food chain.

The political climate was not favorable for building a packing plant. Environmental regulations would have added greatly to the cost if a facility had been approved. Consequently, the Murphys dropped the idea. Instead, they decided to stick with what they knew best, raising hogs, and to diversify by moving into other locations. They chose to expand in Iowa, long the leading hog state, where feed costs were lower than in North Carolina. Also, more processors were present, and marketing options were greater. The Midwest was still in the agricultural recession of the 1980s, and opportunities were available.

RANDY STOECKER, a graduate in agricultural economics from Kansas State University, had been a leader in Pig Improvement Company (PIC), which developed a major innovation for the industry. PIC had established the concept of segregated early weaning, which called for removing the pigs from their mother 17 or 21 days after birth. Apparently the maternal immunity from the sow peaked at 14 days, and the quicker the piglets could be moved to the nursery, the more advantageous it was. The Murphys were the first large operators to adopt that concept commercially.

In 1987 Stoecker was employed by the Murphys and named group vice president in charge of production of midwestern operations. One of the first actions after Stoecker was employed was to purchase Plainview Hog Company, of Sioux City, Iowa. Plainview had 83,000 feeder pigs on inventory with contract finishers in a large variety of facilities, many of which were old barns not well adapted to protecting the pigs against the weather. In addition, inferior or obsolete feeding equipment was permitting too much feed waste. Both of the above were responsible for a low feeding efficiency. The Murphys knew they could correct the equipment and the buildings to provide better conditions for the hogs. Initially, feed consumption was 3.9 pounds of feed to 1 pound of gain. By improving the ration, it was reduced to 3.3 pounds. Better feeding equipment reduced the requirement to 3.0 pounds, and later, when proper confinement buildings were constructed, the feeding ratio fell to 2.7 pounds to 1 pound of gain. That represented an increase in feed efficiency of 31 percent. Better feed efficiency is important economically. At the same time, it greatly reduces the environmental problem, because the more efficiently a hog utilizes the feed, the less manure there is. After the corrections were made, the Murphys' records proved that they could produce hogs 4¢ a pound cheaper in Iowa than in North Carolina because of lower feed costs.

However, the Murphys experienced an unexpected surprise in the Midwest. Contract farming was first used in the 1920s in watermelon production in the southeastern states. Watermelons were a volatile crop requiring a rapid rotation to control disease. In the 1940s contracting started in the poultry business because of market volatility, and by the 1960s contracting covered almost 100 percent of broiler production. By 1981 contractual integration covered almost 32 percent of all agricultural sales, but somehow Midwesterners did not understand contracting. "In fact, it almost appeared as if they saw contracting as something evil." In North Carolina, small-scale farmers who needed supplemental income looked upon contract production as a viable alternative to doing off-farm work or even exiting from farming. But that was not the case in the Midwest, even though many farmers were groping for opportunities to survive.

After Stoecker realized how strong the emotions were against contracting, he reasoned that it would be best to take a "go easy" approach. Care was taken to select a good independent contractor, and only one confinement facility was constructed the first year. That provided good positive results, and 10 contracts and facilities went into effect the second year. The younger, more progressive farmers understood the advantages and were quick to accept contracts. By April 2000 about 900 finishers were under contract in Iowa alone. Other companies had similar success, and by then more than 65 percent of the new confinement buildings being constructed in the state were by contract producers.

———◆———

OSCAR WENHOLM, JR., of South Dakota, is a midwestern hog raiser who felt that the business environment had changed and that farmers should be willing to adapt to the new conditions. He and his wife were at the point where they were going to sell their hog equipment. After more than 30 years of feeding hogs, they felt they were not being well rewarded for their effort. They realized that they could no longer "afford to farm that way." Then, Wenholm started contracting with Murphy Farms, "the best business partner I can imagine," and things improved for the couple. Wenholm has a 3,300-hog finishing site, which means he can finish more than 6,000 hogs annually, and uses 55,000 bushels of corn and 600 tons of soybean meal. In 1998 South Dakotans, in an emotionally charged contest, voted to ban corporations from owning livestock. Fortunately, existing contract growers were grandfathered in and can continue to operate.

———◆———

ONE OF THE BIGGEST OBJECTIONS of those who opposed the entry of Murphy Family Farms into Iowa was that "corporate factory farms smelled." The Murphys had experienced the same campaign in North Carolina, so they were better prepared to confront the problem. Being aware of the different climatic conditions of the Midwest and after working closely with Iowa State University, the Murphys decided that new contract producers should install Slurrystores for aboveground manure storage and handling. A Slurrystore is made of glass-lined steel plates bolted together to form a large, circular, aboveground tank with a cement bottom. Above-ground storage, instead of lagoons, was necessary to prevent seepage into underground water. To reduce odor, chopped barley straw is blown into the Slurrystore and floats on the manure, forming a crust. This becomes a bio-cover, which has minimized odor problems "except for those who protest because of a mind-set." When the Slurrystore is ready to be emptied, an agitator pump thoroughly mixes the barley straw with the manure to provide the maximum benefit when it is injected into the soil. The Slurrystores have virtually provided total protection against environmental problems and reduced the odor problems. At the same time, applying the manure on the soil makes the system cost effective. The Murphys and Iowa State Research and Development (R&D) are working on developing microbes that will consume any offensive gases.

ACQUISITIONS continued. In 1988 Murphy's became the nation's largest hog producer and, based on total sales, ranked twenty-first among privately owned companies in North Carolina. The following year Murphy's achieved forty-ninth ranking among the largest farms in the nation. In 1991 Murphy of Missouri was established specifically to develop sow farms and nurseries. The 50,000 sows in Missouri were expected to supply feeder pigs first to the contract finishers in Iowa and then to those in North Carolina, Oklahoma, Illinois, South Dakota, and eventually Utah.

Murphy's hoped to continue partnering with independent contractors just as it had since 1964. The company helped contractors obtain financing and gave advice on construction. The producers furnished land, labor, and facilities but assumed no risk. In return they received an established price for pigs raised based on performance. By the mid-1990s more than 600 grower partners were working with Murphy's, and not since 1975 had there been a bad loan among the contractors. In North Carolina, the company ran a mix of company and contractor facilities; in Missouri, the nurseries were 100 percent contractor owned; and in Iowa, finishing was done 100 percent by contract farmers. The company had one of the nation's largest R&D facilities dedicated to applied research in nutrition, genetics, environment, and equipment. All the knowledge gained was available to its business partners.

BY the late 1980s Pete Murphy's Quarter M Farms was responsible for more than 20 percent of the hogs produced under contract for Murphy Farms. In 1989 Wendell Murphy's son, Wendell H. Murphy, Jr., nicknamed Dell (hereafter Dell), founded DM Farms, made up of

6,000 acres. With facilities for 16,000 sows, it finished 325,000 hogs annually. Bryan Allen heads the sow service for DM Farms. Allen's wife, Allison, has finishing facilities under contract, as do both her parents. Dell and his sister are partnered with a contractor on a series of farms with more than 50,000 sows. In 1990 Joyce Murphy Norman established a contract finishing operation that she called BAZ after family members.

THE COMPANY eventually had nine elevators within a 100-mile radius of Rose Hill, with a total holding capacity of 2,957,829 bushels. They were used for assembling corn and soybeans for the mills. The Register Mill, founded in 1962, was expanded and, by 1990, produced 4,000 tons of mixed feed weekly. On March 19, 1984, Norman Murphy purchased the Rose Mary Mill, at Rose Hill, for $1 million. That purchase represented 100 times more in dollars than the note Wendell wanted him to cosign in 1962. The Rose Mary Mill initially had a weekly capacity of 6,500 tons, but in 1992 it was upgraded to 10,000 tons and produced four grower rations. In January 1994 The Chief Mill, named in honor of Norman Murphy, was opened at Rose Hill. It had a weekly capacity of 25,000 tons and produced eight different rations. At the time of its dedication, it was reputedly the world's largest mill. The Chief has a capacity of 1.6 million bushels and receives three 65-car trains (700,000 bushels) of corn each week from Ohio and farther west. Another 250 semi loads of soybeans, mostly from the southeastern states, arrive weekly. In addition, fat and trace elements are needed to complete the scientifically mixed and pelleted ration. Each of the 35 company semi feed trucks holds 22 tons and hauls four loads during each of two shifts five days a week, delivering 30,800 tons to the hog facilities. A staff of 19 works in two shifts in The Chief Mill to produce that output. In addition, the company has feed mills in Algona and Manning, Iowa, each capable of preparing 10,000 tons weekly of a total mixed ration.

In 1993 Murphy's, Smithfield Foods, Carroll Foods, and Prestage Farms created Circle Four Farms on 50,000 acres of scrub land at Milford, Utah. This was a totally integrated hog facility, with a feed mill and with breeding, farrowing, nursery, and finishing units for producing 2.5 million hogs annually for an accompanying packing plant. The facility was located in the western United States to put it

The Chief opened in 1994. The completely computerized facility enables 19 workers on two shifts to manufacture more than 1,000 semi loads of a total mixed ration every five days.

closer to the new markets in the Pacific Rim nations. The Beaver County, Utah, area receives only 10 inches of rain annually. This enables most of the liquid waste to evaporate in the desert air. The solids in the lagoons are treated with microbes for eventual use as fertilizer. Utah State University established a swine management program with internships at the facility. In June 1998 Murphy's sold its 45,000 sows in Circle Four to Smithfield Foods.

◆

WHEN Wendell Murphy was asked what was the most difficult obstacle he had faced in business, he replied:

I do not remember ever having any insurmountable problem until I came up against the environmental problems starting after all the publicity in 1995. Since 1995 I have spent nearly all my time on political and environmental considerations. I cannot run my business. I have to rely on others to do that job. . . . They [the media] are out to get us. They have tarnished our reputation, and that hurts us lots more than any money it cost us.

The media accuse us of being immoral, unethical, and illegal, especially regarding my 10 years in the legislature.

Murphy was particularly disturbed because all the time he was in the legislature (1982–1992), he did not recall that the media, and particularly the Raleigh *News & Observer*, had written anything negative about the way he voted while serving. It was not until 1995, well after he had left the North Carolina senate, that a series of articles was published. "When the media do an article, they do not have to prove their point."

Walt Cherry noted that the pork industry in North Carolina had to employ a public relations director to release factual information to counteract all the negative publicity put out by the media and by environmental and public action groups. "They know they have to ride the emotional aspect; they do not have to prove their points."

One of the most baffling problems the hog industry faces is that the technical specialists in conservation recommend that cities build lagoons for their sewage, but they will not defend lagoons for animal waste. Cherry added, "If farmers do it [spill from lagoons], it's bad, but cities can do it." He cited the number of discharges from hog lagoons and the number from municipal lagoons, which clearly indicated a dual standard discriminating against agricultural lagoons.

After the industry outgrew the historic pasture-type hog raising and the animals were housed in confinement buildings, the waste disposal problem became important. The farmers involved knew that animal waste had significant fertilizer value and obviously wanted to apply it to their best advantage. One reason confinement facilities were disbursed was to give ample acreage for each unit. This minimizes the cost of transporting the manure and maximizes its value. Murphy's has planted 13 varieties of trees on 5 different sites to use the manure on tree plantations. The most rapidly growing trees are ready for harvest in 7 to 10 years, after which a new crop is planted.

A waste monitoring system was established that links the more than 700 hog farms to the central computer. The level at each lagoon is recorded at the central office, and if any monitor indicates that a lagoon level is too high, trucks are dispatched to that site to remove some of the manure. If a lagoon at any contract grower's site is filled, that contractor cannot be restocked with hogs until the problem is remedied. Murphy's is serious about how the manure is disposed of and has established a zero-tolerance policy. In at least one case when

Murphy's was made aware of a waste mishandling problem, it removed all the hogs and the company's property from the contractor's farm.

When Pete Murphy was asked what his most serious obstacle to operating the business was, he said that initially it was financing. Then he expounded: "Today it is actually survival, because of environmental regulations. [Simply] being allowed to stay in business is more difficult than any other problem."

IN 1975, 56-year-old Norman Murphy had bypass heart surgery, but he surprised everyone by coming back "stronger than ever. He was back in the business right away." He remained active until his death in June 1990, less than two months from his seventy-first birthday. His demise led to a restructuring of Murphy Farms, in which he had held 50 percent of the stock. His will left equal shares to Wendell, Joyce, and Pete, which resulted in Wendell holding 66.66 percent of the stock and Joyce and Pete each holding 16.66 percent.

Wendell was only 52 years of age and still very much in charge, but his brother and his sister were well aware that he was showing the signs of pressure. Pete commented that Wendell's biggest problem

The home office of Murphy Family Farms, constructed in 1992 on the south edge of Rose Hill, North Carolina.

was that he "liked to micro-manage." Joyce called him a worrier. Wendell admitted: "I took after my daddy. I was too much of a detail man when I first started, but in later years I have changed." He added that he had a talent for spotting good people and motivating them to join Murphy's. Once they were there, he let them do their jobs because "I did not know the hog business." When asked what his greatest weakness was, he replied, "I was not firm enough when the [executive] committee sometimes was making what I thought was a wrong decision. Generally, the results proved that my instinct was correct."

His sister and his brother obviously were content to let Wendell be in charge. Pete admitted that his best break in life was being Wendell's younger brother and "having him pave the way." His weakness was his temptation to enjoy success. The three got along with minimum friction, which is not typical in many family-run businesses.

Jim Stocker, who was employed in 1973 and was president of the company, a member of the board, and very much a confidant of Wendell, commented, "Wendell felt we were getting a little stale and needed a new leader to get some spark." Stocker, Pete, and Dell made up the search committee for a new president. They selected Jerry Godwin, president of Champion Spark Plug Company. It is their consensus that today the company has the best management team in the industry. In 1995 Murphy Farms became Murphy Family Farms, which encompassed Murphy Farms, Quarter M Farms, BAZ, and DM Farms.

ON LABOR DAY 1999, 37 years to the day that Murphy Farms was founded, a letter of intent to merge with Smithfield Foods was signed. Talks of merging with Smithfield started in the late 1980s because "it was inevitable that we had to be in the packing industry." As time passed it became obvious that getting permits to build a processing plant would be impossible. If the permits could have been secured, the cost of meeting the requirements would have made a plant too costly. Walt Cherry commented that the biggest obstacle to pork production in North Carolina is the lack of good processors. "The political climate has kept processors out. The activists have stopped at least two."

Wendell emphasized that he realized "[it] was too late, because of environmental considerations, to get into the packing industry. We did

not merge with Smithfield because of *money problems*. But this was the time to do it. Total integration will be here in the *not-too-distant future*. The water is running downhill. There are too many synergies for it not to happen."

Wendell was very emphatic that the agreement with Smithfield was a merger, *not a sale*. He stressed that 1996 and 1997 were the two most profitable years ever for Murphy's and that the losses of 1998 and 1999 were not nearly as great as the profits of the previous high years, so the company still had ample cash reserves. This writer made a point of the fact that many media articles stressed that Smithfield had purchased Murphy's. The Murphy people commented that the transaction was a transfer of stock, that no cash changed hands, and that not all of Murphy Family Farms was involved. Wendell holds more than 20 percent of Smithfield stock.

During the downturn in the hog industry from July 1998 to October 1999, many small hog producers looked to the government to step in and help them. Others wanted regulations enacted to stop processors from owning hogs. It is the consensus of Murphy leadership that both of the above would have been backward steps. Today, in addition to its involvement with Smithfield, Murphy Family Enterprises, owned by family members, retains the Register Mill, five grain elevators, several hog enterprises, and extensive real estate holdings, including substantial development property. The more than 1,600 independent farmer contractors and about 1,700 employees know that Wendell Murphy, more than any other individual, brought the pork industry into a new age.

I knew before I met Wendell Murphy that my interview with him would be a pleasant experience. On the Sunday before I met him, I had a conversation with a local clergyman who did not know my purpose for being in the area. He said to me, "The Murphys have done more good for this area than most people will ever know." From that comment I learned that Wendell and his siblings had come from humble beginnings and had not forgotten those roots.

— Hiram M. Drache

Bibliography

Interviews

Allen, Bryan. Rose Hill, NC, February 22, 2000.

Cherry, Walt. Raleigh, NC, February 24, 2000.

Murphy, Harry D. ("Pete"). Rose Hill, NC, February 22, 2000.

Murphy, Wendell H. Rose Hill, NC, February 21, 2000.

Murphy, Wendell H., Jr. ("Dell"). Rose Hill, NC, February 22, 2000.

Norman, Joyce Murphy. Rose Hill, NC, February 22, 2000.

Parker, Harry C. Chinquapin, NC. Telephone interview, May 8, 2000.

Stocker, Jim. Rose Hill, NC, February 22–23, 2000.

Stoecker, Randy. Ames, IA. Telephone interview, May 2, 2000.

Miscellaneous

Johnson, Charles. "Wendell's Way: America's Hog Magnate Started with Nothing and Says You Could Have Done It, Too," *Top Producer*, Mid-September 1992, 8–11.

(Raleigh) News & Observer, December 2, 1995, January 7, 1996, August 3, 1997.

Successful Farming, Vol. 96, No. 1, January 1998, 22.

Wenholm, Oscar, Jr. "Contracting Saved My Farm," *Successful Farming*, Vol. 98, No. 4, March 2000, Special Bonus Page.

NAPI
(Navajo Agricultural
Products Industry)

Culture Versus Economics

While flying into Farmington, New Mexico, in anticipation of interviewing the management team of a large irrigated agricultural operation, all I could see was rugged rock outcropping in an otherwise barren desert. I wondered, Where can anyone farm in this country? Especially 110,630 acres of irrigated land! The next day a friend from North Dakota, now living in Farmington, and I drove to the headquarters of NAPI (Navajo Agricultural Products Industry) located south of Farmington on New Mexico Route 371. After finding the headquarters, we surmised that the farm should not be far. We headed south and east on an excellent hard-surfaced road. Suddenly we crested a hill, and there ahead of us lay a large plain dotted with irrigation towers.

We stopped in awe, for here, at 5,600 feet elevation, was a desert plateau waiting for water to produce a bountiful harvest. I thought, What potential, a farmer's dream, 10,000 acres in one block under virtually controlled conditions. Satisfied with our initial observation, we pointed our car west toward Shiprock, New Mexico, and then back to Farmington. That evening all I could think about was what an exciting experience it was going to be to learn about what, when finished, would "reputedly be the largest single-site farm under irrigation in the world."

— Hiram M. Drache

Part of the last leg of the 43-mile main canal, which, in its journey from Navajo Dam, crosses the desert with the use of five tunnels and nine laterals. (Photo by Andy Rohall)

BEFORE learning what the past and present management personnel have to say about NAPI, it is important to know something of the geographical and historical setting. In the period after the Civil War, our nation experienced a rapid growth. Western lands were being settled by immigrants and eastern Americans who sought to fulfill their dreams of owning land as provided by the Homestead Act of 1862. The tremendous growth in agricultural production provided the nation with the abundance of food, fiber, and other farm products needed to develop its urban industrial society. As settlers moved west, they soon overran land that Native Americans had called home for centuries. In the ensuing struggle, the natives were forced onto reservations.

One of the largest reservations was to become the home of the Navajo Nation. It contains 16 million acres—25,000 square miles—in the Four Corners area where Arizona, Colorado, New Mexico, and Utah meet. A treaty dated 1868 between the Navajos and the U.S. government provided that besides the land, there was also to be assistance for farming. In 1908 the benefits were enhanced when the

Supreme Court decreed in the Winter's Doctrine that the Indians had rights to waters touching their reservation in any way. It was presumed that Congress intended that the Indians should have sufficient water to satisfy the needs of the reservation and that the federally reserved water rights dated to the creation of the reservation regardless of whether water had been used.

In 1914 appropriations were made to the Bureau of Indian Affairs (BIA) to provide for the stipulations of the treaty. The Snyder Act of 1921 authorized the BIA to upgrade existing (Navajo-built) irrigation systems. The position of the Indians relative to irrigation systems was further enhanced on July 1, 1932, with the passage of the Leavitt Act, which provided that repayment of construction costs associated with government irrigation projects on Indian land is deferred as long as the land remains under the ownership of the Indians.

During the droughts of the 1930s, the Navajos experienced major losses to their cattle herds. They asked for a 200,000-acre irrigation project to compensate for their livestock losses. In 1945, in conjunction with the above, the BIA proposed the 100,000-acre Shiprock project, which supposedly the Bureau of Reclamation (BOR) supported. The purpose of the Shiprock project was to teach the Navajos about updated practices of irrigated farming.

The next activity that moved the project ahead did not take place until 1957. That year the state of New Mexico granted the Navajo Indian Irrigation Project (NIIP) 630,000 acre-feet of water for irrigation, power, and domestic purposes. The NIIP became jointly funded by the BIA and the BOR.

During the next five years, the Navajo Dam was constructed. This was followed by the authorization of the NIIP and San Juan–Chama Canal Project. An appropriation of $135 million, in 1961 equivalency, was to be used to create an infrastructure to irrigate 110,630 acres as well as to serve water contracts for municipal and industrial uses. The avowed purpose of the project was to provide employment for 18,000 individuals on family farms and related enterprises.

A vision statement written at that time read: "The collective vision . . . saw NIIP as a method of economic development for the Navajo Nation and the Four Corners Region. The large 'agricultural park' was to be used as the foundation for integrated industries to create jobs using a renewable resource for the depressed area of the Southwest." The BOR restudied the project and concluded that the plan was not good, nor was the choice of lands to be served. The revised cost estimate as of January 1966 was $197 million.

In 1966 a joint BIA-BOR state committee reevaluated the project to "optimize the benefits to the Indians from the water allotted to NIIP." At that point the NIIP stated that its objective was "to bring the highest economic returns to the Navajo consistent with optimum and desired levels of living and community life." The land was to be developed at the rate of 10,000 acres per year from 1971 through 1981. In 1966 the tribe was still determined to establish individual family farms, but by then total projected employment opportunities were reduced to 8,800. This was a realization that labor-intensive agriculture was no longer in vogue, but the guiding philosophy of the project was that agriculture should continue to be a focal point of the Navajo Nation.

However, as time passed, the thought of allocating the land to 1,120 individual farmers in parcels of less than 100 acres each no longer seemed feasible. Too few of those potential Navajo farmers had the necessary expertise to properly manage an irrigated farm on even that limited scale. Fewer still had the means to raise the capital needed. Even more significant was the fact that Navajo youth were now becoming better educated, and once they realized the opportunities available to them elsewhere, they were reluctant to return to the reservation to work either for the government or in agriculture. The Tribal Council abandoned its plans for individual farms and on May 11, 1967, decided in favor of forming a tribal corporate enterprise— Navajo Farm Enterprise. This entity was to be managed and operated by the Navajos, possibly along the lines of a commune. Somehow the idea of a commune did not seem proper, and on April 16, 1970, the Tribal Council created Navajo Agricultural Products Industry, which was structured along corporate lines, and assumed financial responsibility for its operation.

The establishment of NAPI marked a real turning point for the NIIP because it reflected a change in the philosophy from the traditional agrarian life style of the Navajos to a more modern type of agriculture. A New Mexico State University Experiment Station was established nearby to work with the Navajos in determining the most feasible farming practices, and an earnest effort was made to get the project moving.

AFTER graduation from college with a degree in agricultural engineering, Albert Keller took a position with the BIA. In 1970 he was assigned to the NAPI project at Shiprock, which entailed overseeing 2,000 acres of irrigated alfalfa and a complementary livestock operation. This line of agriculture was similar to what the Navajos had practiced, and it proved to be a "very successful operation." The appropriation ceiling for the project was increased to $206 million.

IN 1973 the BIA called for preparation of an environmental impact statement and developed a comprehensive plan for the development of the irrigation system. The Navajo Tribal Council gave NAPI complete jurisdiction over NIIP real property. One of the major decisions made then was that all the land should be watered with center-pivot irrigators (an all-sprinkler system rather than the commonly used system that needed to be manually moved). Center pivots required less-experienced individuals to operate them and reduced water needs. It was estimated that 330,000 acre-feet of water annually would be ample. The Office of Management and Budget (OMB) agreed to the more costly but more efficient center-pivot system and revised its estimated costs to $281.4 million.

In 1975 New Mexico State University made an agreement with the Navajo Community College System to assist in training managers, supervisors, and technical personnel for the proposed NAPI operation. In November the first water was released, and on April 10, 1976, the first water was delivered to the 10,000 acres in Block 1. Everything was looking up for the NIIP and, hence, NAPI. From then until 1980 was the NIIP's most favorable level of federal appropriations—about $24 million annually for design and construction. Progress proceeded rapidly, and in 1978 the first water was delivered to the 10,000 acres in Block 2, followed in 1979 by the arrival of water to Block 3. In 1978 and 1979, respectively, work commenced on Blocks 4 and 5.

From 1976 through 1978, NAPI had hired several Navajo general managers but then fired them because they were not prepared to handle startup problems. After only three years of operating Block 1 and one year of Block 2, the cumulative operating losses were more than $19 million, nearly $500 per acre per year. The sizeable debt caused widespread alarm in the Tribal Council, which backed down from its firm stand on Navajo preference for all available positions. With the

Main canal and headquarters (right center) of Block 1, with a view of part of the 10,000 acres that were opened to irrigation April 10, 1976. (Photo by Andy Rohall)

beginning of the 1979 crop year, it employed Ball Agricultural Systems, Inc., to manage NAPI. Ironically, despite the accumulated losses, an elaborate office complex was constructed at that time.

In 1980, after the new administration took over in Washington, D.C., it requested a reduction in all domestic spending for the next eight years. This stopped most construction on the NIIP project. That request was followed by a study called for by the OMB to determine the validity of the NIIP. An inspector from the department recommended stopping the project and, after consultation with those working with the Endangered Species Act, prevented work beyond Block 8. As a result of that action, the BOR reduced its Farmington staff from 100 to 5 and considered closing its office there.

In conjunction with the above activity, the BIA employed Boyle Engineering to investigate the status of the NIIP and NAPI. The Boyle study suggested that block development slow down to provide time for stabilization and maturation of NAPI's agribusiness activities. An independent examination by the Department of the Interior (DOI) declared that NAPI was a viable enterprise but needed to improve its management and that the NIIP should continue after reorganization. In 1982, after the DOI report, Ball Agricultural Systems, which had incurred major losses, was discharged from its management contract. Albert Keller, who was still employed with the BIA, was requested by the Tribal Council and the NAPI board to do the firing. The next Monday Keller was in Washington, D.C., where he received a reprimand from the Assistant Secretary of the BIA. The agency did not look upon his activity with favor, but Keller had endeared himself to the Navajo leadership.

The DOI then recommended that the BIA and the Navajo Nation absorb the accumulated losses of NAPI and contribute funds for additional working capital. In 1984 the NAPI debt was restructured, with the BIA and the tribe each contributing more than $30 million. In the meantime, a joint task force from the BIA and the BOR determined that NAPI still had sufficient potential to cover costs at that time. It stressed that NAPI should emphasize production rather than concentrate on being an employment agency. It specified that its cropping program should be aimed at crops that would provide a solid foundation for vertical integration and that the Navajo Nation should provide capital for such a program.

In 1984 a Presidential Commission on Indian Reservations made the following recommendations to the President:

That there be a modernization of tribal government to separate executive, legislative, and judicial powers to stabilize non-Indian concerns relative to sovereign immunity;

That there be a reorganization of trust responsibilities to allow for 99-year leases of land so it could be used as collateral;

That tribal enterprise be privatized to separate the council's governmental functions from its for-profit businesses;

That the Tribes have authority over the state and federal governments to solve the problem of dual taxation;

And that the Tribes be given the same position as local and state political units to issue industrial revenue bonds.

As of February 2000, very little had been done relative to implementing these recommendations except some minor changes regarding dual taxation, specifically on the sale of motor fuels. A provision was made that the tribe be permitted to issue industrial revenue bonds for construction of a potato processing plant.

The tribal leaders were being forced by the financial institutions to make a change. They realized that they could not continue under the leadership of Ball Agricultural Systems but were at a loss as to where to turn for management. In 1983 and 1984, under successive managers, heavy losses were sustained, which, in 1985, necessitated a restructuring of the accumulated debt previously referred to. In December 1984 Albert Keller retired after 30 years with the BIA and was asked by the chairman of the Tribal Council if he would take over as general manager effective January 1985. He asked for and was given assurance that he would have a free hand in management.

His first challenge was convincing First Interstate of Arizona to lend NAPI operating money. The cash shortage was so serious that he felt compelled to have his 10 top staff members personally be responsible for their vehicles, which they paid for on a monthly basis until funds became available. Keller reinforced the management staff by employing Don Hunt, a potato specialist, and Charley Higgins, an agronomist. Gerry Anderson, a professionally trained animal scientist, who had been with the Ball group, stayed on to manage the feedlot. Anderson's backup manager was LoRenzo Bates, whose degree was in range science and who had been working with NAPI's cattle program since 1975. He spent eight years under Anderson and gained the respect of Keller as one of the up-and-coming Navajos. Those five—Keller, Hunt, Higgins, Anderson, and Bates—gave NAPI a professionally trained management staff, and they made a difference.

In 1985 Keller placed more emphasis on high-value crops, particularly potatoes, which eventually were increased to 6,000 acres. That helped to turn the tide, and a positive cash flow was established. Losses were reduced to $600,000 in 1985 and were cut in each successive year until 1988, when NAPI posted its first ever profit—a net of $361,190. An accounting change in 1989 stretched that year to 14 months, which may not have made a great difference, but in any case, a net profit of $4.9 million was recorded. That was followed in February 1991 with the 1990 annual report that indicated a profit of $1.5 million, making three successive years of profit.

IN THE MEANTIME, in 1988, after an audit by the Inspector General of the DOI, it was recommended that the project be terminated, with compensation to the Navajo Nation for losses. The DOI felt that costs were excessive and questioned the availability of water. It was believed that water would be more valuable for uses other than farming. This was especially so since the final 40,000 acres of the project would require additional energy to pump water to those fields, which would increase expenses. At that date the total investment in the project exceeded $300 million. The NAPI board of directors responded with a revised comprehensive plan for completion of the NIIP. Its report concluded that "completion of the project can generate significant social, economic, and environmental benefits." The BIA challenged the DOI's negative report and in 1990 engaged the Southwest Research and Development Company to restudy the project. The Southwest report demonstrated a positive cost/benefit analysis and supported the BIA's decision to complete the NIIP program. From that time, with the BIA's backing, the Tribal Council and the NAPI board have continued the development of the farm and related enterprises.

After the above action, the BOR restaffed its Farmington office in preparation for continued construction. Funding was sustained at about $25 million annually for continued development, but there were insufficient funds for operation, management, and on-farm development to keep pace with construction.

IN 1990 tribal politics came into play, and politics were of primary importance as far as the farm was concerned. During the time that Keller was general manager, Peter McDonald was elected council chairman and was later sent to prison for "taking tribal funds." During the 1990 campaign for tribal council delegates and chairman, Peterson Zah had promised that, if elected, he would put Navajos back into management positions. Keller realized that despite having excellent relations with the Navajos, he would be replaced. The track record of NAPI during his administration was immaterial.

LoRenzo Bates, current general manager, commented on the campaign's outcome:

> Tribal politics began to interfere with business decisions. The Nation decided to terminate all non-Navajo management personnel. This was probably the tribe's biggest mistake because there were not many trained Navajos to take over. The Nation just was not ready to take over management responsibility. For-

tunately, crop prices were good, which made up for management mistakes. But then the federal government began to question the NAPI project because it was losing so much money.

As soon as the election was over, Keller knew his time was limited, and during his final months, he was not able to be effective. He understood the culture and the politics involved and was not upset over his termination and that of his hired staff in August 1991. Except for Bates, Tsosie Lewis, who was trained in agronomy, and Pierre Dotson, who had a master's degree in business, there were no tribal members to provide the needed leadership expertise. That small core represented the first generation of trained Navajos who faced the task of developing the know-how and then teaching the next generation.

On May 26, 1992, Secretary of the Interior Manuel Lujan, a native of New Mexico, issued a memorandum to several agencies of his commitment to Chairman Peterson Zah of the Tribal Council to move construction of the NIIP without undue delay. This included construction of the Burnham Lateral (canal), the first of the facilities needed to provide water to Blocks 8 through 11. The development continued, and in 1997 NAPI devised a grand strategy to use the resources of the NIIP in a joint venture to capture a larger portion of the profit stream and increase employment opportunities. The Nation was to provide $10 million in equity to assist in forming a partnership with R. D. Offutt Company to build an $85 million potato processing plant. Subsequent to the above action, NAPI "proclaimed" a break-even in operating income for fiscal years 1996–97 and 1997–98.

KELLER had recommended that Bates be appointed to the general manager's position, but Bates was reluctant "because of the politics involved." The leaders then picked Tsosie Lewis, a 19-year veteran employee and a Navajo with a college degree. In the $4\frac{1}{2}$ years from the time Keller left until Bates was appointed in January 1996, Lewis and three other persons served as managers, two in the final year.

LoRenzo Bates had been raised on the reservation. His parents were employed by the BIA and lived on a small farming operation specializing in cattle. When the Tribe and the BIA gave land-use permits, his parents received 75 acres, of which 50 were irrigated. Theirs was one of the first farms on the canal system. In later years they increased their operation by leasing five sections of land chiefly for pasture.

After Bates graduated from Navajo Mission High School in Farmington, his goal was "to get as far away from the farm as I could because all it [the farm] was, was constant work and drudgery." He wanted to go to San Jose Community College in California, but his parents did not want him to go "that far away." He enrolled at Dixie Junior College in St. George, Utah. He said of the college: "I really liked it. It was a smaller place much to my liking. I was one of only three Native Americans." Each summer Bates returned home and worked on the farm.

After graduating from Dixie he transferred to Utah State University but soon dropped out because of low grades. He took a job at the Navajo coal mine, where he was well paid but was unhappy because the night shift interfered with his "night life." It did not take him long to realize that there was "no chance to go up the ladder" unless he finished college. He returned to Utah State and in 1974 received a B.S. degree, with a major in range science.

He learned that the BIA agency at Shiprock was looking for range scientists. A Comprehensive Employment Training Act (CETA) program was in place to train people for work on the NAPI project, which had just started. Bates was the only applicant out of 30 to be selected. His first assignment was as a ranch laborer, which paid $2.65 an hour.

In 1975 he was employed by NAPI as a trainee, with primary emphasis in handling the cattle herd. This led to his involvement in the 12,000-head NAPI feedlot. After a few years Bates was asked to head the feedlot, but he declined and continued to work under manager Gerry Anderson. During that time he learned about the politics involved and saw "what the general manager had to do." As he gained confidence he found himself challenging Anderson, but he realized that he would not have survived the general manager's job during those years. After Tsosie Lewis was removed in 1992, Bates became farm operations director. All the time, he was learning more about tribal politics. He credits Keller with being a "real political mentor, and I followed his [Keller's] advice."

When Bates became general manager in January 1996, he found the core of a Navajo management team on deck but needing "to be organized and given direction. We had to improve efficiency, get our yields up, and set some long-range goals." He feels fortunate that he has survived but always wonders how to anticipate what the 88 council delegates are thinking. Some of the senior council members have been allies. Bates particularly cited Morris Johnson as influencing him as to "how to play the game" but all the time realizing that the council had to be made to understand how to run a business.

Bates is fully aware that the federal government has expended $450 million on the project, most of it on the canal conveyance system. The federal government had dragged its feet, which put the project about 25 years behind schedule and increased the cost. Part of the reason was that NAPI had not made money during most of its years. The program was not working as a means of employment, and the Nation had failed to capitalize on opportunities that were available. The commodity price decline caused a $4 million operating loss instead of the projected break-even in 1997, which "made the politicians look at us as a big black hole." Bates took that as a real setback. In the 1999 tribal election, 50 percent of the delegates were changed, which meant the education program about NAPI had to begin anew.

NAPI comes under criticism because it does not have to pay real estate taxes or water costs. This gives it a real financial advantage over producers who have heavy costs for both. However, until NAPI gets the total 110,630 acres developed, it must spend about $800 per acre from operating income to prepare new land for production. Currently it has 60,000 acres under cultivation. This explains some of the heavy losses incurred. NAPI cannot collateralize the 60,000 acres of developed land, which management feels is worth $3,000 an acre. NAPI's operating line with Wells Fargo Bank is secured by the crop, so it must be repaid each year. The real estate debt is held by the Navajo Nation, so management can take comfort in the fact that it is secure with its land base. However, NAPI has to contend with a governing group that does not understand agriculture and has little interest in it. This indicates a changing attitude of the Navajos in contrast to their presumed heritage as an agricultural society.

U NDER general manager Bates, NAPI management is divided into the following divisions: Farm Operations; Financial; Operation Maintenance of the Canal System; Marketing; Livestock; Human Resources & Safety; and Facility Maintenance.

Left to right, in front of the NAPI headquarters building: Buddy Benally, Director of Marketing; Lorenzo Bates, General Manager of NAPI; and Stephen Lee, Director of Farm Operations. (Photo by Andy Rohall)

Stephen Lee grew up on the reservation, but his parents were not involved in agriculture. Lee started college with the intention of going into medicine but, after one year, decided college was not to his liking. In 1977, at age 20, he applied to work at NAPI. His first job was moving the original irrigation pipes—the labor-intensive system prior to the purchase of center pivots. Next, he was promoted to irrigation parts person and, in 1980, to irrigation repair person. Later, he became assistant director of farm operations, and in 1996 he succeeded Bates as director of farm operations.

Buddy Benally also grew up on the reservation, where his parents lived on a four-acre irrigated farm. In high school he joined the Future Farmers of America (FFA) and became interested in agriculture. After high school he worked on the NAPI pilot program located on Block 7. This consisted of about 2,100 acres of irrigated pasture and a cow/calf enterprise. After working there for about six years, he was encouraged to go to New Mexico State University. He wanted to work in animal science, but after two years his advisor convinced him to switch to crops. His master's degree project involved a study of growing 2,000 acres of Russet potatoes for the fresh-pack market. In 1983 he completed a master of science degree, with a major in agricultural economics, and rejoined NAPI.

Lee has 138 permanent full-time employees on the farm. Each year, starting in late February, he begins seasonal hiring and continues until about 300 are employed by April 10. This crew works until planting is finished, about the end of May. In late July hiring starts again, peaking with a crew of 500 in early September. Fall is the wet season at Farmington, and if the weather does not cooperate, the harvest crew can be on duty through December. The 1999 harvest lasted until January 10, 2000, when the last corn was combined. Most of the seasonal workers are "repeaters from previous seasons," which reflects the 55 percent unemployment rate on the reservation.

One of Lee's biggest challenges is securing equipment operators, for tribal regulations mandate that all drivers of moving machinery must have valid drivers licenses. This is a risk management precaution, because many of the workers have lost their licenses due to driving violations. An ongoing problem is workers who remain for two weeks "just to get a check to make a car payment." The turnover rate is lower today than it was initially when so much manual labor was required. High turnover and high labor cost made it necessary to install automation wherever possible, or the farm could not have survived.

Benally has a staff of 40 in marketing, of whom 2 work in the alfalfa pelleting plant; 4 are engaged in the bagging plant, which bags shell corn and/or alfalfa pellets into 80-pound bags for retail sales; 30, including several women, are in the fresh-pack potato department, which involves both bulk and bagged shipments; and 4 are on the marketing staff.

Because departmental managers must account for their departments' budgets, they would like to see staff reduction.

There is too much fat that needs to be trimmed. Lots of facilities could be eliminated, but there are too many of our workers who expect this will go on. . . . [W]e have to face the world. One of the ways to reduce payroll would be to have more cross training, because too many of the workers think they only have to do one thing.

Despite startup problems, NAPI has a historical average of 170 bushels of corn per acre, achieving a peak yield of 199 bushels on

18,000 acres in 1997. About 80 percent of the average production is contracted prior to planting each year. Manure from the feedlot is generally spread on ground to be planted to corn because corn is the best crop for securing weed control. NAPI's 9,000 acres of alfalfa annually produce about 54,000 tons of high-quality hay, 60 percent of which is contracted to dairies. One dairy has purchased at least 10,000 tons annually since NAPI started raising alfalfa. Hay that does not make the quality expected by the dairies is used in the NAPI feedlot. Wheat is used for alfalfa seed-down and is cut in the dough stage. It is dried and baled for retail for cow herds, sheep, or horses and commands a price equal to that of top alfalfa. Baled wheat is mar-

Eighteen balers of a custom operator baling alfalfa on a 640-acre irrigated farm in Block 2. Looking toward Angel Peak, altitude 6,988 feet. This crop was sold to large dairies. Note the 13 center pivots in operation in the background. (Photo by Andy Rohall)

keted in May and eases the cash-flow needs at a time when there is little or no other income. Pinto beans have consistently been one of most profitable crops, averaging about 1,800 pounds per acre. To gain economy of scale, NAPI prefers to stick to one variety of bean and plant about 8,000 acres per year when it can secure a good contract price. NAPI constructed a plant specifically for edible beans, where they are cleaned and stored for sale to food processors.

Over the years NAPI has grown many high-value specialty crops, but the most consistent high-value large-acreage crop has been potatoes. To date NAPI has been limited to raising only 6,000 acres. Some of these potatoes are contracted to Frito-Lay, and most of the rest are sold for the fresh market, which has a "limited window" of sales between the harvest seasons of competing areas.

To reduce the full-time staff on the farm, NAPI leases its tractors, combines, trucks, and vehicles. This secures state-of-the-art technology and at the same time eliminates the need for keeping its shops open year-round. NAPI does not need skilled mechanics because the leasing company replaces at once any equipment with a major problem. This has virtually eliminated down time. All the alfalfa harvest is contracted out to a custom operator, who does the swathing, baling, stacking, and hauling. NAPI feels this is the best way to assure timely harvest.

Once NAPI realized the potential of potatoes, it knew its best chance for long-range consistent profit would be through value-added (processed) potatoes. Since 1984 it has attempted to entice processors, particularly R. D. Offutt Company, to build a plant for processing NAPI potatoes.

I**T IS AT THIS POINT** where culture clashes with economics. The tribal leaders are concerned about putting people on the payroll, but the managers must try to make their departments profitable. Tribal leaders expect the managers to maximize profits, "but they will not change their tradition to make that possible." Reality and heritage also come into conflict with the fact that once Navajo youth complete high school, "they do not want anything to do with agriculture, especially if it involves manual labor." Part of that problem is caused by the fact that NAPI workers start at $7.50 an hour and the youth know that power plant and mine workers start much higher. This is where the generation gap appears greatest. NAPI management experience has been that if a high school graduate comes to NAPI, it will be only long enough to receive training before moving on to another job. This is true even though NAPI has a good fringe-benefit program in place.

Another stumbling block in attempting to get high school graduates to return to the reservation is that outside of NAPI, "9 times out of 10, the only other job opening will be a governmental position, and that is not a good option." Profitability because of value added by processing makes a potato processing plant essential, for it would provide year-round jobs at a higher skill level than seasonal farm work. Further education tends to lead the Navajos away from the reservation because the youth gain a whole new perspective on what the world has to offer. In 1999 the Navajo Nation offered 4,800 scholarships, but only "about half of them were given out." Youth who have not finished high school are looked upon as not being aggressive and not having high standards about their work. Getting them to perform according to a set standard is a real management challenge.

WHEN Albert Keller left the general manager's position, he cautioned the board of directors that governmental and internal politics should not dominate the economic decisions that had to be made or the potential for profitability would diminish. He suggested that individuals be recruited for the board who had no political or personal interest in NAPI. Keller understood the potential and the problems, but he had no power to change the culture.

In addition to NAPI, the Navajo Nation has several enterprises that have produced a flow of income. The Navajo Tribal Utility Authority delivers the electricity generated by the two large power plants that use coal mined on the reservation. The Navajo Indian Construction Authority builds sewer lines, water lines, and roads on the reservation, where it has no competition. It has 638 authority grants. Navajo Arts and Crafts sells the jewelry made by members of the Nation. The Navajos own radio and television stations on the reservation. Several community colleges and shopping centers, as well as distribution of the reservation's oil and gasoline, generate income. Tourism also provides a profit. The Navajo Forest Products industry has a monopoly on the timber on the reservation. All these enterprises have the potential to generate a greater profit, but external influences often handicap their operations, not unlike NAPI's experience.

The farm management people contend that the farm operating programs make money most years, but excessive external overhead causes losses. LoRenzo Bates, when asked to identify his biggest challenge, replied, "It is that I do not have the freedom to do what I want to do like an independent entrepreneur." But he has learned to put his head on the pillow and leave the problems behind, or he could not survive as the general manager.

Bates knows the stumbling blocks to bringing expertise and capital onto the reservation. They are threefold: double taxation, Navajo preference when it comes to employment, and Navajo sovereignty, which makes an outsider's investment on the reservation very risky. Potential outside commercial ventures have attempted to hurdle the barrier with little luck because they are expected to accept major compromises that make it difficult to do business in a profitable manner. Bates is concerned that if further processing cannot become a reality for its potato production, the farm, even with no taxes or water cost, will at best be hard pressed to make a profit. The full-time employment has been reduced from 300 to 186, and the seasonal labor force has been cut from more than 700 to about 500. He emphasized "our water is an entitlement, not a right, so if the water is not used on a regular basis, political powers beyond the reservation might decide that there is a more urgent need for water elsewhere [and] we could lose it." He added, "Water has never been a priority of the Navajo Nation because [the Nation has] always taken it for granted."

When asked what the alternatives could be if the BIA or the Nation decided that NAPI could no longer continue, Bates suggested that maybe the 110,630 acres and facilities should be rented to an independent operator. Then, Bates could concentrate on upgrading and replacing the water delivery system, for without water the lease would have little value. There is the possibility that NAPI could pursue going into poultry production and processing. It could expand the fresh-pack operation for potatoes or possibly expand into other vegetables. But, in all cases, niche markets would have to be found.

Which will be the winner—tradition and culture or economics? Will NAPI continue to drift and lose money, in keeping with one former director's comment that "It was never intended to make money"? Or will the BIA/BOR decide that saving the $450 million investment will require the skill of an entrepreneur?

Bibliography

Interviews

Bates, LoRenzo. Farmington, NM, January 10, 2000.

Benally, Buddy. Farmington, NM, January 11, 2000.

Keller, Albert L. Farmington, NM, January 11, 2000.

Lee, Stephen. Farmington, NM, January 11, 2000.

Miscellaneous

DAAAKEH NITSAA: The Large Farm. A National Commitment to the Future: The Navajo Indian Irrigation Project. 1987.

"Desert Oasis," *American Vegetable Grower*, October 1999, 18–20.

Harrison, Christie Ann. *NAPI: A Navajo Success Story.* University of New Mexico, May 1997.

NAPI (Navajo Agricultural Products Industry) Annual Report, 1989, May 1990; *1990*, May 1991; *1991*, May 1991; *1992* and *1993*, no dates. (No annual reports are available for subsequent years.)

Navajo Indian Irrigation Project: Chronology. Bureau of Indian Affairs, Bureau of Reclamation, and Navajo Agricultural Products Industry, January 19, 1999.

Navajo Pride at Work. Navajo Agricultural Products Industry brochure. No date.

Joe, Mike, and Sue Naumes

"Change—The Freeway of Business"

JOSEPH PETER NAUMES (PETE) was born June 24, 1873, in Chicago. In 1907 he left the "Windy City" for Hood River, Oregon, where he established an orchard to produce and sell fruit. He prospered until the late 1920s, when he fell on hard times and lost everything. Pete wanted to remain in the fruit business and in 1929 became manager of the car-lot shipping department of the Pacific Fruit and Produce Company at Medford, Oregon. Medford was well known as the center of the productive Rogue River Valley pear industry.

Pete Naumes was an entrepreneur at heart and was undaunted by his earlier failure. During the 1930s, when the opportunity presented itself, he ventured again and developed Naumes Orchards of Oregon, Inc., to produce apples and pears. Next, he formed the Associated Fruit Company for marketing his crop as well as the crops of other growers. He had a good reputation in the Rogue River area as a businessperson and as an innovator in the orchard industry. One of his proudest achievements was winning a gold medal from the Oregon Horticultural Society for the best box of apples. Numerous awards and innovations would be accredited to Naumes in the decades ahead.

WILLIAM JOSEPH NAUMES (JOE) was born October 16, 1910, at Hood River. Like most farm children, Joe worked at his father's orchard as soon as he was old enough to do so. In 1928 Joe entered Santa Clara University, but when his father hit on hard times, he dropped out to earn money. In 1932, armed with his earnings and some scholarship help, he reentered Santa Clara, graduating in 1934.

Immediately upon graduation Joe took a position as a fruit inspector, a job he held until late 1937, when he secured employment that proved to be a real asset in the years ahead. Joe had taken Latin and Spanish at Santa Clara. This gave him the opportunity to apply for a position as manager of several fruit packing houses for a fruit distributor in Argentina. For the next 20 months he supervised Spanish-speaking laborers and perfected his spoken and written Spanish to where he could write documents and deliver speeches in Spanish with ease.

In 1939 he returned from Argentina to become a field man for the pear industry headquartered in Chicago. Then he became a buyer for the F. M. Ball Bartlett pear cannery. In 1941 he entered the Navy, and on November 1 of that year he married Frances Annette McCormick. Shortly after the war started, he was transferred to Pearl Harbor, where he served as a supply officer for the duration. He held the rank of Lieutenant Commander at the time of his discharge in 1946.

T HERE was never a doubt in Joe's mind about returning to the fruit business. His father had always given him moral support in all his activities in the business, and Joe probably hoped his father would take him in as a partner, but that was not to be. Pete Naumes had earned his way through life and apparently felt the next generation would profit most if it did the same. He was not responsive to the idea of taking his son in as a partner. There is some speculation that Gene Thorndike, of the First National Bank of Oregon, in Medford, may have discouraged a father-son partnership. In any case, in 1950 77-year-old Pete Naumes sold the Associated Fruit Company and its well-known Paradise Brand name to other parties.

In 1946 Joe Naumes, armed with only limited savings from his service days, was taken into partnership with Steve Nye to form the Nye-Naumes Packing Company. Nye was well established in the business and wanted a partner. Joe, age 36, was well versed in the fruit industry and suited Nye's needs. Banker Gene Thorndike apparently felt comfortable with the Nye-Naumes arrangement. He told Joe, "By all financial standards I should not lend you money, but I will because I think you are going to make it."

The Nye-Naumes Packing Company was a totally integrated firm. Nye managed the orchard, and Naumes operated the warehouse and recruited growers to provide additional fruit to make the packing company more profitable. This was one of the earliest locally owned packing firms in the Rogue River Valley. Next, the partners developed the NANPAK label, which has continued to this day. When it came time to erect a new packing house, Joe bucked tradition and built a ground-level house. He had seen forklifts in operation in the Navy and decided that such machines could make a ground-level packing house feasible. The facility was the first of its kind in the area.

The firm prospered and soon farmed more than 1,000 acres. It packed all its own fruit and that of several other farmers. The partnership's orchard was considered sizeable for those days. In the early 1950s Nye and Naumes expanded again, when they joined with Paul Culbertson to purchase and operate the orchards and packing facilities of the Crystal Springs Company, owned by the Spatz Brothers. Next, they acquired an interest in the Rogue River Valley Packing Corporation, a pear cannery.

J IM SEMPLE came to Nye-Naumes in 1949 as a packing house supervisor and worked up to the position of superintendent of the packing plant and cold storage. Semple knew both Pete and Joe Naumes and recalled that Joe was very much on his own, in no way relying on his father. Semple continued: "Joe was a self-made natural leader. He was so good with people that I think he would have been a fine priest. He was filled with futuristic ideas." Semple related how the ground-level packing house using a Hyster (forklift) led to the use of pallets and cardboard boxes for handling pears and apples.

Joe seemed to be everywhere in the orchards or on the packing house floor. Semple related Joe's often repeated phrase that the owner had to plant his footsteps in the soil if he expected his people to follow him. The bracero program was still in effect, and Joe used the management skills he had honed in Argentina and in the Navy. Semple noted: "The Mexicans were so pleased to have an employer who both spoke and wrote their language so well. He liked people and it showed, for everybody liked him." Semple said that Joe had the ability to lead without getting under your skin.

Labor has always been the major cost of running orchards, representing 70 percent or more of operating expenses. Because virtually all of the harvesting and much of the pruning involve manual labor done while on ladders, finding workers is difficult. At Medford, pruning and grafting require a crew during five fall and winter months. Spraying starts in February and is carried on throughout the growing season. During the spring months, artificial heat is often necessary to prevent the crop from being harmed by frost. Irrigation starts in June and continues into September, overlapping the harvest season, which commences in August and ends in October, depending on the variety of fruit.

The Nye-Naumes Packing Company was fortunate to be able to employ bracero workers for orchard work until 1965. During those years most of the permanent workers in the packing house and cold storage still came from the local area. In 1958 Bill Kusel became an employee of Nye-Naumes when the company bought the Anders Frink orchard. He supervised crews in the orchard and recalled the difference in the work ethic of various groups. During the pruning season, when braceros were not available, the crew was made up of "winos," who were picked up each morning and delivered to their quarters each evening. Kusel recalled that it was necessary to have at least 35 employed so that 20 would be working each day. "So many had hangovers." While pruning was in progress, a smaller crew of year-round employees sprayed the trees and disked the soil between the trees. Hand-hoeing was done around the base of each tree as a precaution against mice.

The workers under the bracero program left Mexico in May, as soon as the peak labor season started, and worked north until the season was over in October. Then they returned to Mexico. Kusel said:

> These were good people to work with and so reliable, but it took patience. They understood that they were part of the foreign aid program for Mexico. Many of them were interested in some venture in Mexico and came north to earn extra money to get it going. When the braceros came for harvest, they were formed into crews of 20 at the beginning of the season, and when harvest was over, the same 20 were still there. They wanted the bonus that came with working the full harvest. . . .

[Joe] always had time to visit with the field workers, which they liked because he spoke to them in their language.

THE NAUMES FAMILY experienced a turning point in 1963 when Steve Nye decided to retire. None of the Nye children were interested in the business, and Joe realized this was a golden opportunity for his family. He had often told his family that Steve Nye was the perfect partner. He and Nye had a very respectful and trusting relationship. Over the years they developed a totally integrated firm that made them leaders in the industry.

Joe's son, Michael (Mike), said, "It was 1963 when we made a family decision to buy out the Nye interest. . . . I was a junior in high school, and Dad decided to invest a half million dollars to buy the Nye stock. We had a good pear crop, and it helped to fund the entire purchase."

Once Joe became sole owner of the firm, he traveled throughout the world searching for new ideas. As a result of those travels, he adopted interplanting, double rowing, and trellising to enhance quality and maximize production of the orchards. He worked closely with researchers to develop red pear varieties. He was one of the pioneers in the Rogue River Valley in replacing smudge pots with wind machines to aid in the battle against frost.

Kusel recalled the major changes he saw in the orchard industry over the years that were particularly beneficial in reducing the demand for labor. The first change was the shift from gravity irrigation pipes to low-rise stationary sprinklers. The aluminum gravity pipes were 25 feet long and had to be moved every 12 hours, which meant working in muddy soil while changing to the next row. The work was usually done by a five-person crew. The company was allowed $2^7/_{10}$ acre-feet of water annually, so lots of pipes had to be moved. Switching to stationary sprinklers made more efficient use of water and eliminated the drudgery of changing pipes twice a day.

The next big change came in methods used to guard against frost damage. Whenever frost warnings were signaled, the smudge pots had to be lit. About 25 smudge pots were required to protect each acre of orchard. At full heat, each pot used a gallon of fuel per hour. On the average, one person could handle 200 pots, which provided protection for 8 acres. The company had about 2,000 acres of orchard by then, which meant a great expense in both labor and fuel cost, even with fuel at 8¢ a gallon, the price when Kusel first started. Juan Moreno, who began working for the company in 1969, commented that he would never forget his first year, when he worked 45 nights in a row lighting and filling smudge pots. Some nights the pots were started by 8:00 P.M. and on a cold night would use up to 10,000 gallons of fuel. Sometimes students were let out of school to help during cold spells.

In 1973, when the energy crisis hit and fuel rose to $1.25 a gallon, it became prohibitive to continue using smudge pots. Despite the initial capital cost, the company completed the switch to overhead sprinklers and wind machines. A wind machine could protect 10 acres and required a minimum amount of labor and only 5 gallons of fuel per hour of operation.

The next innovation came in assembling the harvested fruit. Up to that time each worker placed the harvested fruit in 48-pound lugs positioned in the tree rows. Tractor-drawn trailers were slowly pulled through the orchard, and the lugs were loaded onto them. To speed up harvest, bins holding 24 lugs each were placed near where the picking was done, and the filled lugs were placed in them. When the bins were filled, a forklift moved them to a waiting truck at the edge of the orchard that hauled them to the packing house, where they were unloaded by a forklift. A forklift driver and a truck driver replaced several people who formerly loaded and unloaded the fruit manually.

By the 1970s the two major challenges remaining to reduce labor requirements were pruning and harvesting. By then research had started on developing smaller trees, making those tasks possible from ground level. Hand-harvesting large trees required resetting ladders as many as eight times to get all the fruit. Pruning required the same procedure. As smaller trees were developed, the tree population per acre increased. Trees formerly 25 feet apart and quite high were now planted in double rows and only 8 feet apart, permitting 800 trees per acre. Because the new trees were much smaller, most of the fruit could be picked from the ground, and only three ladder settings were needed to reach the highest fruit.

JOE realized that very few local people wanted to work in the orchards. Even students no longer applied for summer work. The bracero program was scheduled to end in 1965, and securing labor would become more difficult with each passing year. He turned to the Mexican families who worked for him to recruit other workers. He knew that the migrant families had cliques that watched out for each other. They knew where jobs were available and the best places to work. Apparently the migrant families fell into three categories: those who did not want to settle down, preferring instead to follow the harvest each year and then return to Mexico for the winter; those who wanted to earn enough working in the harvest fields each year to get their own business started in Mexico so they could settle down permanently there; and those who wanted to get permanent work and settle in the United States.

Juan Moreno recalled that his father started with the company in 1964 and finished his working career there before retiring in Mexico. Juan had only three years of elementary school because, as a child, he had accompanied his parents to the harvest fields. In 1969, after several years of odd jobs, 19-year-old Juan started working for the Naumeses on a permanent basis. His is a touching story of someone who has come up the economic ladder by having a good work ethic and remaining with one employer.

Juan's first year's experience with the smudge pots was related previously. When asked for his recollections since 1969, he commented that when he first started, some crews were a combination of locals and migrant workers, whereas other crews were mostly Mexican families. Each crew consisted of about 20 individuals. Some of the Mexican families made their homes in the area. Others went back to Mexico each year but most of the time returned to the same farms each season.

After working for the company for 15 years and having done every job in the orchard, Juan was asked to become a crew leader. He was 34 years old, had been an exceptionally reliable worker, and had good people skills. Joe had taught Juan all the things he needed to look for regarding tree health and soil condition. When asked to describe his experiences as a crew leader, Juan said:

> We used to have lots better workers—everybody came to work every day. You started the picking season with 20 people in each crew, and you ended with the same number. [The workers] were more mature then. These people came back every year and kept nice camps.

Today you start with 20, and before the season is over, you have worked with 60 to 80. Some days you have 7 not show up, and some days you have only 10 or 12 all day. Sometimes they quit during the day. Many times they show up at the office after a couple days' work and want their pay. They often leave for another job that pays less. I don't understand it.

Juan explained that on one variety of fruit, a fast picker can pick six bins a day at $12 per bin, while a slower picker can sometimes pick only two bins. On another variety the pay is $14 per bin. In all cases a $2-per-bin bonus is paid if the workers stay to the end of harvest. He lamented that some people who did pruning under his supervision "did not even learn to do a good job of pruning. It's the same thing over every year—teaching—teaching—they do not learn." In his final comment on work habits, he expressed these feelings: "I like irrigation season best because we do not have too many people, so [there are] not so many problems. During harvest you are looking for people every day."

After his parents retired in Mexico, Juan and his wife and younger children visited them every two years until his father died. Since then they have visited his mother annually. Juan is proud of his son who worked for the Naumeses in the summers and as soon as he graduated from Medford High School became a full-time employee. Within a short period he was sent to refrigeration school, and at age 22 he is in charge of refrigeration engineering. Juan smiled when he said, "He gets really good pay."

IN 1970, after 21 years with the company, Jim Semple, superintendent of the packing plant and refrigerated units, resigned because he realized that family members soon would be coming aboard and would assume the "good positions." He had always wanted to own an orchard, and this was the time to make the change. Semple purchased and operated a small but productive orchard and worked as a consultant for Terra International.

Semple guessed correctly, for in 1971 Joe's nephew, Pete Naumes, joined the firm. In 1972 Mike came on board. Mike had graduated from Santa Clara in 1968 and then served 19 months in Viet Nam as a chaplain's assistant. After his tour of service Mike entered Cornell, and in 1972 he completed his M.B.A. with "an agricultural emphasis." He stressed that his experience in the service gave him insights into persons of all ranks and positions, which proved valuable to him in business. Mike professes to be shy but said his exposure in the military did more to help him overcome his shyness than his six years in college.

Mike and his sister Susan (Sue) both recalled that their parents were great about encouraging the children to do whatever they wanted. But Sister Mary Pat Naumes, who never had any interest in the business, said, "Father encouraged me to follow my interest, which led to taking vows. However, in retrospect, our parents did things to create an interest in the company." She gave the following example. Having the family together for the evening meal "was a must." Sister Mary Pat said, "Father made a point to bring information about the company to be discussed at dinner. It was intended to keep us all interested in the business." Sue said, "The dinner discussions developed an interest and family pride about the company, which may have been what our parents were hoping for."

Succession is a major problem for most family businesses, and despite the myth of the family farm, this is particularly true in agriculture. In the case of Joe Naumes, he probably wanted his children to have an experience different from his, and he "had a marvelous way of leading them." Semple and Kusel both commented that Joe was very deliberate about instilling a good work ethic in his children. Both put in considerable time hoeing mice-prevention strips around the trees. At age 11 or 12 Mike and Sue were expected to get up at 5:00 A.M. to change the irrigation pipes. Mike was 12 the first time his father took him along to the lawyer's office, where he listened to the two men discuss vital matters. It was about then that he first "got the idea" that he might want to join the business.

By age 13 Mike was in charge of pruning and picking crews. At age 16 he was a crew supervisor and was confronted with a sit-down strike because picking conditions were poor. Mike realized he dare not turn to his father for advice on how to handle the matter, so he asked Bill Kusel for help. When Mike was 17 he decided he definitely wanted to become part of the company.

In the fall of 1971, when the head of the packing house quit, Mike left school to take over and had a "baptism of fire." Several machines in the packing house caused problems. Mike said, "I am no mechanic." Bad weather and an early frost caused decay problems that were compounded because the sorting staff was weak and the packing staff was strong. Mike said, "I was caught in a real educational experience. Dad knew the problems I was having, but he let me work through them. He was busy with his end. I was really glad when I could return to graduate school."

In the fall of 1972 Mike joined the company for good, but he was not sure what his role would be. He recalled his early feelings:

> I had trouble adjusting to working with people who could not comprehend even the simplest direction. It seemed I got the jobs no one else wanted. Dad relied on me to work in insurance and government regulations. I liked that. Dad was a total person; he did everything in the company, but it outgrew him. He was very precise and . . . a great idea person.
>
> During the 1972 harvest season, I ran the packing house in Medford, and Tom Spatz ran the Crystal Springs house. Then, in November our sales manager had a heart attack. I was forced to take over Oregon Pears Sales, which was run out of our office. Dad asked me to do it. I knew lots of the people because I had traveled all over the country on sales trips with Dad. It was a natural takeover—I grew into the job.

Floods that destroyed the north-south rail lines created the biggest challenge in Mike's sales job that fall. The situation provided a real indoctrination into selling because there was no indication when the rail lines would be back in service. At that time the Naumeses shipped 95 percent of their produce by rail and 5 percent by long-haul trucks. Suddenly they were forced to shift entirely to trucks. Only a couple of orders were lost, however, and those because not enough trucks were available. Unfortunately for the railroads, they never regained most of their former business. The railroads wanted to haul 2,700 lugs per car, and the chain stores wanted semi loads split among several stores for quick turnovers. Trucks had the advantage of being able to deliver to the East Coast in five days, whereas the railroad required two weeks or more. By 1998 trucks moved 95 percent of the Naumeses' production.

When Mike joined the company, the business was still strictly centered in the Medford area. The firm consisted of the Naumes Packing Company; Growers Refrigeration Company, Inc., a cooperative owned by many orchardists; Medford Pear Company; an interest in the Rogue River Valley Packing Corporation, a pear cannery; Mary

Mac, a 60-acre orchard; Crystal Springs, five orchards jointly owned with the Culbertson family; and Naumes Orchards, an 80-acre family orchard owned with the Albertson family, who also owned orchards in the Hood River area, and Murray Albertson (no relation to the aforementioned Albertson family), a speculator in the business from Boston.

Left to right: Laura (Mrs. Mike), Mike, and Sue Naumes, with Laura and Mike's triplets, Cynthia, Sean, and Joe. Laura is company vice president. She is a foremost researcher in Integrated Pest Management and is in charge of chemical operations. Mike is president, and Sue is secretary-treasurer.

SUE NAUMES, the second of Joe's children to join the firm, first became involved with the company when she rode with her dad to count and record how many lugs of fruit were being harvested. Her father then reported to the packing houses how much fruit would be coming from each of the orchards. At about age nine she was given a hoe and shown how to pull and grub weeds about 2 feet around each tree. As a young girl Sue was also known for picking "shiners"—pears missed by the picking crew. That was how she earned some of her first spending money. At age 11 or 12 she joined a crew that changed irrigation pipes twice a day seven days a week. Her next step was working with crews of teenagers thinning out the smaller and defective pear trees. When harvest came she picked fruit with the same group.

Sue had no idea what she wanted to do when she entered Santa Clara University, but everyone encouraged her to go to law school. While in college she spent time at the state legislature. She related that experience: "It was a real eye-opener for an idealistic college kid to spend a year there." In 1970 she graduated with a degree in political science. The notion of getting into politics was still in her mind, so she entered law school. There she enjoyed tax work and property laws and decided that she did not want to pursue a career in politics. "I had no conscious turning point about going into the family business, but I saw that Dad was giving Mike the tough decisions and that he seemed to thrive on dealing with them. This appealed to me."

In 1974, when Sue received her law degree, she joined her father and her brother. Her father suggested that she start by reading the last few years of correspondence so she could learn what had been taking place. Next, he sent her to the packing house to learn how that operated. Then, she was transferred to Growers Refrigeration, which Nye-Naumes had started for tax reasons, to become acquainted with the workings of the cold-storage facility as well as to meet member growers.

IN 1973 the orchard had grown to 1,200 acres, and the company was incorporated as Naumes of Oregon. About the same time, the company purchased Medford Ice and Cold Storage, a railroad ice warehouse that was no longer in use. That purchase was fortunate, for shortly after that the Crystal Springs packing house, which was owned with the Culbertsons, burned. The Culbertson-Naumes partnership was not functioning smoothly, and this proved to be an excellent time to separate. In late 1974 Sue was assigned to do the preliminary insurance work for settlement of the Crystal Springs fire. Then she was put in charge of the Melrose packing plant in Medford, which the Naumeses had acquired. The Melrose plant had the first weight sizer, which enabled more accurate and uniform sizing to fill the shipping boxes to the correct weight.

In early 1975 the Rogue River Cannery, a one-product, low-tonnage facility, failed. The four largest suppliers, including the Naumeses, purchased the cannery and got involved in the concentrate business. Next, the Naumeses severed their connection with Oregon Pears Sales Cooperative because the cooperative's sales personnel did not understand the business. Purchasing Medford Ice and Cold Storage, dissolving the partnership with the Culbertsons, buying the cannery, and severing ties with Oregon Pears in a short span of time was a major turning point for the company.

AFTER the above steps were completed, Joe sensed that Mike and Sue were capable, so it was time to expand the business. He understood that firms in production agriculture had to align with each other to develop power in the marketplace and that they had to gain that clout without cut-

Bosc pears being run through the automated/computerized packing lines at the Melrose packing plant in Medford, Oregon. Each year 2 million 44-pound cartons are packed at this plant, enough for about 2,000 semitrailer loads.

ting each other's throats in the process. The Naumeses entered into an agreement with a puree product producer to guarantee a reliable outlet for some of their production and were ready to step out.

In early 1976 they purchased three parcels of land totaling 500 acres planted to Bartlett pears and walnuts in the Yuba City, California, area. At that time, this was considered a large holding, especially of Bartlett pears. The operation had been owned by C. E. Sullivan, and none of the Sullivan children were interested in the business. Mike recalled that at $3,000 an acre, the transaction came to more than $1.5 million. This seemed big, but his 66-year-old father was poised to move. Mike said, "Dad seldom had trouble financing. His motto was always let your banker, lawyer, and priest know your problems, and then they will feel you are aboveboard with them. He was cautious and never moved until he had all bases covered."

With the purchase in California, the Naumeses gained on their competitors because they were able to harvest a month earlier, enabling them to market fresh products over a longer period. This was important, for it gave them contact with their customers for more months a year.

California was basically a frost-free area, but fire blight necessitated spraying the trees every third day while they were in bloom. The Bartletts were so susceptible to the blight that a shift was made to Fiji and Granny Smith apples and to Bosc pears at Yuba City, but the disease remains a million-dollar-a-year problem there. In 1977 Sue and her cousin Pete were sent to Yuba City to manage that operation. When harvest was over at Yuba City, Sue transferred to Medford to manage the warehouse during harvest.

Sue noted that the Yuba City purchase stimulated her father to new heights. She said, "Dad was the visionary, always looking for new deals, and he left the details to us. He kept pushing us and was always in the background observing how we handled each challenge."

In 1978 Joe found what he had been dreaming of. He had known about the DiGiorgio operation for many years. It was a publicly traded company that, at its peak, had more than 40,000 acres, but because of a lack of family interest, the largest stockholders had gradually moved out of farming. During the harvest of 1978, the produce from Yuba City was hauled to Medford. This was not good economy, and the Naumeses decided they needed a packing house in California. When Joe learned that the DiGiorgios were not having a good crop, he contacted them. Three weeks later, on July 10, 1978, Joe offered $8 million for the last remaining 2,500 acres of the DiGiorgio farming operation in the Sacramento Valley, at Marysville, California.

During negotiations the DiGiorgios had flown the Naumeses around the orchard and even to Medford in their airplane. Sue noted, "Dad, at age 69, was eager not only for the DiGiorgio operation but also for an airplane. He realized what an asset it would be to our business." In one transaction the company more than doubled its operation. The DiGiorgio holdings consisted of the Dantoni and the New England orchards, which, when combined, made up the world's largest pear orchard. The acreage also included quince, persimmons, walnuts, canning peaches, and alfalfa. In addition, each orchard had a packing house and a refrigerated storage facility. Mike and Sue both went to Marysville to manage the operation and to assess the person-nel. After a few weeks it was decided to make Steve Maxey, the DiGiorgio foreman, overall general manager. From that point on, Sue made weekly trips to Marysville, and Mike came down for harvest and for special problems.

In February 1979, the company purchased its first airplane, a Cessna 414, which seemed to broaden Joe's horizon. For several years he had sought to spread the risk by expanding the geographical base to improve the company's odds with the weather and to diversify in a greater variety of fruits, as well as by going into concentrates. Joe thought that exposure over a broader area would help the firm shed any provincial inclinations and enable it to pick the best ideas from each area farmed. Mike and Sue readily agreed. Joe reasoned that if he went north to Washington, the harvest season could be prolonged even more.

IN 1980 the company acquired the Chelan Apple Company, of Chelan, Washington, which consisted of 400 acres, plus the 250-acre apple orchard of J. K. McArthur and Son, at Entiat. The com-bined orchards consisted of 500 acres of red delicious and golden deli-cious apples, 100 acres of pears, and 50 acres of nectarines and cher-ries. Both properties had packing houses and refrigerated storage large enough to provide service to the 24 other growers in the area.

With the latest addition, the company had 5,000 acres of orchards in three states, making it the world's largest producer of pears. The company name, Naumes of Oregon, Inc., became obsolete with the expansion into three states, so in 1980 it was changed to Naumes, Inc.

The Naumeses observed that smaller independent growers with a single product and a short harvest season were finding it difficult to survive. They realized that to protect themselves in the competitive global economy, they had to expand and to improve their long-range marketing strategy by having a greater variety of fruits available over a longer season.

When Mike started with the company, the harvest season ran from late August to early October. Only eight years later, it extended from June through October. The packing houses operated for from 8 to 10 weeks, and after the Washington purchase, their season was extended up to 28 weeks. More important, fresh fruits were now marketed year-round, and the Naumeses could provide buyers with mixed lots of pears, apples, and cherries. In 1981, their first full year after the Washington purchases, they marketed 1 million boxes (18,000 tons) of fresh pears and 350,000 boxes (7,000 tons) of fresh apples. In addition, they sold 14,000 tons of pears for canning and 3,000 tons of pears for juice.

In 1985 they purchased the 1,000-acre Rogue River Orchard near Medford, which consisted of 600 acres of established orchard and 400 acres that were being planted. The Rogue River Orchard Company had been owned by the Harvey Mudd family since 1929, but family members no longer had any interest in it. In 1987 the 800-acre Feather River Orchard near Marysville, California, was repossessed by an insurance company. The Naumeses acquired it and merged it with their existing operations there. Feather River Orchard contained prunes, which gave them an entry to the Pacific export market. In 1988, to maintain management control of the extended operation, they purchased their second aircraft, a Piper Cheyenne. To maximize the use of the planes, a schedule was prepared at the beginning of each month, which was altered only for emergencies.

A view of Rogue River Valley with Medford, Oregon, in the background. In the foreground is the thousand-acre Rogue River orchard. The slender white towers are wind machines used to prevent frost damage.

During the 1980s the company purchased several other smaller orchards in the Medford area that were going out of business because they were too small to be economically viable. The Naumeses had foreseen that orchards not only had to increase in size but also had to

integrate or be closely allied with value-added firms to survive. In a 1985 speech, Mike stated that most of the smaller orchards in the Medford area would fall by the wayside and that the three or four survivors would produce more pears than ever. The Naumeses were determined to be one of the survivors and by 1987 had 7,000 acres of totally integrated facilities in three states. With the exception of walnuts, cherries, and canning peaches, they packed, stored, and marketed all their production. At his death from a car accident in 1987, Joe Naumes was acclaimed one of the world's largest fruit growers and the world's largest producer of Bosc and Bartlett pears. Unlike the great majority of those in agriculture, he saw a son and a daughter firmly entrenched in the estate he enjoyed creating.

NATIONALLY, much of agriculture experienced economic problems during the 1981–1986 period, but Naumes, Inc., and most western fruit growers were not adversely affected. Instead, this is when they added several thousand acres. The world economy was booming, and interest rates were down, as was the value of the dollar. This greatly aided exports. During those years the greatest challenges the industry faced were environmental problems, a shrinking labor supply, new diseases, and changes in marketing. None of these were surprises, but they all compounded the stress on the already hard pressed smaller producers.

In the 1970s the annual operating cost per acre varied from $1,100 to $1,300. By the mid-1980s it had risen to $1,800 to $2,000. The cost of establishing an acre of orchard increased from $5,000 in the 1970s to $10,000 a decade later. In the 1980s a decision was made to shift to Bosc, Comice, Seckel, Forelle, red, and Asian pears and to Granny Smith apples. Changing trees was a slow process, but it was necessary as a way of responding to the market and as a means of diversifying.

To cope with the ever-shrinking labor supply, the Naumeses continued working with plant breeders to produce smaller trees that would make picking and pruning both safer and less labor intensive. Double-rowed orchards were established to increase the number of trees from 70 to 400 per acre to 800 per acre. Production continued to improve. A company rootstock nursery was established to hasten development of new varieties. The new orchards were developed on posts and wire trellises to maximize the bearing surfaces of the trees.

Under the Tatura system of trellising, tree population stepped up to 1,100 trees per acre. Grass was planted between tree rows to shade the ground, to make it more solid to work on, and to reduce the need for tillage between the trees. Every tree variety has different requirements for spacing and density to gain the best return per acre, but the answers could be found only through experimentation, which took time.

I N 1991 integration and diversification continued with the addition of juice/concentrate plants in California and in Washington. In 1994 the 600-acre Pateras ranch was added to the existing operations at Chelan. The new varieties on the Pateras ranch aided the Naumeses in diversification as well as in gaining further economies of scale. In June 1992 a juice concentrate business was secured at Marysville. Two years later the bankrupt Endurance Fruit Processing facility, a concentrate plant at Wapato, Washington, was acquired at a cost of $3.5 million. The plant was formerly owned by the Yakima Nation and provided work for about 40 people. It became a profit center after the initial problems were overcome.

The 600-acre Pateras ranch orchard near Chelan in central Washington. This orchard is on a mountainside about 1,200 feet above the Columbia River.

A news release by the *Fruit Grower* after that purchase declared Naumes, Inc., the number one grower of apples, cherries, pears, and peaches, as well as a producer of chestnuts, persimmons, pomegranates, prunes, and walnuts. In all, the company was responsible for more than 90 varieties of products, up from only 8 varieties of pears in the 1970s. In addition, it processed 130,000 tons of apples, pears, kiwi, cherries, plums, and peaches, making it one of the largest juice concentrators in the nation. The firm's total dollar volume had increased from $8 million in 1976 to $60 million in the mid-1990s, with projections for $200 million within a decade. About one-third of the total dollar volume came from the production sector.

Naumes, Inc., was totally integrated from orchard to sales. To control the quality and preserve the identity of its fresh sales, the company adhered to a strict policy of packing only produce grown in its own orchards. However, in California it contracted 7,000 tons of Bartlett pears to SnoKist for canning. Naumes juice plants rely on produce from independent growers, where run-of-the-mill quality is satisfactory. *The Packer* reported that the company hoped to increase its sales of packed specialty products from 2 million boxes in 1998 to 4 million within a decade.

In 1997, when the company purchased the 500-acre Earnest Orchards, of Medford, Mike stated that Naumes, Inc., hoped to shed some of its smaller orchards that were not economical to operate. He said that OSHA regulated the moving of sprayers from one field to another, which created an added cost. However, Oregon land-use laws made it virtually impossible to shed uneconomical parcels. The number of growers in the Rogue River Valley had declined from more than 200 in 1960 to 20 in 1995, of which 5 integrated firms controlled 92 percent of the tonnage.

Some of the remaining growers wanted to leave the business, but they could not find buyers because their units did not blend in with those of existing operators and land-use laws prohibited rezoning for uses other than agriculture. In 1997 the three largest producers in Medford had combined sales of $400 million and provided employment for 5,000 people. Naumes, Inc., held 65 percent of the acreage in the Medford area, making California second in area and third in sales of the three states in which it operated. However, Medford was the company's home office from which it did its accounting, financing, and fresh marketing.

When asked what was the minimum unit for survival in the orchard business, Mike replied, "A thousand acres totally integrated. It takes that much production to support a viable packing unit, but it is not enough to support a marketing unit." He pointed out that most of the remaining 20 orchards were 500 acres or more but were dependent upon the 3 largest firms to market their products. It is likely that within five years, only 10 of the small producers will still be in business. Mike explained that many small growers with 10 to 50 acres were able to continue in Washington because more of them were involved in cooperatives and the state had more independent commercial packers. The latest trend, however, is toward vertical integration, which will continue to encourage larger production units.

Part of the reason for the decline of the smaller growers is that buyers often prefer to purchase from totally integrated producers. Tradition has it that in the Rogue River Valley area, in a seven-year cycle, there is one year of hail, one year of frost, one year of low prices, and four years in which to recover.

The good fortune of the 1980s came to an end in the 1990s, when three consecutive bad years took their toll in the valley. In 1993 hail struck just before harvest; in 1994 the crop was short because of poor pollination, and prices were low; in 1995 the crop was good, but prices hit a 20-year low. The Naumeses left 1,500 tons of unpicked fruit, which proved to be a good decision. Because of the surplus, much of the crop had to be re-packed. On the average, the growers lost $500 to $1,000 an acre during each of those years. Fortunately, a very short pear crop in 1996 caused prices to bounce back to $280 to $410 a ton for No. 1 canning pears. The supply was at a 40-year low, and many canneries received only 50 percent of their normal tonnage.

GLOBAL MARKETING, refrigeration, and processing have all worked together to take traditional "one country one year" shortage factors with major price cycles out of the picture. The fruit growing industry, like other sectors of U.S. agriculture, is dependent upon exports to survive. The industry needs to export at least 35 percent of its production to be profitable. The problem is compounded by the fact that most fruits can be stored for only 9 months, except some apples, which will keep for 14 months. In 1998 a worldwide glut of all fruits occurred. At the same time, expansion using European and U.S. capital was taking place in the Southern Hemisphere. This expansion continues. To keep up with the ever-growing global economy, international partnerships will be formed, whereby producers will link up with their counterparts in the Southern Hemisphere and sell to each other in the off-season. With fruit, as with most agricultural commodities, the United States is the supplier of last resort, which causes our producers to be the first to suffer and the last to recover. These forces changed the fruit industry drastically in the 1990s. Naumes, Inc., has factored global marketing into its long-range plans. Mike said, "We favored NAFTA because we need it. In the long run it will be better for everybody. The problems in the short run are perceived inequities."

ONE OF THE MOST NOTICEABLE CHALLENGES in marketing has been the consolidation of buying power. Ten chain stores have a real grip on the food industry, but the computer and the consumer have also fed the process of change. To keep pace with the changes in marketing, producers and packers have no choice but to become involved with large marketing groups. Tony Freytag, director of marketing for Naumes, Inc., commented that alliance with national marketing agencies is a necessary extension for producers and processors. Freytag noted that many of the smaller growers do not feel any urgency to change their ways of marketing, and they will continue the long-range trend of "falling by the wayside." He said, "We have to be big enough in our thinking to see globally and not spend time fighting change. You never get to the end of the road of change—the freeway of business—it just keeps going."

The Naumeses, having been in the business since 1946, are well established and known nationally. They have been innovative on the production side and have tried to anticipate consumer demands. Buyers for the chain stores have been difficult to deal with because they follow computer guidelines "and do not seem to know the product like the old-time produce man, nor do they understand the problems of the producer." The Naumeses have an advantage by being totally integrated. They market under three names—Cream of Chelan, Snow Crest, and Naumes. They sell nine flavors of juices in single strength or concentrate and will also custom blend. They are one of 70 growers who have a strategic alliance with C. H. Robinson, one of the oldest and largest produce distributors in the nation, for much of their marketing. A recent sale to Northwest Airlines of 4 million apples (138 to a 40-pound carton, totaling 29,000 cartons and weighing nearly 600 tons) is an excellent illustration of the need to meet unique market demands. The requirements of the airline necessitated costly specialized equipment and diligent line workers. Eye appeal is a major concern of the chain stores and of big customers such as Northwest, even though the product may not be what the ultimate customer wants.

THE COMPANY'S 8,000 acres of orchard, 7 packing plants, and 2 concentrate plants require a work force of more than 500 year round and more than 2,500 in peak season. Labor represents 70 percent of annual operating cost. Mike noted that as long as he has been in the business, securing workers during peak season has been an ongoing problem. Each year the government becomes more involved, especially from a regulatory standpoint.

Mike believes that three possible solutions may exist. One solution may be to return to some kind of guest-worker program under which workers live in company housing and go back to their homeland when the peak season is over. The Naumeses have housing for 30 permanent and 200 temporary employees in the Medford area and for about 500 workers in California, where the state imposes an annual fee of $50 per worker to open the houses. The second solution proposed by some in the industry would be to expand production to foreign nations and export to the United States. The third solution would be to mechanize more, but the biggest challenge there is harvesting for the fresh market, where bruises must be avoided because appearance is so critical. Harvesting for canning and juice does not present the same problem. Mike wonders if the labor situation will have to deteriorate to the point that crops will rot in the field before significant notice will be given to the shortage. This awareness would come only after the shelves are bare.

The Naumeses have experienced that few workers, regardless of the wages, want "the hard labor of agriculture" and that seasonality makes it even less desirable. Therefore, they are forced to rely on migratory Mexicans. This further complicates the labor problem, for the law requires that before employers can hire, they must require two sources of identification from every worker. However, the employers have no way of knowing how valid the references are. The Immigration and Naturalization Service (INS) can enter the plants at any time.

If the INS determines that the identification a worker used is not valid, the employer is liable, even though the company followed the letter of the law. Mike stated that it cost Naumes, Inc., a million dollars each time the INS made a search. He said that consumers are probably unaware of, or indifferent to, such events, but they must bear the final cost.

In a recent year there was an attempt to organize the farm workers in the Medford area. At a meeting for this purpose, it came out that the biggest problem the workers had was with field-level supervisors. Most of the field supervisors were former field workers who had been promoted but had not received proper training. Ironically, the two largest growers in the area sent 50 people to the meeting, and the other 22 growers sent a total of 6 people. Mike's conclusion was that apparently the smaller growers did not think training supervisors was important, which further compounds the problem for the industry.

The Naumeses want to maintain their reputation as producers and marketers of fresh fruit, and they will continue along those lines as long as they can secure an adequate labor supply. If they have to reduce acres to correspond with the shrinking labor supply, they will put more emphasis on increased production per acre. In recent years, however, they have become more interested in the concentrate business because, by using computers to monitor costly equipment, they can operate a concentrate plant with eight people per shift.

To IMPROVE YIELDS and minimize labor, the company has increasingly relied on chemicals. That presents a new dilemma, which Mike addressed as follows:

> I'm often amazed by people's paranoia to pesticides. The truth is that the synthetic materials we use are in such small doses and are so diluted that they are very safe. Fruit grown organically is subject to much higher levels of toxicity. Organic growers are allowed to spray with any naturally occurring substance. The naturally occurring substances are substantially more toxic than the synthetic ones we use.

Mike pointed out that the loss of valuable existing chemicals is especially costly because new ones are slow in coming to the market due to the expense of getting them developed and certified. He specifically cited the loss of a fungicide used to combat decay organisms while buyers have a zero tolerance for decay. It cost $2 to re-pack a box of fruit, in addition to the loss of fruit that could not be re-packed. He noted, "Survival is hopeless for the small operator in such an environment. Larger operators at least have staffs to deal with escalating government regulations, with administrative rules, and with the proliferation of agencies. This all takes time and money."

Laura Earnest (Mrs. Mike) Naumes, in charge of the company's chemical operations, related that pears are one of the most susceptible fruits to insects, which attack both the fruit itself and the trees. She stated that fruit can be affected by growers in adjoining orchards who prefer the organic method and do not use chemicals. Their trees and fruit become infested, and the orchardists who use chemicals have no way of protecting themselves from the organic orchards next to them. An even worse plight comes from orchards that have been abandoned and are invaded by disease and insects, leading to damage in surrounding orchards. In severe cases, active orchardists are at risk of losing their export market, for which they must certify that the fruit has not been exposed to specific disease or insects. Bill Kusel, who works with Laura Naumes in the chemical operation, said, "We are in the Chemical Age, and we will have to learn how to handle it or we will not have an ample supply of good fruits."

The company also uses biological control by applying techniques that reduce the male insect's ability to sense females, thereby curtailing breeding. The Naumeses used that technique on 500 acres with good results, and they will be applying it to other orchards because it has proven very cost-effective. In the Medford unit alone, which is their smallest unit, they have 17 certified spray operators. These people are at work throughout most of the year, which gives some idea of the cost of the battle against disease and insects.

From an economic standpoint, the regulatory aspect is serious and costly to the industry. The Food Quality Protection Act and other regulations cause some chemicals to be removed from the market and inhibit companies from developing new ones. Laura said, "We can only hope that technology will keep improving to overcome the objections so production can remain viable."

THE NAUMESES look to the future fully aware that the global economy will play an ever-increasing role in their operation, but they
feel that this will not impact what they do any more than the cheap-
food policy of the United States, under which the American farmer
has worked for decades. Gyrations in the market are no more harmful
than regulations. Mike frequently stated that governmental harass-
ment has caused more distress to the industry than nearly any other
factor. He said:

> Too many regulations . . . too many forms . . . fines if the
> forms are not properly filled out. The government does not seem
> to understand that you cannot determine the direction of agri-
> culture by legislation. The state [Oregon] restricts the sale of
> land that is zoned for agriculture. You cannot sell under any
> conditions, because economics are not part of the consider-
> ation. This defeats the purpose of trying to preserve farmers.

Mike continued: "OSHA, instead of being a business advocate and
a champion of safety, seems to have as its main goal to generate as
much revenue as possible to substantiate its existence."

Long range, the Naumeses believe that exports will be expanded.
They and the industry will get rid of unprofitable grades and sizes.
Technology will provide the answer to more innovative, automated,
and less expensive packaging. Higher tree population and better tree
count and health will improve predictability of production. Global
information about weather and markets will help planning. Marketing
will increasingly become the key to success, so closer attention has to
be paid to the constantly changing desires of the consumer. In that
respect the company aims to increase fresh-cut, frozen sliced, dehydrated, and puree products. It is searching for compatible high-value
crops that will lengthen the harvest season so year-round employment
can be offered to more persons. The search for ways to increase production per acre will continue. Mike closed his dissertation on future
plans by stating that the company needs to employ a "topnotch professional leader," which would allow Mike to spend more time doing
visionary work. This would be a safeguard in case none of his three
children has any interest in the business.

Mike's final comment about the future was that he is concerned
"that society does not appreciate the value of agriculture. This is probably more true in California than in Oregon or Washington. The aver-

age citizen in the number one agricultural state [California] has no concept what agriculture is."

Bibliography

Interviews

Freytag, Tony. Medford, OR, November 19, 1998.

Kusel, Bill. Medford, OR, November 17, 1998.

Moreno, Juan. Medford, OR, November 18, 1998.

Naumes, Laura Earnest (Mrs. Michael). Medford, OR, November 18, 1998.

Naumes, Michael D. Medford, OR, November 16–19, 1998.

Naumes, Sister Mary Pat. Medford, OR, November 19, 1998.

Naumes, Susan F. Medford, OR, November 16–19, 1998.

Semple, Jim. Medford, OR, November 17, 1998.

Miscellaneous

"Future of Agriculture." Speech by Mike Naumes, April 17, 1995, presented to Jackson County Chamber of Commerce.

"Keep Up with Market Demand," *Western Fruit Grower*, March 1987, August 1994.

Medford Mail Tribune, November 19, 1980.

Naumes, Mike. Speech to the Oregon Bankers Association Convention, February 20, 1997.

———. Speech to the Oregon Land Title Association Convention, June 24, 1997.

Oregon Business, June 1993.

The Packer, March 23, 1998.

Successful Farming, May 1989.

Yakima (Washington) Herald, June 27, 1994.

Leonard Odde

"I Didn't Know It Would Be So Easy"

WHEN farmers or ranchers in the Dakotas and other Plains states are asked how large their operations are, they often reply in quarter sections, not acres. Such responses are due in part to the fact that it takes more acres to generate a given income in these states than in states more favorably endowed for crops.

Fortunately, the four-wheel-drive tractor and the air seeder, which were pioneered on the Plains, and the 60-foot toolbar planter for beans and corn can till large areas at minimum cost. This enables one person to operate several thousand acres with limited hired help.

A June 1 picture of part of the nearly 10,000 acres of corn planted by 36-row Kinze planters, with 20-inch rows no-tilled into spring wheat residue. Each planter unit is capable of planting 500 acres per day.

Although the farms are large in acres, their gross incomes are considerably less per acre than in more productive areas. With few exceptions, government payments are a larger portion of farmers' incomes here than in most other regions. Those payments have served to perpetuate many farms that otherwise would not be economically viable.

The more-than-40,000-acre Odde farm and subsidiary enterprises are an excellent example of efficient use of the limited labor supply on the Plains. Located on the western edge of the Corn Belt in east central South Dakota, L & O Acres is one of the most dynamic farms in the Midwest and on the northern Great Plains. Corn, soybeans, and sunflowers are cropped in this predominately small-grain area in various combinations. Odde saw the potential of diversification and made it work for him.

ARNE ODDE was seven years old when he arrived in the United States from Norway in 1890. His parents moved to the South Dakota frontier and settled along the northern border just east of the Missouri River near the village of Herreid. In 1918, at age 35, Arne married Ragna Toriseth, who had immigrated from Norway in 1901. Neither had gone to school past the eighth grade, and both had spent their time since school working on neighboring farms or for their parents. By May 1930, with the birth of Leonard (Len), the Oddes had a family of five girls and four boys.

The Odde farm consisted of 320 acres, which even in good years could not produce more than enough for a subsistence living. Eight horses, needed to provide the power, consumed about one-third of the total production of the farm. Odde purchased a 15-30 McCormick-Deering tractor in the 1920s, but, as Len explained, "When the bad years came, there was no cash for fuel, so the tractor was not used. Dad refused to borrow for fear of losing the farm." The 10-cow milking herd "generally supplied enough milk for the family plus a little to churn into butter for table use, but there was never any to sell." The chicken flock was large enough to supply family needs. The Oddes also had 15 sows that each produced 2 litters a year, with about 6 pigs per litter. "We ate lots of salt pork and pork gravy with boiled potatoes, milk, and bread." Pork was the major source of income, which in the dry years, with accompanying low prices, was as little as a $1,000 a year. The income did not improve much until the 1940s.

Although on most farms the oldest children stayed home to help with the farm work, that was not the case in the Odde family. The eight oldest children went to high school, but Len, the youngest, did not go beyond the eighth grade in country school. When the opportunity arose, the Odde children worked off the farm. They all lived at home while they attended high school even though they worked out. After high school, only the oldest brother, who was "Dad's chief helper, stayed on the farm. All the rest went into other work." Len was 14 when he first "worked out" wherever he could get a job. Most of the time he received a dollar a day. He remembered the work was not very steady and only when farmers really needed help in busy times. During World War II his oldest brother went into the service. A second brother had died, so Len was compelled to stay on the farm.

In 1947, after his oldest brother returned, Len took a job with Anderson Welding Supply in Aberdeen, South Dakota, for $20 a week. He said, "I couldn't wait to get away from the farm. I was sure I would never farm again." He was 17 when he left the farm to start his new job, which he recalled was hard work but "a snap compared to farming and with much shorter work days." He remained at that job for the next five years and by then earned $50 a week. In the meantime, he married Beverly Larson, from Columbia, South Dakota, who worked as a secretary in Aberdeen. From 1952 to April 1954, while Len was in military service, Beverly remained at her job in Aberdeen. When his military service ended, Len rejoined Anderson Welding Supply. He became a route salesman and earned $400 a month.

Len decided to move on and in 1957 interviewed for a job with Williams Hardware Company, of Minneapolis. He became a route salesman on a straight commission. His territory consisted of the northern counties of South Dakota from the Missouri River east and into Minnesota as far as Fergus Falls. Most of his traveling was still on gravel roads. He called on hardware stores, blacksmiths, welding shops, and small manufacturers. He made "good money, and it was a great confidence builder."

Two of the accounts he called on while traveling for Williams Hardware led to the next phase of his career. The family lived at Wahpeton, North Dakota, where Beverly was employed, and Len became involved with individuals who were starting to manufacture specialized farm implements. They wanted to expand and in 1963 invited Len to invest and be one of the three original incorporators of Wil-Rich Manufacturing Company. He became vice president of marketing. The new company could not afford to have him on the road full time, so he continued his sales job and sold Wil-Rich products on a moonlight basis.

In 1967 Eugene Dahl, an officer in Melroe Manufacturing Company, at Gwinner, North Dakota, whom Len had learned to know while selling for Williams Hardware, offered him a job as materials manager. Len took a cut in pay to work for Melroe because he wanted to get off the road so he could spend more time with his family. However, he continued to sell for Wil-Rich. By 1969 Wil-Rich was doing well enough to require his services full time, so he left Melroe.

The double-piston lift was Wil-Rich's first really successful product, and it put the company on solid ground. Len was president of the firm during 1970 and 1971, but he did not like the work. He wanted to do something else and offered his stock to his partners. Wil-Rich was later purchased by Lear Aircraft Company.

After Len left Melroe in 1969, he kept in contact with Eugene Dahl, who had become chairman of the board of Steiger Tractor Company. As soon as he sold his interest in Wil-Rich, Len was employed as vice president of marketing at Steiger. His timing was perfect, for the agricultural boom caused by the dollar devaluation and the Russian decision to buy grain skyrocketed tractor sales.

Len and Beverly both came from very modest homes, so they carefully shepherded their newly gained funds from the sale of the Wil-Rich stock and from Steiger commissions. In 1973, before land prices started to rise, Len purchased 2,000 acres in the Westport area north of Aberdeen. The chief reason for buying there was that it was near Beverly's home and she liked the area. According to Len, his intentions were to "farm a few thousand acres and enjoy a nice, easy life. As soon as the children were grown, Beverly

Leonard Odde at age 63, two decades after he purchased a few thousand acres and planned to "enjoy a nice, easy life" even though at age 17 he was sure he "would never farm again."

took a job at St. Luke's Hospital. We each had our own business. She never became involved in farming because we agreed that one boss is all you need."

Len wanted to remain with Steiger for a few more years and farm as a sideline. That all changed in 1978, when he resigned because he did not like what was taking place in the company. He had no way of knowing what was ahead, but after 1980, Steiger's good fortunes eroded as the farm economy weakened. The six years of agricultural recession, which were so disastrous for Steiger, proved to be a boon for Len's farming operations. In retrospect, had he known what was ahead, he probably would not have entered farming when he did.

While marketing for Wil-Rich and Steiger, Len had the opportunity to work with a cross section of farmers in the United States and abroad. He related what he had observed:

> On those jobs I watched many successful farmers. I quickly grasped this was not the kind of farming I had seen when I was growing up on the farm. I realized that there were real opportunities in agriculture if one started with a large enough base. I also understood that those who were doing the best were those who kept expanding on a regular basis. In 1973 there was only one four-wheel-drive [tractor] in operation in Brown County [South Dakota]. I saw lots of land on which farmers were not doing well because they were too locked in on small operations. Many of them had inherited [their farms], but they were not willing to take the risk to expand by renting additional land.
>
> What happened next was amazing—the availability of land was unbelievable. It was so easy to add acres. The next thing we knew we doubled our operation. Leon Schaunan came to us and asked if we wanted to rent his farm. That gave us 4,000 acres total, and we had the backbone of our operation. I had learned from observing the successful farmers we sold Steigers to that they were making 25 percent return on their money if they did a good job of managing. But I realized they were risk takers.
>
> I had acquired a considerable amount of Steiger stock while I worked there and made a fair profit when I sold it, but it would have been a mistake to have used that to buy land. I became a believer in renting when I saw what the topnotch operators were doing. If you buy land, it loads you down with

long-term debt and does not leave you with any flexibility in financing. Renting greatly enlarges your capital base. Next year [1998] we will be operating over 40,000 acres, of which 4,000 acres are owned by us. Our average land cost is much less than if we were paying interest, taxes, and principal. Every acre that you don't own frees you of dollar obligations.

From 1973 to now [1997], our rents have tripled, but so have most other costs, so it is relative. We found it easy to add land because of retiring farmers or those who were discontinuing farming but were happy to find a safe home for their land. When I started I never realized this would happen, and I didn't know it would be so easy. The local people saw that we did a good job of taking care of the land, and they came to us when they had land to rent.

Not every year has been easy. My first tough year came in 1976, when the area suffered a drought. Our production was basically zero. I personally visited every landlord. One of them rode the combine with us for an hour, and the hopper was not filled. We stopped combining. I got every landlord to agree to wait until 1977 to collect the 1976 rent. Fortunately, I was still employed by Steiger, and we were able to keep the trucks rolling.

Len explained how he adapted to meet the needs of the business, most of which he had learned by observing what successful farmers had done. The land he farmed was not easily available to any railroad line. He did not want to tie up funds for large long-term storage facilities, so in 1974 he purchased his first double-trailer semis. Originally, most of his grain was hauled directly to the Minneapolis or Mississippi River terminals. To protect himself from market fluctuations, he practiced hedging and forward contracting.

Trucking was a key to the success of his operation. Although the return on investment was not as good as in farming, trucking complemented farming and was essential to Len's side ventures in the machinery business. After the crop failure of 1976, the trucks were pushed hard to generate as much revenue as possible, for that income was critical to survival. The trucks hauled for elevators from areas that had crops to sell. The truck fleet grew to meet the needs of the farm, and by the early 1990s it contained 12 double-trailer rigs. The fleet is operated independently of the farm as L & O Acres Transport,

Inc. When not needed to drive trucks, the employees do farm or shop work.

Len never intended to spend his time doing field work. Initially, he was involved with his job at Steiger, and by the time he left that position, the farm business was large enough to command his full attention in management. He prides himself for having built a good record-keeping system and knowing what his costs were from the time he started to farm. He never wanted to be in the position of farmers he knew who were not doing well but would not sit down to figure out what was wrong. He had seen enough of that while growing up on a subsistence operation.

In 1978 Greg Odde joined his father on the farm. Greg admits that he had never given much thought to farming and was not encouraged by his parents to enter the business. However, while working on the farm during the summers, he became fascinated by all the technology and rapid changes in the industry. This gave him an opinion of agriculture different from what he had heard from so many of his friends whose parents farmed. Now Greg could understand why his father was so driven to keep going even though he admitted he was not particularly excited about the business. Greg shifted to an agronomy major to prepare himself for the farm. After graduation his chief duty was securing and applying herbicides, pesticides, and fertilizers. In addition, he was responsible for the custom work of those three lines.

In 1973, when Len purchased his first land, he noted that little fertilizer and few chemicals were being used locally. But neighbors observed what he did and his results, and before long they came to him to inquire about his doing custom work. Len appreciated the chance to spread his cost of owning the equipment, but from the start he determined he would not take more custom work than he could handle with the machines he needed for his own farming. Len and Greg both agree that the use of fertilizer and chemicals is the best way to increase yields and to save land while maintaining profitability.

In 1975 Todd Ochsner, a nephew who was still in high school, started working on the farm on weekends and during vacation. Todd had spent summers on his grandfather's farm since he was 11 years old and observed, "My grandfather did not worry about making money; he just worried about surviving." In 1976 his grandfather could not afford to have Todd work for him, so Todd turned to Uncle Len. The Odde farm was suffering from drought, but it had the perfect job for a high school lad who wanted to stay in shape—picking rocks.

Todd graduated from high school in 1979 and started college part time, with plans to major in finance. He drove trucks for Len in his spare time. After one semester he quit school to drive trucks full time. He made $800 a month, but he was motivated to do better, so he decided to return to college.

Len realized Todd's potential and challenged him. Len was the Wil-Rich distributor for South Dakota and Nebraska and after 1980 was also the Steiger distributor for South Dakota. He offered Todd a job selling the equipment of those companies to dealers. Todd was allowed a draw and received a commission. The $800-a-month truck driver made $45,000 his first year, and more the second year. Then he met with reality, for in 1983 the full brunt of the agricultural downturn hit, and selling became "really tough."

Todd Ochsner said of the next couple years:

> It was the best education I could have received. I learned how to read a financial statement. There were more divorces, farm write-downs, reduced land values that cut the net assets to the bone. I was really educated about the business of farming. However, even in those down times, I observed that farmers are basically optimistic even though they talk negatively. I learned that most farmers want to hear the positive, and that is the message you have to tell them. The farmers who felt the most locked in were those who took over from their parents. That experience made me realize what I like about Leonard—he is always positive and never gives up. He does not tolerate negativism.

The distributorships with Wil-Rich and Steiger served Len well, for they provided some diversification from total reliance on crops. Len had weathered the energy crisis and the drought in 1976, but he said he was unprepared for the political moves. He explained: "The Carter embargo really put the blocks to us because it put the brake on exports." In Len's opinion, the embargo made the years 1981 through 1986 worse for American agriculture than they would have been if exports had not been so negatively impacted.

Ironically, what was a near disaster for many farmers proved to be a boon for Len. Every year during the early 1980s, farmers came to him and asked if he would rent their land. By then, Len had forgotten about farming a couple thousand acres and having a "nice, easy life." He was too motivated to take life easy and rose to meet the new challenges. His accurate records proved that each time he rented more land, the per unit cost was lowered, so he seldom refused an offer to rent land, whether it was a 40-acre parcel or 2,000 acres.

In 1983, when so much land was being offered by area farmers, the Hommelscheins, of Germany, approached Len. They were in the market for land and liked the way Len farmed. The Hommelscheins suggested that they would buy 2,500 acres if he would agree to farm them. Later, they came out to the field where Len was overseeing some work and asked if he would farm another 3,500 acres if they bought them. Without hesitation he accepted the offer. Later, they bought another 1,000 acres, making a total of 7,000 acres. This made the Hommelscheins Len's biggest landlord. Then, South Dakota passed a law limiting future purchases by foreigners to 160 acres. Len understood the emotions involved and why the politicians reacted to the clamor. He regretted that turn of events but was happy that the Hommelscheins were grandfathered in. His closing comments regarding his German landlord: "They are such good people to work with. I see them once a year, but it is always a pleasure doing business with them." With a solid 11,000-acre base, Len had his per unit cost of production pared down, but as he added more land, the cost continued to decline, so he was comfortable about further expansion.

Len was always on the lookout for ways to improve the efficiency of his farming operation. He maintained his connections with the farm machinery manufacturers through Far Better Farm Equipment, a free-standing division created to handle farm equipment for resale as a distributor and/or retailer.

In 1980 Wil-Rich developed its first air seeder. Because Len was a distributor, he tested it on his farm. As expected, the new machine had flaws, but Len was quick to grasp its economic potential. The air seeder provided for reduced tillage, which saved moisture and enhanced the chance for better yields besides being environmentally friendly. It also saved fuel and time. The air seeder was another step to more efficient and economical farming. It was the ideal tillage machine for large-scale grain farmers in the Great Plains and the Prairie Provinces of Canada.

Again, the depressed farm economy of the 1980s presented Len with an unexpected opportunity. Concord, Inc., of Fargo, one of the original air-seeder manufacturers, was caught in the farm economy downturn. Concord representatives approached Len about potential involvement, which led to his becoming the Concord dealer for South

Dakota. Over the next nine years, Far Better Farm Equipment sold 408 new air seeders, which was both an agronomic and financial success for L & O Acres.

When Len first started to farm, he purchased his tractors and combines, but he did not have a fixed program for replacement. Because his equipment was new, large, and well maintained, he willingly sold any machine whenever it was to his advantage. That practice changed as the farm grew and more equipment was needed. To avoid tying up large amounts of capital, he turned to leasing his four-wheel-drive tractors. By 1997 he was leasing eight four-wheel-drives, one more than he needed if everything worked well. But he is a stickler for getting the work done in a timely fashion, and, like his late good friends, David and Ivan Miller, he willingly paid the extra cost to have backup power.

Timing is even more critical in harvesting than in planting. To speed that job, the nine company combines, each with 36-foot direct-cut heads, are supported by six to eight combines of a custom harvester. Len recalled the time when storm clouds threatened and his crew operating the nine company combines harvested 5,000 acres of

Five of the leased fleet of 9390 Steiger four-wheel-drive tractors equipped with fertilizer tanks and hooked to 36-row Kinze no-till planters. Odde learned that every time he added acres, his per unit cost of production was reduced. Equipment such as this made that possible.

50-bushel wheat in 4½ days. The last of 225 double-trailer semi loads left the field just hours before a storm hit. As with his spraying and fertilizer-spreading equipment, Len uses his harvesting equipment to do custom work for others, but only after he has finished harvesting his own crop of wheat, beans, corn, or sunflowers.

Because the truck fleet was used for other than strictly farm business, it became a free-standing division called L & O Acres Transport, Inc. As previously stated, profits from trucking were generally not as good as those from farming, but in a bad crop year, the cash flow generated by the trucks was important. In 1993 the value of the truck fleet was enhanced considerably when a 2.2-million-bushel terminal elevator capable of loading 54-car unitrains was purchased. The elevator was acquired "at a bargain or fire-sale price" from a construction company when the original investors, who planned to develop a sunflower processing plant, defaulted. Len was in a strong financial position and able to take advantage of a totally unexpected opportunity.

The facility, located at Ipswich, South Dakota, became a free-standing division called L & O Acres Terminal, Inc. The terminal was not a part of Len's long-range plan, but he was not one to pass a good

The 2.2-million-bushel L & O Acres terminal located along the Burlington Northern / Santa Fe main line near Ipswich, South Dakota. The siding is being enlarged to handle 108-car unitrains, which would enable 22 million bushels throughput.

opportunity. The farm's production provides only a small part of the terminal's total annual volume, so it cannot be justified solely as a farm elevator. However, because it is a unitrain facility, it provides an improved market for area farmers and is a profitable value-added venture. The challenge ahead is whether the rail siding should be increased to handle 108-car unitrains, which would enable up to 10 turnovers annually for the terminal.

As of 1997, L & O Acres and associated divisions had 30 full-time employees: the farm had 10, including office personnel; the truck fleet had 12; the terminal had 3 or 4; and the shop and the dealership had the remainder. Farming always has priority, but employees are shifted to other positions depending on the immediate need. Len has experienced a very low turnover of labor. In 1997 no employee had less than five years tenure with the company. The truck drivers "make the best money, but they also have the toughest work schedule."

Len credits his good fortune with help to the fact that the people like the "atmosphere of working here." He was quick to point out that being close to Aberdeen (17 miles) was a definite advantage for both hiring and living purposes. It was obvious to him that farmers in areas remote from good communities have a more difficult time accessing qualified workers. Leonard Odde is a people person who enjoys his association with his employees. He maintains tight control of overall activities and "leads good people." In his opinion, he generally has better qualified workers today (1997) than he was able to secure in the 1970s. "We could not do the job we are doing if it were not for the good quality of people working here."

When asked what he thought was the future of farming, Len replied:

> I don't think we have a handle on that yet—nationwide or globally. I think the traditional father-son operation is the most serious problem facing agriculture. It is a real problem because most farms of that type are too small [to support the next generation]. The only thing I know for sure is that most farms will have to continue to grow in total business volume. Personally, we cannot stand still. We have good leadership, and our operation will continue to grow. I am not sure just what the next move will be. Expanding our land base is the easiest way and would cause the least problems. Possibly we should expand the

terminal. A third direction might be to expand the dealership with several short-line companies.

When Todd Ochsner was asked to give his thoughts about the future of the farm, he commented, "Planned growth is essential to the operation. We have managed our growth well. I am convinced that if we ever decide to pare down, we would be better off to sell."

The company has $75,000 invested in computers to do an accurate job of record keeping. Len is not convinced that computers do any better job than he once did with less technology, but he concedes that he does not want to return to pre-computer days. The immediate challenge to L & O Acres is that it is in an agricultural area with several strong competitors for land, whether renting or buying. Len has 35 landlords to negotiate with, which is not a problem until other farmers get aggressive for land. He recalled the active bidding after the good prices of 1996, when many farmers quickly forgot the long-range reality of agriculture's cycles.

As of 1997 Len was pleased that the government supposedly was withdrawing from agriculture. He realized that a free market would probably be subject to greater variations but that in the long term there would be an overall improvement. He is apprehensive, however, about the current policy, for he fears there will be continued intervention because the government does not want to stay out of farming. His reasoning is based on the fact that it is politically important to have a cheap-food policy.

At the time of interview, 67-year-old Leonard Odde, although suffering from Parkinson disease, was still very much in control of L & O Acres, a private holding, and its four subdivisions. His dream of farming a couple thousand acres and having a "nice, easy life" quickly faded because he "never realized it would be so easy to expand." A farm boy with an eighth-grade education, who at age 17 never wanted to see the farm again, gained his schooling and confidence in the school of hard knocks. Len has wonderful people skills and is an excellent salesperson. He has the ability to make major decisions without hesitation. He instinctively knows what to do and has made very few mistakes. He has never stopped looking over the horizon, and even though his farm has grown far beyond the few thousand acres, he would be the first to say it really was easy because he did it so naturally.

When asked to comment on how agriculture has treated him, he replied, "Agriculture has paid me better than anything else I have done or could have done. I have enjoyed it more than anything else I have tried. The potential for agriculture is great."

In his sales career throughout the United States and several foreign countries with Melroe, Wil-Rich, and Steiger, Len's eyes were opened to the potential in farming as carried on by innovative entrepreneurs. In that respect, probably his two best role models were David and Ivan Miller, now deceased, of Wahpeton, North Dakota. The Miller brothers each created an outstanding farming operation of nearly 30,000 acres with virtually no help from their parents. Anyone who knows Leonard Odde knows that he will enjoy farming to his last ounce of energy.

Bibliography

Interviews

Ochsner, Todd. Westport, SD, September 16, 1997.
Odde, Greg. Westport, SD, September 16, 1997.
Odde, Leonard. Westport, SD, September 15–16, 1997.

R. D. Offutt

*"The Entrepreneur Has to Have a Special Work Ethic
That Other People Do Not Understand"*

RONALD D. OFFUTT, JR., never thought of himself as poor, but when it was time for him to enter college, his parents told him that he could not count on any help from home. Ron (for clarity, father and son will be designated as Ronald and Ron, respectively) did not know that his father owed $250,000 on personal notes and on accounts payable. Neither did he realize that the mortgage on the farm, which had cost $16,800, was now $30,000 in addition to the mortgage on the machinery. His father pleaded with creditors John Scott, Frank Thompson, and Paul D. Jones Co., for seed potatoes; Jack Bahlmer for trucks and cars; the Blacks, of Michigan, for onion seed; Sam Setter for fuel; and others for lenience so he would not be forced into bankruptcy. Ronald, the eternal optimist, promised that he would repay them all with interest when he recouped.

The future looked bleak for young Ron after high school graduation until his Moorhead (Minnesota) High School football buddy, David Johnson, came out to the farm with Jake Christiansen, Concordia College football coach, and encouraged him to enroll. Ron was tendered an athletic scholarship and a job in the laundry of the Physical Education Department to supplement his scholarship.

When Ron graduated from college in 1964, his father had pared his obligations to the unsecured creditors down to $100,000. Ronald had managed to do this by double-farming; that is, besides his Minnesota business, he raised 320 acres of potatoes in Missouri each year. He planted in Missouri in February and harvested in July.

In March 1964, just before graduation, a mentor told Ron about a 200-acre farm at Baker, Minnesota, that was for rent that spring. It was a challenge for Ron to convince his father to rent that farm; but what a break it was, for a good crop and high prices resulted in a bonanza. Ronald paid all his unsecured creditors, was current with his secured creditors, and had a net worth of $84,000. From that day on, father and son rode the roller coaster to success. Even in his wildest dreams, Ronald D. Offutt, Sr., could not have imagined that 3½ decades later, his son would be the nation's largest potato farmer.

IN 1928 Bert Offutt, age 41, was killed in a car accident, leaving his widow, Laura; daughters, Pearl and Frances; and sons, 17-year-old Luman and 11-year-old Ronald. Bert Offutt had specialized in potatoes on his 700-acre farm at Orrick, Missouri. He also brokered in seed and table-stock potatoes in a territory stretching from Oklahoma to the Red River Valley of the North. After Bert's death, Laura Offutt took charge of the business and relied on her oldest son, Luman, who quit college to become his mother's chief helper. In the meantime, Ronald attended Kemper Military School and College. After graduation he returned to the farm to raise potatoes. From 1935 to 1937, Luman, who liked cattle, was a cattle buyer in Argentina for Armour and Company. In 1938 he returned to take over the farm and became a very successful cattle feeder, a career he pursued until his retirement. In 1940 Ronald Offutt and Lida Swaford, a coal miner's daughter, were married. Three years later the young couple, with two children, Betty Lou and Ron, struck out on their own.

In his potato transactions Ronald had the good fortune to become acquainted with John Scott, of Gilby, North Dakota, one of the premier potato growers of the Red River Valley. Scott found 320 acres near Inkster, North Dakota, which the Offutts, using money inherited from Bert Offutt's estate for the down payment, purchased and farmed during 1943 and 1944. The relationship between the two men developed into a lifelong bond. Both were strong-willed, visionary individuals, and they became true pals.

Scott was financially successful and willingly helped people he liked. Ronald became one of the greatest recipients of Scott's generosity. In the two years the Offutts were at Inkster, Scott not only taught Ronald how to raise potatoes in the Red River Valley, but he also lent him machinery to raise them. Ronald was highly motivated but could seldom implement his ideas because he was not a good manager or record keeper. For many years, when he did not have money or credit to buy seed potatoes, Scott staked him.

In 1943, their first year at Inkster, Ronald and Lida Offutt had an extremely good crop, but in 1944 the situation was just the opposite. Lida could not stand the isolation of the Inkster area and wanted to be "closer to civilization." The poor crop was enough incentive to encourage the young couple to move. They had nothing to show for their first two years of farming other than a small profit on the sale of the farm plus the original down payment money. They used all they had to make a down payment on a 240-acre farm three miles south of Moorhead, Minnesota.

In spite of the agricultural prosperity of the postwar era, Ronald did not do well. His goal was to raise 400 acres of potatoes each year. To get an adequate land base for proper rotation, he purchased some cheap land a few miles east of Moorhead that was water prone and never proved profitable. During the lean years that followed, the aforementioned seed potato producers, especially John Scott, provided Offutt with seed "on credit."

When the time came, Betty Lou and Ron attended rural Clearview School three miles south of the farm. Because all the students' fathers were farmers, it was natural for Ron and the other boys to discuss farming in their free time. Ron recalled that he became fired up about farming when he was nine years old and drove truck in the field during the fall grain and potato harvest. He even skipped school during potato harvest. The next year his father purchased an IHC Super "C" tractor with a two-row cultivator, which Ron said was "my tractor." In the seventh grade, he stayed out of school for five weeks to help with the potato harvest.

When Betty Lou entered high school, Ron was ready to start the eighth grade. He was transferred to Moorhead Junior High, and they rode the same bus. When potato harvest started, he protested bitterly because he was determined to skip school and drive potato truck again, but his parents had other ideas.

The school bus route went through Thea Gullings' farm north of the Offutt farm. Ron loved the Gullings land. One day he found out it was for rent and begged his father to rent it. The 410-acre Gullings farm was excellent land. Ronald, ever short of capital, hesitated, but his optimism and Ron's pleadings convinced him to plunge ahead. Ron recalled that was the first time he begged his father to rent a farm. The second time, mentioned above, was in 1964 regarding the

200 acres at Baker. Ronald's progress improved after he rented the Gullings farm, which he operated as long as he farmed. The Gullings land and the 1964 crop at Baker proved to be turning points in the Offutts' farming career.

RON said that if his teachers during his early school days had known about attention deficit disorder, they would probably have believed he had it because he was always so restless and competitive. He recalled that his father encouraged him "all the time." An early memory that left a lasting impression on Ron was winning a softball bat for reciting the most Bible verses in vacation Bible school. He said, "That really sparked me. I think it was a real confidence builder and helped my competitive spirit."

After he started school at Moorhead, his farming was relegated to weekends and during the summer. But school took on a new meaning when he became involved with sports—football and wrestling—and vocational agriculture, which led to participation in the Future Farmers of America (FFA). His vocational agriculture teacher, Leo Maattala, knew the right buttons to push to motivate his eager student. Getting A's in agriculture was such a good experience that it created a desire to do better in his other courses. He became a good student.

In his first week of football, Ron met Paul Horn. Horn said of that meeting: "He was about twice as big as I was and real tough looking. I was the quarterback, and Ron was the center." Little did the two realize then that their meeting would be the beginning of a lifelong friendship and eventually a business relationship.

During Ron's high school and college years, his father assigned him the task of picking up day laborers, many of whom were winos, to hand-hoe onions and to work in the onion and potato harvest. Ron said:

> I vividly remember going into town to pick them up and taking them to the farm. We paid them 75¢ an hour and had to pay them at the end of each day so they could drink at night. Each morning I had to make the rounds for more recruits because we had about 70 percent turnover daily. This made a deep impression on me: I never wanted to live that way.

The Offutts were devout people and planted lasting values in their children. Ron reminisced:

> Dad instilled a high moral character, almost as if he were preparing me for the business. He stressed that if you owe people money and can't pay them, talk to them. He drove home the adage that your word is your bond. He was such a visionary, and that rubbed off on me. He allowed me to make mistakes, and if something we tried did not work, he was my picker-upper and would not let me get depressed. It really helped me when he let me talk him into renting the Gullings farm and the Baker farm, especially considering the way those deals turned out.
>
> We were the first in the area to cut seed potatoes and deliver them in bulk to the planter. That was a real cost cutter. Later, Dad allowed me to leave the Red River Valley and go east to the sand lands to raise irrigated potatoes. That was another turning point. There is no doubt that the constant lack of financial success instilled in me the undying desire to succeed. I think when my father died [in 1983 at age 66], he had very complex feelings about me. He was extremely proud of what we did together, but he was jealous, probably because he had not done it. I'm sure it caused him frustrations.

Ron was all business when he was in college and established a solid academic record. He excelled in football and wrestling, was married, and worked on the farm every minute that he was not occupied with college activities. Early in life Ron realized that he wanted to farm, and because his family was close-knit, he assumed he would be in partnership with his father. He recalled that as soon as he graduated from Concordia, his father asked him if he wanted to work for him. Ron replied, "No, I want to be your partner." It took a couple days for Ronald to grasp the idea, but once he agreed, he did so wholeheartedly and established the farm name R. D. Offutt & Son. Ron admitted he was probably a little presumptuous, because he expected his father to carry his note for the portion of debt he would be assuming. Ronald showed good judgment and made his son sign for half the obligations with the financial agencies. Those obligations included a line of "really marginal machinery, half the cost of 480 acres of land, and a potato warehouse."

Then the unexpected happened. In 1945 Ron's grandmother, Laura Offutt, had purchased two sections of land at Hunter, North Dakota. Now that Ron was graduating, the family members at Orrick offered those two sections for rent to Offutt & Son. Ron originally expressed disappointment. He explained to his mentor that he realized he would eventually farm, but before he became immersed on such a large scale, he wanted to spend time working in nonfarm businesses to gain some "street smarts." How many 22-year-old farm boys who wanted to farm, staring at an opportunity to take over two sections of prime land, would have reacted in that manner?

In 1965 the partnership rented the two sections of land at Hunter on a one-third/two-thirds basis. Ronald did not have good records, so the first challenge for Ron was to apply his college economics and business training and set up a record-keeping system. In 1966 the basement of the family home was partitioned to create space for an office. That year the Offutts rented another full section, which gave them 2,570 acres of good land on which they could rotate potatoes. This provided for economy of scale for the contemporary technology. With a bigger crop coming than their warehouse could handle, they turned to Production Credit Association (PCA) for a $48,000 loan to build a 33,000-hundredweight (cwt.) potato storage facility at Glyndon, Minnesota. Ron remembered, "Father, always the eternal optimist, was happy to step out and expand."

In early 1967 a series of nine Quonsets in which the government had stored grain, located on 10 acres of land near Glyndon, came up for auction. Ron borrowed the money from PCA to purchase the land on which the Quonsets were situated. When the day came to auction off the Quonsets, he announced that anyone purchasing them would have to remove them at once. There were no bidders, and Ron purchased all nine, again borrowing the full purchase price from PCA. Each Quonset was capable of storing 13,000 cwt. of potatoes. A big crop was on the way, so three Quonsets were immediately modified for potato storage.

The next big event of 1967 came in November, when Grant Mattson, the John Deere dealer at Casselton, North Dakota, informed Ron that he planned to sell his business. Mattson was a pioneer in leasing farm machinery. The Offutts were his biggest leasing customer. They were expanding rapidly and understood the advantages of leasing over getting further encumbered with debt for new equipment.

Mattson had taken a liking to Ron and encouraged him to buy the dealership. He suggested that if someone else purchased it, that person might not have the necessary capital to lease machinery on the scale that the Offutts required. This caused a mild panic with the partners because they were not sure they could get the financing they needed for new equipment. The financial situation was not helped by a large 1967 potato crop that flooded the market and "took the sails out of the price."

Buying the dealership was Ron's project, and he had to do some creative financing to come up with the $100,000 that Mattson wanted. The warehouse built at Glyndon in 1966 was sold for exactly what it cost, which freed up the $12,000 the Offutts had paid off on the mortgage. Ronald agreed to sell 160 acres of the 480 acres east of Moorhead, which freed up another $36,000. The next $20,000 came from Charley Laitner, a machinery salesperson in Missouri, who became a partner with Ron and the first manager of the dealership. A Moorhead banker lent $20,000 on a personal note. After exploring every possible source of cash, the 25-year-old "entrepreneur" was forced to turn to his grandmother, who reluctantly provided the final $12,000 on a short-term note.

The original John Deere dealership at Casselton, North Dakota, purchased in 1968 by R. D. Offutt.

Ron's reason for purchasing the dealership was that he could raise money more easily for a new venture than he could to sustain the Offutts' already highly leveraged farming operation and buy the necessary equipment to operate it. That desperate financial situation made it necessary for him to learn how to maximize the dollars available to him. In March 1968, after five hectic months of searching for financing, he thought he was ready to take over the business. He commented on what took place next:

> I actually bought $87,000 worth of parts, tools, shop equipment, and furniture. This gave me $13,000 to start operating. I rented the building from Grant for six years, after which I was to pay $150,000 for it. Grant wanted me to have the business, and that is why he was willing to rent rather than sell it. When Grant found out I had only $13,000 in cash for operating capital, he got serious. He took me out to five farmers who were big customers and had cash. He convinced them to prepay their parts expenses for the year. This gave me another $20,000. Our next step was to create a separate rental company, which we called Farmers Equipment Rental. We arranged for local financing for 12 tractors and 10 combines. The farm rented several of the tractors, which eased the farm's need for cash. The dealership was successful from the start and was a key to getting us on a roll.

Ron soon learned that if the dealership was to be successful, he had to be involved in the management. This meant he had to divide his time between farming full time and running the dealership, which took a toll on his personal life. He did well in the dealership but soon realized that he was not a detail person. He liked "painting with a wider brush," and he found that freedom in farming. Those who knew him understood that he was every bit the visionary his father was, but he was much better at implementing and managing.

Now R. D. Offutt & Son farmed 2,570 acres, and again finances were stretched to the limit, but the Offutts realized that opportunities still were to be had. Alert area farmers were looking for alternative crops. Per capita potato consumption had hit a low in the 1950s, but then potato chips, french fries, granules, canned potatoes, hash browns, shoestrings, and other processed products increased dramatically in popularity. Per capita potato consumption rose sixfold by 1965. The more alert Red River Valley farmers saw potatoes as an opportunity but did not have the expertise or the equipment to get into the business. This gave the Offutts another opening.

In 1966 the father-son team entered into the first of many potato growing partnerships. They partnered with Ernie Oberg, a contemporary of Ron's, a hardworking individual and good farmer who had good land. The Offutts furnished the expertise and the specialized equipment, and the proceeds were divided on the basis of inputs. That partnership is still in effect at the time of this writing. The partnership relieved the Offutts of having to seek more land, and because they were established brokers, they benefited from marketing the crop.

Ronald was well connected in the potato industry, which had a fairly tight "good-old-boy network." He continued some of the relationships his father, Bert, had initiated in the 1920s and earlier. His chief market was with Guys Foods, of Kansas City, Missouri, and a firm in Denver.

In the 1960s, french fry and other processors were still located in large cities so they could deliver the fries directly to the consumer. After refrigeration was perfected so french fries and hash browns, particularly, could be kept frozen during shipping, processing plants were built at the point of potato production. This greatly reduced the cost of transportation. The Offutts seized the opportunity to fill a void in the upper Midwest potato business.

The next major partnership came about in 1969, when the Farmhand Company, fronting for Bill Burnap, approached the John Deere dealership at Casselton in search of someone interested in growing potatoes under irrigation. Burnap, a Texas oil producer, owned more than 1,300 acres in the Becker and Big Lake area of Minnesota and was looking for someone to joint venture with him. The Offutts entered the venture, which included three quarters (480 acres) of land with three walking center-pivot irrigators. With Burnap's guarantee, they were able to borrow at the St. Cloud PCA to buy a half interest in the venture. The first year resulted in an average crop. The chipping potatoes did not do well, but the Kennebec potatoes were just what the french fry people wanted. The partners were encouraged, and in 1970 planted four quarters of Kennebecs under irrigation for the french fry processors in Minneapolis, Chicago, and Michigan. After one year's experience with the walking irrigators, they switched to the newer wheel-supported center pivots. The Offutts were launched into furnishing potatoes, grown under irrigation, for the french fry industry, and Ronald earned brokerage on every potato sold.

The Offutts expanded rapidly at Big Lake and were always able to get money when other farmers could not because of their ties with Burnap. The partnership with Burnap lasted until March 1977, when he wanted to sell his farming interest to raise cash for developing a chain of gas stations / convenience stores. Fortunately, for the first time ever, the Offutts had the asking price—$400,000. That enterprise included all 1,300 acres that Burnap owned there.

BACK in the Red River Valley, the Offutts consolidated their operations and prepared themselves for the next move. In February 1967 Ron relocated his office from the basement room of his parents' home to a 12-by-12-foot space in the mall at Glyndon. Grace Wiedemann became his first secretary and full-time bookkeeper. He realized that if he wanted to continue expanding, he would need to improve his record-keeping system. Ron said, "We were making money in all our enterprises, but still we were outrunning our demand for capital because of rapid expansion. Working under the premise that you always get money one way or the other, I had to get ready for the next step."

In 1970 Ronald Offutt & Associates, a straight partnership, was established. The partnership with Ernie Oberg had worked so well that other farmers were interested in becoming involved. An individual partnership was established for each operation so that each was clearly a distinct entity and there was no concern about cross-filing for financial or tax reasons. Elmer Oberg, Warren Wear, Wayne Seter, Lester Anderson, and Allen Richards all formed individual partnerships. The real reason for those partnerships was to raise money. The Offutts needed to develop the rest of the Quonsets into potato storage facilities and build another very large storage building on the site. Ron added:

> Because Father could get the contracts for potatoes, we were able to show the finance people our production contracts. All the risk we were taking was production risk, not market risk.

These partners were looking for another crop. We did all the work and furnished all the specialized equipment. We were doing "A"-level marketing but only "C"-level production, but we always got our potatoes sold. Every time we increased sales, the brokerage income grew.

The Offutts were fortunate that they had organized Offutt & Associates when they did, for shortly after that, the "financial roof caved in." Norwest Bank, which had been the financial agency for the dealership at Casselton, "pulled out because our record keeping was not good enough to satisfy it." Ron continued: "I had to raise money at Clay County State Bank in Dilworth and at Moorhead PCA. All the partners of Offutt & Associates were also financed at Moorhead PCA. The Moorhead branch had to overline with the Minneapolis office of PCA." Minneapolis PCA was not comfortable with the Offutt operation. Its termination will be described later.

The Offutts were on a roll, and 1971 was no exception. Besides the John Deere dealership at Casselton, Grant Mattson had a Chevrolet agency. He knew that Ron could not afford to buy both businesses, so he sold the Chevrolet agency to Walt Sanders. But Ron told Mattson that when he could afford it, he would "put it [the combined dealership] back together." When his partner, Charley Laitner, left and profits were sufficiently good, Ron unwisely decided to buy the Chevrolet agency. However, it was not large enough to justify full-time management, and by 1974 it had consumed all the capital he had put into it. Consequently, he disposed of it. He reminisced of that venture: "First, I wanted to put it [the car agency] back together with the John Deere dealership when I could afford it. Second, I liked being in the automobile business so I could buy my vehicles wholesale. Those were both the wrong reasons."

During the years of struggle with the car agency, Ron had a stroke of good luck. Betty Lou and her husband, Larry Scott, decided that they did not like life in California and that they wanted to return to the Midwest. Ron asked Scott to take over management of the Casselton dealership, and he accepted. This was during a high period in agriculture, and with Scott's good management, that dealership became a star performer in the struggling Offutt empire. Scott made an excellent record in the business and became a partner.

The second venture for 1971 was in what Ron does best—producing potatoes. Dick Bucholtz rented the Offutts' three quarters of sand land with pivots at Perham, Minnesota. Ron said: "Father had more sales than we had potatoes, so I had to get more land. This was my first experience with sand land, and for the first time, I realized that irrigated sand land was superior to the Red River Valley for raising potatoes. I kept this knowledge to myself. The Perham area was perfect."

The Offutts' timing was perfect too, for just then a large amount of land was coming out of the Conservation Reserve Program of 1956, or what is generally called the Soil Bank. The land at Perham was not productive without irrigation. Most of the owners could not afford irrigators or, like most farmers who participated in the Soil Bank, had no desire to continue farming. Generally, those farmers had not done well financially, and the Soil Bank was a graceful way for them to ease out of farming. They were mostly older farmers who either could not afford to drill wells and install irrigators or had no desire to do so. Most of the land had been idled for the length of the Soil Bank contract (generally 10 years), and some that had previously expired was being used to graze turkeys. Of the first 40 farms the Offutts purchased, not one farmer was displaced. Every one either had already retired or was working elsewhere.

The land was available at reasonable terms, and Perham was located within economical hauling distance of the base of the Offutts' operations at Glyndon. During 1972 Ron secured options on 10 quarters of land at Perham. He recalled:

I had no idea how I was going to buy or develop those 1,600 acres. Then, through a mutual friend, I was introduced to John W. Swenson, who not only owned several thousand acres but also had great financial resources. I remember the first time I visited with Swenson alone, in the fall of 1972. At the second meeting, he talked partnership. Swenson asked, "Would you rather be on the Perham prairie with 10 or 15 circles and financing yourself or be in a partnership with 100 circles and no money problems?"

This led to my greatest disagreement with my father, because he did not like Swenson, but Father deferred to me. During the winter of 1972–73, we put in more than 20 pivots. We could acquire all kinds of land for $300 an acre. Swenson and Marion Gill, an employee who knew land and how to deal, purchased several thousand acres on contract-for-deed. In the spring of 1973 we planted about 1,200 acres. That fall the potatoes were all trucked to Western Potato in Grand Forks [North Dakota] and to Harry Kay, Northern Potato, in Minneapolis, for french fries.

Potato prices were good in 1973, but because of the huge Russian grain purchases, so were the prices of most other agricultural products. Potato prices, however, did not increase as much as the prices of other farm products. Thus, a national movement was started to organize potato growers for collective bargaining. The leaders of the movement reasoned that this was a good time to stand firm against the processors because, if need be, they could shift to other crops that were profitable. Potato prices were about $1.75 a cwt., and the growers wanted $5.00.

Ron, like his father, was inclined to be a "self-help person" and not an organization fan. He went to one meeting of the growers and decided he wanted no part of the bargaining movement. The Offutts continued fulfilling their sales contracts in the fall and winter of 1973 for $2 a cwt. Swenson could not understand why they were doing that when the open market for potatoes was $5.

Living up to the terms of the 1973 contracts paid off in the spring of 1974. The processors offered $4 a cwt., and the producers group held out for $5. The Offutts went to all their traditional buyers and sold a million cwt. of potatoes at $4 a cwt. Then, they were faced with a new challenge. They had only 2,500 acres of land for potatoes, and they needed another 1,000 acres to produce enough to fulfill the contracts. Larry Monico, a Park Rapids farmer and partner of Dick Buckholtz, walked into the Offutts' office and asked them if they wanted to partner on 600 acres of potatoes. Next, Vern Hagen, a prominent Red River Valley potato grower and long-time Offutt friend, led them to John Ojtonowicz, who agreed to raise 400 acres. This was Ojtonowicz's first experience with producing potatoes, but he did a good job and remained in the business.

A SIGNIFICANT TURNING POINT occurred at this juncture. Ron reminisced:

One night at the Elks Club, Paul Horn approached me and told me I really needed a strong accountant. I had Phyllis Kline, who I thought was doing a good job, so I was not so sure. Father was protective of Phyllis, a widow with a son, and wanted to make sure she had work. Paul recommended Al Knoll, who was plant manager for Ready Italy, which was owned by Jenos, Inc., a Jeno Paulucci Company. To protect myself I hired Al on a six-week trial basis.

It wasn't long until I realized that he brought professional accounting skills to our organization. He had the ability to project and forecast. He had credibility with First Bank. And he had the savvy to piece a deal together. My father learned to respect him, and Al established credibility with him. Al had the ability to put the brakes on both of us. He was an excellent person for us to have to talk to any third party. He was just the person we needed to run a tight ship.

Al Knoll was employed in April 1974. He recalled that Ron not only did not tell his father, but he moved Al into Ronald's space in the 20-by-36-foot office building at Glyndon, which the Offutts had constructed in 1970. When Ronald came to work, he was startled to see Al sitting in his office. He immediately stepped into Ron's office, where "some discussion ensued," after which father and son shared an office.

Al looked at employment with the Offutts as an "opportunity to establish some systems." He said, "I soon earned my stripes by organizing the books well enough to satisfy Travelers [Insurance Company] and secure a loan of $1.5 million." Part of that loan was used to purchase the second floor of the FM Building in Moorhead the summer of 1976. The new office did much to enhance the company's image and made it more accessible to the public. The Glyndon office was then used as the truck dispatching and farm operations office.

To PREPARE for the huge anticipated crop, Offutt-Swenson, Inc., and Perham Potato Farms had to build a large potato warehouse that would cost a million dollars. According to the partnership agreement, Swenson was to handle the financing. The understanding was that he would arrange for long-term financing in the partnership name, but for some reason Swenson went short term. In March 1975 the bank called the partners in and asked the Offutts to match Swenson in the up-front financing necessary for a long-term loan.

Swenson knew that the Offutts had no cash, and the Offutts knew that Swenson knew that. This caused them to realize that the partnership was not going to work. Supported by their newly employed accountant, Al Knoll, by attorney Earl Colburn, and by Vern Hagen, the Offutts offered to buy Swenson out for $2 million and assume $2.5 million in debt. Or, Swenson could buy them out. Swenson was not a potato grower but was taken in by the high prices of 1974 and jumped at the latter opportunity. Then, the Offutts discovered that was what Swenson wanted when the bank gave Swenson a letter of commitment on the spot. Ron smiled as he recalled, "This was the first time in my life I ever had big cash."

BECAUSE 1974 had been a good year, for the first time Ronald, age 57, and his wife, Lida, took the winter off and went to Arizona. They returned to Arizona in the fall of 1975. They were still there when the sale to Swenson was consummated. Ron called them in Phoenix and told them they had $2 million net. The Offutts figured they had no value in the Perham operation, so whatever they received was net. Ronald's response was, "That's great for you, Son." Ron sensed that his father was not sure what to expect and perhaps felt that he was on the way out. Ron immediately told his father that half the total was his because they were still partners in the potato business.

Laura Offutt, Ronald's mother, had once stated, "Ronald can stand debt, but he cannot stand prosperity." Ron says that he inherited his father's ability to tolerate debt but was surprised at what happened next.

Ronald lived up to his mother's statement. In the spring of 1976, on their way home from Arizona, Ronald and Lida stopped at a café in Dalhart, Texas. Ronald learned that Winrock, the 21,000-acre irrigated Winthrop Rockefeller farm and feedlot, was for sale for $12 million. Ronald had always been fascinated by the cattle business, and Winrock was just the challenge he needed. The property had been on the market for a long time, and Rockefeller was anxious to sell. The Offutts apparently made the first serious offer. It was only a matter of time until they were the owners of Winrock, including the equipment and the cattle. The million-dollar down payment was made with the agreement that the mortgage be against the real estate only. This freed them to use the chattel property (livestock and equipment) as security for the 1976 crop expenses.

UNLIKE his 59-year-old father, who quickly invested his proceeds in Winrock, 33-year-old Ron wanted time to think about the best way to lever his million dollars from the sale of the Perham operation. He purchased a certificate of deposit at First Bank, where he did business with a college friend, David Birkeland. That proved to be a good move for several reasons. The Minneapolis office of PCA apparently had never liked the Offutts' involvement in the Perham operation and wanted to be paid out. Ron borrowed against his CD to pay off the Minneapolis loan. This is when Burnap, the partner at Big Lake, was bought out.

RON likes growing potatoes more than any other activity, and in 1974 he had a chance to test his ability again. The Offutts contracted to produce 200,000 cwt. of potatoes for the Pillsbury plant at Grand Forks at $2.50 a cwt. and 400,000 cwt. for Western Potato Company at $3.50 a cwt. after they were sorted. The contracts provided for shipment directly from the field, so they did not have to plan for storage, and they expected prompt payment. They could produce potatoes for $2.00 to $2.50 a cwt. and were satisfied to make 50¢ a cwt. on 250 cwt. per acre. This was their first experience working with Western Potato, which operated out of Little Falls, Minnesota. Unfortunately, Western was in financial difficulty and could not pay for the potatoes delivered. By the summer of 1975, Western owed the Offutts about a million dollars. Ron realized that the Western receivable was basically his net worth. Western had managed to pay off some of its bill but did not have money to harvest its crop at Little Falls. The Offutts moved their equipment and their entire harvest crew to Little Falls and harvested almost enough to satisfy the amount due.

Recovering nearly a million dollars in receivables made 1975 a good year by a "squeaker." Was it a bad omen for what was to come? Ron liked growing potatoes on irrigated sand land, so in the spring of 1976, when a 1,200-acre farm came on the market at Spring Green, Wisconsin, he took the challenge. At the same time, he purchased a

3,000-acre potato farm at Stevens Point, Wisconsin, on a contract-for-deed. He recalled:

> With the Spring Green farm, we reentered the market with potatoes for chipping. With the Stevens Point farm, we were going to produce potatoes for the Ore-Ida french fry plant at nearby Plover. It was the first time we had ever planted potatoes without a sales contract. That was our big mistake, for the Ore-Ida plant was not ready in time to take the crop. We lost a million dollars selling on the open market.

But the Offutts' troubles were not over. In November 1977 a severe blizzard at Winrock upset the cattle on feed, resulting in a million-dollar loss. The wind was so bad that sand took the paint off the leeward side of a just delivered tractor and cut wooden fence posts off at ground level.

Meanwhile, in 1977 the Offutts built potato houses at Becker and Park Rapids, Minnesota. The potatoes stored in the Park Rapids plant were under contract to the Simplot plant at Grand Forks. That winter the Simplot plant burned. Simplot invoked the act of God clause and did not purchase the potatoes. About the same time, the potatoes in the newly constructed Becker warehouse, which were contracted to Ore-Ida, started breaking down because of hollow heart. Thus, they did not meet specifications and had to be disposed of. Fortunately, the potatoes at the Park Rapids warehouse were available to fulfill the Ore-Ida contract. Ron recalled:

> The combined losses from the feedlot and the Wisconsin potato farms, along with the loss of a full warehouse of potatoes with hollow heart, put us into the mode of selling assets. This is what we did to recoup. We were against the wall and immediately went to the Ore-Ida people and asked to be relieved of our five-year contract to deliver potatoes from the Stevens Point farm. Ore-Ida accommodated us. We got rid of that big Stevens Point warehouse and farm as well as the Spring Green farm.

Fortunately, rising land prices in the late 1970s enabled Ron to sell both Wisconsin farms at a profit. But he learned a valuable lesson that he said prepared him for the financial downturn that took place in the agricultural economy in the 1980s. Burnap was no longer his partner at Becker / Big Lake, which meant he had to raise capital to run that farm. He turned to Frank Kasowski, a Casselton farmer with superb managerial skills, and established a partnership. Kasowski had cash for machinery, other inputs, and an airplane that enhanced on-the-spot management of the widely separated farms. That partnership proved very beneficial to both parties and was still going strong at the time of this writing.

IN 1973 Ronald made his first sale of potatoes to Jeff Brooten, owner of Potato Processing Company, in Atlanta, Georgia. He had known Brooten for many years but had not previously done business with him. The Offutts trucked potatoes to Atlanta until 1977, when they purchased Brooten's business. The Offutts wanted to expand their market for potatoes now that they had access to all the land they could farm. But they were looking ahead and wanted to learn more about processing potatoes. By purchasing Brooten's firm, they could do so with the least risk. The plant was old, with outdated equipment, but the company had a large cash reserve and an excellent reputation operating under the well-known and respected Dixie label.

The Offutts operated the plant from 1977 until January 1981. They learned about the manufacturing process and a little about marketing french fries, which was their goal. But, as mentioned above, the potato processing industry was going through a revolution, and the small, urban-centered plants were being replaced by new, larger plants built where the potatoes were produced. It was not the Offutts' intention to make the Atlanta operation a permanent part of their business.

However, to integrate the business better, they decided to get into trucking so that they could supply the Atlanta plant with potatoes from the Minnesota farms. To utilize the trucks better, they secured a warehouse for carpeting at Dalton, Georgia, where carpet from area mills could be assembled for backhauls to the Midwest. They soon learned that their management skills lay in producing, processing, and marketing potatoes, not in the carpet and trucking businesses. The furniture dealers were caught in the inflationary cycle and were suffering under big accounts receivable. They could not pay for the carpets that the Offutts delivered. Instead of having profitable backhauls, the Offutts suffered severe losses, and within three years they discontin-

ued the carpet business and sold the truck fleet and warehouse. Ron reflected on the experience: "There were two or three times in my career when I should have been totally buried but made it. This was one of them."

IN SPITE OF THE LOSSES in the automobile venture at Casselton, Ron was still enamored with the idea of owning a dealership. In late 1974, Kiefer Chevrolet Company, of Moorhead, a long-time successful agency, was for sale. Offutt assessed his opportunity and sensed that it was a large enough business to justify good management. Bob Campbell was employed as manager and did an excellent job of running the business and making money. Three years later the Muscatell Chevrolet agency, in sister city Fargo, North Dakota, became available. The success with the Kiefer agency prompted Offutt to buy Muscatell. The much larger dealership overwhelmed management, and his third attempt at owning an automobile agency was liquidated within six months.

About the same time the struggle with the Muscatell agency was taking place, David Birkeland, of First Bank, came to the office. He suggested that money was going to become costly and that Offutts ought to retrench. Birkeland pointed out the problems ahead and advised that it might be best to liquidate wherever they could. Ron recalled that what was suggested was not to his liking, but he followed Birkeland's advice. Within two years interest rates were 15.5 percent. Birkeland's advice was on the mark.

IN APRIL 1979, to learn the pitfalls and advantages of integration, Ron and his mentor met with Earl B. Olson, founder of Jennie-O Foods, who had successfully integrated the production, processing, and marketing of turkeys. Olson proved to Ron that processing was more profitable than production but that it was essential to be integrated because of the synergies involved. When Ron came away from that meeting, he related to his mentor that Olson had told him what he wanted to hear and convinced him that his biggest opportunity for growth, diversification, and stability in the potato business was to build a processing plant.

Because his greatest concentration of potatoes was in the Big Lake / Becker area and because that was near the large Twin Cities market, Ron was all set to build at Big Lake. He spoke about his plans to his mentor, who suggested that he consider Park Rapids, the county seat of Hubbard County, because that county was traditionally a high unemployment area. Ron was familiar with Park Rapids because he was already growing potatoes in the nearby Osage and Perham areas. After studying data presented by his mentor, he decided on Park Rapids and broke ground there in the fall of 1980.

Ron received a $4 million HUD grant because he was moving into an impacted area. He had another $7 million from Travelers Insurance Company arranged for, but Travelers specified that the plant had to make 10,000 pounds of french fries per hour for 48 hours before the company would make the loan. Time was running out on the loan commitments, and Ron had not secured short-term financing. Art Greenberg, a Red River Valley potato grower, entrepreneur, and long-time friend of Ronald and Ron, came to the rescue. Greenberg offered to call a mutual friend, Pete Taggares, of Pasco, Washington, who was one of the nation's largest potato growers and processors and was always interested in new ventures. Taggares flew to Park Rapids to assess the situation and asked Ron if he wanted a partner. Ron and Taggares joined hands and formed Chef Reddy. But Taggares insisted that the plant had to be larger to gain economy of scale. He had a good financial statement plus some strong backing, so with his involvement the partners secured another $10 million from First Bank to complete the larger plant.

Because Taggares was familiar with producing french fries, he was made managing partner, and Ron was the sole source of potatoes. By September 1981 the fries were in cartons, and by August 31, 1982, 44 million pounds of finished product had passed over the loading dock. Production increased each year through 1985, when 130 million pounds of finished product requiring nearly 300 million pounds of raw potatoes were prepared. The association with Taggares was a major turning point for Ron.

RONALD surprised his son in 1976 when he used his million dollars from the sale of the Perham operation to purchase Winrock. When Ron's mentor told him that the water table in the Ogalala aquifer was dropping, which might be a bad omen, he replied, "Dad likes cattle and is so excited about the operation." Winrock had great potential, but when the Offutts had a profitable corn crop, the cattle lost money, and when cattle prices were good, they had a short corn crop. In 1981, because the cost of irrigating was excessive due to a prolonged period of hot, dry winds, the Offutts began to have doubts about Winrock's long-range prospects.

In January 1982 the Offutts learned that John Hancock Life Insurance Company was repossessing 5,000 acres of land at Perham that Swenson had purchased from them in 1975. This was at the time when the french fry plant at Park Rapids was coming into full production. More potatoes were needed to keep it going. It became obvious that because Winrock had not lived up to its expectations, it should be sold to free up funds to purchase the land at Perham from John Hancock. Both transactions took place in 1982.

A FRIENDSHIP that started on the football field in the mid-1950s culminated in a business association in March 1986. Paul Horn majored in agronomy and business administration in college and then earned a master of science degree in agronomy and horticulture. Paul's parents farmed land adjacent to the Offutts' property, and the families were friends. After Paul joined his father on the farm in the late 1960s, Ron suggested that the Offutt and Horn farming operations merge. Paul's father declined because he thought it was "too great a risk." At age 37 Paul injured his back and through 1984 underwent six operations. He was in constant pain and had to quit farming. People close to Ron Offutt know he has a big heart when it comes to a friend in need. True to his character, he invited Paul to come up to his office, then in the Dakota Bank Building in Fargo, but he had no assignment. Al Knoll, who Paul had recommended to Ron, was the accountant and responsible for finances. He was assisted by Don Jemtrud and by Phyllis Kline, who was the bookkeeper and payroll person. They and Ron made up the office staff. The 200 employees who operated the farm and four machinery dealerships made up the rest of the company personnel.

Ron was asked by Howard Bisonette, a potato expert from the University of Minnesota, to travel to China with him. Ron asked Paul to join them. After observing conditions in China, they realized that a french fry plant was not feasible but that a flake plant would be practical. Potato flakes would be a good contribution to the Chinese diet. The three thought they could be involved in such a plant until they came face to face with Chinese politics. After that trip, Ron and Ernie Oberg agreed to rent Paul's land. Paul's health improved, his pain subsided, and in March 1986 he formally joined the R. D. Offutt Company and was put in charge of the farm operations. This freed up Ron to look for more opportunities. Ronald Offutt was no longer living, so Al Knoll and Paul Horn became shareholders with Ron in the R. D. Offutt Company.

THE OFFUTT-TAGGARES arrangement went smoothly until 1987, when Taggares started working on another venture with a major food processing company that detracted from his duties at the Chef Reddy operation. That venture involved a french fry plant in Nebraska. Ron realized that if that deal materialized, he would be bought out or he would have to buy out Taggares. If Ron was bought out, it would be the end of his midwestern potato farming as he had structured it. It was imperative that he buy out Taggares. Ron went to First Bank and explained why he wanted money. He pointed out, however, that if the bank made the loan, Taggares would no longer be guaranteeing the Chef Reddy plant. First Bank was not willing to make the total loan needed to buy out Taggares.

Ron was forced to look elsewhere and turned to the St. Paul Bank for Cooperatives (The Bank). Within 10 days The Bank agreed to make a $27 million loan to refinance the plant and pay for Taggares' equity. The loan was finalized in March 1987, and the partnership was terminated at the end of July. In the meantime, Kathy Taggares, Pete's daughter, sold 130 million pounds of finished product to Burger King and other food-service companies. This gave Ron cash to carry on. Taggares and Ron parted friends because they had always been honest with each other. Several years later that relationship was consummated in a major transaction in the lives of both men.

SOON AFTER the settlement with Taggares, Universal Foods, of Milwaukee, approached Ron expressing interest in buying the Park

Rapids plant. In November 1987 Universal leased the plant for five years, with the right to buy at the expiration of the lease. The lease was written so Ron remained the sole-source supplier. This gave him a strong bargaining point in case Universal should decide to buy at a later date, because he did not want to be left without a market.

The Universal lease freed Ron up for other ventures, which came along in rapid succession. In the spring of 1989 he purchased a 58 percent interest in the Barrel-O-Fun potato chip plant at nearby Perham. Kenny Nelson held 25 percent, with the rest of the stock owned by four others, including Paul Horn and Al Knoll. At the time of purchase Barrel-O-Fun had gross sales of $10 million, and by 1994 that figure was $28 million. But Ron began selling far more potatoes to Frito Lay, a competitor, than to Barrel-O-Fun, which made the financial connection with Frito Lay of greater value than ownership in Barrel-O-Fun. To avoid a conflict of interest, Offutt, Horn, and Knoll sold their stock to the Barrel-O-Fun management team.

During the summer of 1989 Ron purchased the Pillsbury potato flake plant at Grand Forks, North Dakota. This plant had been closed, and all the machinery removed. But he wanted it as an outlet for lower-quality potatoes produced at Park Rapids that could not be used in french fries or hash browns. He purchased used equipment and restarted the plant, which was a success. In the early winter of 1996–97, that plant was expanded, and new equipment installed to increase production. It was barely in operation when in the spring of 1997, the waters of the Red River inundated Grand Forks, flooding the plant. The facility is used solely for production of potato flakes for industrial users. About 30 percent of its production is exported to Japan and Taiwan. The plant contracts for about 15,000 acres of potatoes and utilizes about 1,500 acres from Offutt farms to process 3.4 million cwt. of potatoes annually.

A G-CAPITAL was the third major venture during the late 1980s. Organized in January 1989, Ag-Capital was the brain child of Lee Rosin, of The Bank for Cooperatives, and was actually founded by The Bank. Ag-Capital was capitalized at $30 million and by 2000 had grown to $480 million. It specializes in agribusiness loans and leases, especially in California, Idaho, and Washington, although it has a sizeable portfolio in the Red River Valley. Ag-Capital has proven to be a profitable venture for the company. Having access to capital was a real turning point for the R. D. Offutt Company. After several years of working with The Bank, Ag-Capital turned to Rabo Bank as its equity partner.

I N LATE 1991, after three years of operating the Park Rapids plant, Universal Foods decided that it wanted to reduce its presence in potatoes and paid to get out of its contract. Ron resumed operating the plant. His biggest immediate challenge was selling his production, because he had little experience selling finished products and no sales force. Taggares and Universal had handled that end of the business without any input from Ron. He knew how to produce and process potatoes, but he needed help in marketing them. He searched for a marketing arm and developed a strategic partnership with Lamb-Weston, which managed the plant and sold the products, with Ron retaining full ownership of the facilities.

The RDO frozen-potato processing plant at Park Rapids in 1994. At that time a million pounds of potatoes, nearly all produced by Offutt farms, were processed daily into french fries or hash browns. Potatoes are floated from the warehouse (foreground) underground to the plant.

The relationship worked well, and in 1997 Ron invested an additional $60 million in plant expansion and modernization. This brought the total investment in that facility to $110 million. With the expansion, plant capacity increased from 140 million to 480 million pounds of finished product. McDonald's is the largest single customer of the Park Rapids plant, with the balance of its output being sold to food-service distributors in the East and the Midwest. Although production increased more than threefold, state-of-the-art equipment made it possible to attain that goal by merely increasing the employment from 250 to 480 persons. The plant consumes the production from 24,000 acres located within a radius of 110 miles, mostly south of Park Rapids.

To accommodate the Park Rapids and Grand Forks plants and fulfill the contracts with Frito Lay, potato production was increased to 26,300 acres in Minnesota and 11,000 acres in North Dakota. In the spring of 1998 a strategic decision was made to rent out all the rotation crops so the farm crews could concentrate on raising potatoes. Other farmers could rent company land and qualify for government payments, but Ron could not because the company far exceeded payment limits. Like many entrepreneurs, he preferred to operate without governmental restrictions.

This decision to concentrate on raising potatoes was made after Ron acquired a contract to produce 6,000 acres of potatoes for the J. R. Simplot facility in Grand Forks. He had to expand rapidly to get potato ground, and by 2000 he was raising potatoes on nearly 68,000 acres. His decision was based on the ability to contract all the company raised and on the desire to minimize the need to employ more labor or to own equipment for other crops. Ron likes raising potatoes and excels in doing so. Like most large-scale producers, he found that economies of scale, because of the synergies involved, followed along with expansion. The company no longer transports machinery from one farm unit to another because the individual farm managers prefer to be responsible for their equipment and take better care of it than if it were transported between farms. This has lowered the cost of equipment upkeep, reduced the need for purchasing new products, and eliminated the hassle of constantly moving the equipment. A good program of educating farm managers on the economics of equipment cost was the key.

After the Park Rapids plant expansion was completed and production reached new levels, Lamb-Weston purchased half interest in the facility, which became known as Lamb-Weston/RDO. Offutt retains the exclusive potato production contract. The relationship with Lamb-Weston grew beyond the Park Rapids plant. Lamb-Weston was expanding its plants in Washington and Oregon and invited Ron to grow potatoes for the firm there. At the same time, he developed a similar relationship with Frito Lay and produces potatoes for its plants in New Mexico and Texas.

IN JANUARY 1997 Ron formed a partnership with Milt Carter and Randy Spevak, two growers from Watertown, South Dakota. The partnership was called CSS and farmed 10,000 acres of potatoes in Nebraska, South Dakota, and Texas under contract for Frito Lay. CSS grows only potatoes and rents out the land used for rotation crops.

In March 1997 the CSS partnership became involved with Nonpareil, at O'Neill, Nebraska. Nonpareil is one of Idaho's largest fresh-pack shippers of potatoes. The business at O'Neill is based on 6,500 acres, of which 5,200 are irrigated. CSS's responsibility is to produce the potatoes. Nonpareil manages the plant and its 112 workers. The partners would like to expand production at the plant but are not able to employ the additional 100 workers necessary to fill a second shift. Each day 10,000 cwt. of fresh-pack potatoes are shipped to eastern metropolitan areas, and the adjacent flake plant processes lower-quality potatoes for industrial users.

In July 1999 Ron purchased a 25 percent interest in Winnemucca Farms, near Winnemucca in northwestern Nevada, and in November of that year he purchased the remaining 75 percent of that firm. Winnemucca Farms consists of 10,000 acres of irrigated land, which is used to raise wheat and potatoes. In addition, it has a fresh-pack operation and a potato flake plant that use all the potatoes raised on the farm plus purchases from Idaho producers. The fresh-pack products and processed flakes are sold primarily to metropolitan areas of California, Nevada, and Utah. Winnemucca Farms had a very successful history of raising potatoes and grain, but in later years it failed to keep current with changing technology, which is why it became available.

IN LATE 1998 Pete Taggares offered Ron the biggest opportunity or challenge, depending upon one's outlook, of his career. Taggares owned a lease from the government on 90,000 acres, 27,300 of which were irrigated, at Boardman, Oregon. Taggares was not feeling well and had a premonition that his days were numbered. He offered to sublease that property to Ron for 10 years at a fixed rate, after which time Ron could assume the remaining years of the lease. No less an expert on potato land than J. R. Simplot told this writer that those acres represent some of the finest potato ground found anywhere. Simplot formerly had partnered with Taggares on that property, and they named the unit Simtag. Taggares' premonition proved correct, for only three weeks after he had completed the transaction with Ron, he died. At that time, 36 years after Ron became a partner with his father, who owned 240 acres of heavily mortgaged land, the R. D. Offutt Company had amassed 316,000 acres of land, along with a wide variety of other interests.

A 2000 photo of Ron Offutt, at age 57, the nation's largest potato producer.

The 27,300 acres of irrigated land at Boardman are served by water from the Columbia River at an annual cost of about $125 an acre. An additional 13,000 acres of that property could easily be irrigated. The current rotation consists of 8,000 acres of potatoes, 6,000 of mint, 6,000 of alfalfa, and 5,000 of high-moisture shell corn, with the balance in sweet corn. About half the six crops of alfalfa produced

A satellite photo of center-pivot fields from about 275 miles in space. The colors are from the refractive ability of the crops. The green shaded circles are fields of potatoes. The dark lines are roads, both old and new. Satellite at this height is not yet technologically developed to pick up the infrared problems that can be down at lower altitudes.

each year is loaded directly into containers on the farm and shipped to Japan. The balance is sold to nearby dairies and feedlots.

Again Ron found a partner to further develop synergies and make Simtag more viable. He learned that a cheese plant was coming to Boardman, and he casually mentioned that to a college friend. The friend introduced him to John Bos, owner of the 11,000-cow Maple Dairy, at Bakersfield, California. Bos supplies milk to the Tillamook (Oregon) Dairy Cooperative, a major West Coast cheese manufacturer. Bos and Ron developed a partnership to establish two dairies, with a combined total of 12,500 cows, to produce milk for the cheese plant. Alfalfa is an excellent rotation crop with potatoes, and the land will benefit from the manure provided by the livestock.

Additional synergies are developed by association with the North West Beef feedlot, which is adjacent to Simtag. North West is a custom feedlot that finishes about 35,000 animals annually. It will purchase high-moisture corn produced on 5,000 acres, as well as the off-grade hay that is not acceptable for export or as dairy feed. Manure from the North West feedlot will also be applied to Simtag land.

RON was asked what he might do differently if he could retrace some of his steps. He commented that the acquisition of his first machinery dealership in 1966, which opened so many opportunities for him then, does not seem as appealing today. He added: "I suspect if we had no machinery affiliations today, we could probably be buying our equipment cheaper than we are. We got into the machinery business to help our farming operation, but that is no longer valid. This is part of a changing trend in merchandising."

After 1966 Ron continued to purchase dealerships. In 1998 he combined his holdings of 16 agricultural dealerships with 42 construction and truck dealerships into RDO Equipment, a public company. RDO stock, like that of nearly all agriculture-related businesses in the late 1990s and in 2000, has had only lukewarm reception from the public. The high-technology sectors have taken all the glamour out of any industry dealing with such a mundane commodity as food. That is part of the curse of being in an industry that is still as overcapitalized as American agriculture.

WHEN Ron was asked to describe the major challenges of his first 36 years of farming, without hesitation he replied, "There is more opportunity than there is money. Financing has always been a major hurdle. Some of that has been caused by that fact that I have always aggressively sought ways to expand. But interestingly, in 1998 I would have said the shortage of good people was my major challenge." After a brief pause, he added, "I am not sure that publicly held corporate farms would attract investors to agriculture, because the industry is still too cyclical." He did not allude to the fact that he had solved much of that problem within his own company by specializing in his area of greatest competency—raising potatoes. He took advantage of that ability by capturing a series of production contracts and by being an owner/partner in four potato processing plants.

During recent years Ron, like nearly every farmer, has had several costly bouts with environmentalists. He bluntly stated: "If it were left to the environmentalists, the cost of agricultural production would be such that U.S. agriculture could not compete in the world. . . . Mega-agriculture needs micro-agriculture to [politically help us] create a balance between the environmental extremists and reality. This would allow us to compete in the global market."

When asked to describe his greatest strengths and weaknesses, he was ready, for he was sure he was going to be asked that question. He confidently commented first on his strong points: "I am a visionary with the ability to speak to the vision and to generate enthusiasm in bankers, partners, and employees. I have had the tenacity to see it through the tough times. But [I am] rational enough to recognize when I have made a mistake and to change course. Don't go down with the ship."

His two closest colleagues say: "His greatest strength is his ability to motivate people, but that gets more difficult as the company grows. He understands immediately what will work or what will not. Good leaders have a sixth sense that allows them to see into the future and anticipate what will happen. They also have the ability to pick good management. Since 1976, average yields have increased from 250 cwt. per acre to 480 cwt."

Ron closed his multi-day interview by describing his greatest weakness: "It depends upon the day you ask that question. But it is probably allowing too many people to detract me from the business at hand." He is not the only entrepreneur who feels that way. That recalls the caption to this chapter, "The entrepreneur has to have a special work ethic that other people do not understand." But anyone who knows Ron Offutt knows that he is bothered by others only when he lets them do so.

Bibliography

Interviews

Horn, Paul T. Fargo, ND, July 7, 2000.

Knoll, Allan F. Fargo, ND, July 3, 2000.

Offutt, Ronald D. Fargo, ND, September 4–6, 1997, and May 16, 2000.

Earl B. Olson

"The Impossible Is an Opportunity"

IN 1893 a 19-year-old Swede, Olaf Olson, arrived at St. James, Minnesota, and immediately sought a job as a farm hand. After a few months he traveled to Murdock to join his brother, John, who had settled there 10 years earlier. As soon as he was eligible, Olaf applied for a 160-acre homestead and soon after married Anna Sophie Anderson, also a Swedish immigrant. The Olsons raised three children on the homestead, where Sophie toiled until her death at age 65 and where Olaf actively farmed until his death at age 99. The Olsons raised wheat, corn, and hay to feed their 8 horses, 25 milk cows, and a flock of chickens.

That was the setting into which Earl B. Olson was born May 8, 1915. By the time he was six, Earl was expected to milk six cows each morning before walking one mile to country school and to repeat that task each evening along with other chores. Earl and his siblings did the work expected of them, but it did not take Earl long to realize that farming was not to his liking.

Fortunately for Earl, his parents were interested in education, so after he finished country school, they encouraged him to go to the West Central School of Agriculture at Morris, which was about 35 miles from the farm. His parents were unable to pay for Earl's living expenses while he was in Morris, so he worked at any job he could get. He finally landed regular work with a lumber yard, shoveling coal from train cars on Saturdays and during vacations for 12¢ an hour. It was a dirty and hard job for the slender youth. In 1932, while he was still in school, his parents nearly lost the farm. Crop prices fell to new lows, and without the cream checks and egg money, they might not have held on.

In June 1933 Earl completed the standard three-year agricultural course. He had just turned 18 and was ready to enter the work-a-day world. Olaf was a member of the board of the Murdock Cooperative

Creamery Association and was able to secure a position for Earl. He started at $35 a month and eventually did every job in the creamery except the bookkeeping. The manager had long weekend parties and started "to slip on the job." To keep the operation going, Earl quickly learned how to test cream and keep the related records.

Initially, Earl had an apartment upstairs in one of the business buildings, but to save money and to make sure the boiler was fired on time, he moved to the shower room of the creamery. He was such a sound sleeper that to make sure he heard his alarm clock, he set it on a pie tin so it would make more noise. Later, he purchased a second clock to make doubly sure he would not oversleep.

A 1980 photo of Earl B. Olson in his mid-60s. At age 85 he continues to make daily trips to his office.

His diligence paid off, for he was given an increase to $37.50 a month. The Murdock board recognized that his efforts were keeping the creamery functioning. A few months later the board of nearby Swift Falls Cooperative Creamery decided either to close the creamery's doors or to hire a new manager. In the fall of 1936 the 21-year-old Olson was offered the creamery manager's job at $80 a month plus 1 percent of gross sales.

The creamery's financial status improved under Olson's leadership, and his salary was increased to $100 a month plus the 1 percent commission. At that time, farm-to-market roads still left much to be desired, and trips to town with produce were time-consuming for the farmers, especially those who still used horses. Olson thought the creamery should put trucks on the road to pick up the patrons' cream and eggs. The board saw the wisdom of the truck routes but had no funds to purchase trucks. However, it gave Olson permission to go ahead on his own. His business acumen paid off, for by then he had established a credit record with Russell Hanson at the Benson bank.

He traded his 1935 Chevrolet car as a payment on a new $1,200 ³/₄-ton Ford pickup.

Now Olson arose each morning at 4 o'clock to fire the boiler and then went on the route to collect cream and eggs. The innovation proved to be a hit with the hard-pressed farmers, and before long Olson personally hired a man to make three routes daily. Farmers in the area deserted their cooperatives to take advantage of the pickup service of the Swift Falls creamery, and a second truck was purchased.

Next, he convinced the board to add feed, seed corn, fertilizer, and farm supplies, which the trucks delivered as they picked up the farmers' produce. Olson irked his fellow cooperative managers when he paid higher prices for cream, eggs, and poultry, but the reduced margin on those products was offset by the added income from feed, seed, fertilizer, and farm supply sales. He said that his competitors were too satisfied with the status quo. Olson purchased a third truck, all the time collecting a larger income for trucking as well as the 1 percent commission on gross sales. He said, "I have always had the desire to make money and enjoyed the fun of making a better way of life for a great many people."After he had three trucks, the route business continued to grow, so he contracted with local truckers to handle the new routes.

While working at the Murdock creamery, Olson became acquainted with Dorothy Erickson, who was a clerk in her father's grocery store in the village. By 1938 he was sure enough of his future that he and Dorothy were married. They rented a house in Swift Falls, a community of 75 people. The house, like most rural homes of that time, had no electricity or running water. Within a few months the Olsons purchased the house for $3,500.

Then Olson faced an extreme challenge. He had gone into the freezer to obtain a small block of ice for a customer. On his exit from the cold-storage vault, he was met by a wave of scalding water that his helper had accidently dumped. His legs were so badly scalded that the skin came off when the doctor removed Olson's shoes and stockings. The doctor would not permit him to be on his feet for the next eight months. This was before the days when employer-sponsored sick leave and health and accident insurance became commonplace. However, group policies were being sold to cooperatives, and Olson had convinced the board to carry a policy on the employees. Ironically, the cautious Olson had taken out a personal loss-of-income policy not long before the accident and was able to collect on it. The hospital agreed to install a telephone in his room so Olson could manage the business. The grateful board continued his salary during his disability.

The prolonged hospital stay gave Olson ample time to think about his future. Philosophically he was not a cooperative-minded person, and he realized that he wanted to run his own business. However, he did not have a clear vision of what to do and lacked the financial know-how to start a business. Also, he was in no position to take on a new challenge, for he needed time to regain his strength.

By 1939 the creamery that had been nearly bankrupt in 1936 showed a positive cash flow. Olson convinced the board to respond to member demands and build a locker plant. Locker plants were popular in rural areas because virtually no one had refrigerators, but farmers and villagers needed them for their home-butchered poultry and livestock. The locker business, like all the other ventures, added to the operation's growth, while the cooperative competitors thought, as Olson put it, "we were out of our minds." By now Swift Falls Cooperative Creamery had become one of the larger cooperatives in the area and, more important, was on solid financial footing.

By 1941 Olson felt sufficiently recovered and used his privately purchased loss-of-income insurance money to rent a few acres of land near his house in Swift Falls and purchase 300 turkey poults. The board approved of his moonlighting activities with the understanding that he would purchase the needed feed and supplies from the cooperative. He netted $1 per bird the first year. In 1942 he grew 600 turkeys with the same rate of profit. He repeated in 1943 with 1,000 birds and $1,000 profit. In 1944 it was 3,000 birds; in 1945, 5,500; in 1946, 7,500; and in 1947, 8,000. Every year the profit was $1 per bird. This string of good years climaxed in 1948, when he raised 12,000 turkeys and netted $48,000. Olson gleefully stated that profits from the moonlighting venture of a farm boy from Murdock were greater than the salary of the governor of Minnesota.

Everything was going so well at Swift Falls that Olson thought it was time to spread his wings. In 1946 he partnered with Gilbert Ahlstrand and purchased the creamery at Marietta, Minnesota, which included Nassau Produce, four miles from Marietta. Marietta was 73 miles from Swift Falls, but the two partners devised a schedule of rotating visits that they felt would be adequate hands-on management. The Marietta creamery grew into one of the largest privately owned butter producers in the Minnesota–North Dakota–South Dakota area, but somehow the onsite manager was unable to generate satisfactory profits. In 1951, after learning a good lesson in business management, the partners sold the Marietta-Nassau complex, and Olson looked for another hill to climb.

In 1947 Olson found that challenge when he persuaded the Swift Falls board to erect a turkey dressing plant. In some respects this move could be looked upon as rather self-serving. However, it fit into the basic operations, so the board willingly approved. This was the first plant exclusively for dressing turkeys in the five-state area of Minnesota, Wisconsin, North Dakota, South Dakota, and Iowa and contained the very latest in equipment. In retrospect, it was the start of Olson's rise as the pioneer in the industrialization of the turkey business, first in Minnesota and eventually on a national level.

Twenty growers in addition to Olson signed to raise turkeys for the new processing venture. Once again a new opportunity came his way.

Swift Falls turkey dressing plant, where women with pinning knives (knives with straight, dull blades) removed black pin feathers. The line moved at about eight birds per minute. The birds were then packed in ice (entrails, head, and feet intact) and shipped to New York, where butchers eviscerated them in their shops.

Some of the growers agreed to raise turkeys if Olson would partner with them. They needed his expertise and were willing to provide the labor and most of the necessary capital. This pleased Olson, who was learning to master the art of leverage. His personal turkey growing venture was the start of Earl B. Olson Farms (hereafter EBO Farms), which became the largest privately owned turkey producer in the nation.

The creamery board had never increased Olson's base salary of $100 a month, but the 1 percent commission on gross sales rose with every sideline he added to the creamery's overall operation. The turkey processing business increased his income considerably. He still profited from the three route trucks because the board never saw fit to enter that venture even though it was the key to the creamery's success. In 1948 his earnings from the creamery exceeded $25,000. He also had income from his three truck routes in addition to the $48,000 profit from his sideline turkey business.

By then the Olsons had four children and in 1949 decided to build a new house in Swift Falls. Charles, the eldest, was nine in 1948 and raised four turkeys for his 4-H project. Each year Charles increased his turkey project until 1951, when he raised about 75. His goal was to take a turkey to the Minnesota State Fair. In 1951 one of his turkeys was awarded the grand championship by the judges, but the next day 4-H officials removed the ribbon because Charles was too young to qualify for a trip to the state fair. In 1952 the Olsons moved both their family and their new house to Willmar, and that ended backyard turkey projects.

In February 1949, while attending a convention in Chicago, Olson heard that All State Supply, a produce business that bought and sold eggs, butter, chickens, and turkeys, which the company processed in its plant at Willmar, was in financial trouble. The Swift Falls plant employed 100 in the village of 75, and Olson realized that further expansion there was virtually impossible. He turned to the creamery board, which was interested in All State but felt that buying the plant was beyond the constitutional limits of the association. Olson was not to be denied, and the board agreed that if he would buy All State, the creamery would lease it from him and pay him 1¢ a pound plus expenses for all the turkeys processed. His manager's job at Swift Falls would continue.

On the strength of a potential lease and his job with the creamery, Olson turned to his banker friend, Russell Hanson, for money. The price for All State was $100,000, and Olson put in $50,000. At first, Hanson declined because the bank's loan limit was $15,000. That amount was adequate for typical rural businesses, including farms of that day, but this was not a typical deal. Olson was about to drive away from the bank when Hanson called him back and suggested that they try Midland National Bank, in Minneapolis. The next day the two appeared before the loan board of Midland. Hanson presented the case, and then the two were dismissed while the bank board made its decision to approve a 10-year loan. Hanson also helped Olson secure another $100,000 from Woodmen Life and Accident Insurance Company for operating capital.

Olson was beginning to dream of becoming the number one turkey producer and processor in the number one turkey growing state, but he could not foresee the obstacles ahead. In 1949, after eight years of continuous profit from growing turkeys, the inevitable happened. The $4 per bird profit in 1948 tempted Olson to order 35,000 day-old poults. Other turkey growers also wanted to expand, so he was able to secure only 30,000. In the fall of 1949, just as he was starting to operate All State Supply, the price of turkeys dropped. He lost $2 for each bird produced on his own farm, or a total of $60,000. Entrepreneur Olson was in a bind and immediately turned to the Peavey Company, Super Sweet Feeds, and Cargill for help in carrying his feed bill. Fortunately, they all agreed.

In 1949 annual per capita turkey consumption was only 3 pounds. Consequently, it did not take much of an increase in production to flood the market. The traditional cycle in agriculture was at play, with production rising faster than consumption. The cycle in turkey growing was particularly sensitive because the market was limited to whole birds during a short holiday season. At that time, processing consisted of removing the feathers and blast freezing. Dressed frozen turkeys were then shipped to the big city markets with their heads, feet, and entrails intact. Upon their arrival the turkeys were hung in the butcher shops for the consumers to select the bird of their choice. Once the turkey was chosen, the butcher eviscerated it while the customer waited.

Olson wanted to go to more complete processing, but he also hoped to increase per capita turkey consumption by getting people to eat turkey on a year-round basis. His motive for that was simply to improve the efficiency of his turkey farms and the processing plant. Before he could work toward that goal, he had to get out of the financial bind caused by his losing $60,000 in the turkey market and his raising funds to operate All State Supply.

All State Supply was immediately renamed Farmers Produce Company because Olson thought that would be a more appealing name to his farmer clientele. In the fall of 1949 Farmers Produce opened with eight employees, who, along with taking care of the company's other business, processed only 500,000 pounds of turkey. Olson quickly realized that the plant would be more efficient and could handle a much larger volume if it only processed turkeys, for Swift Falls handled several million pounds by then.

Turkey growers were very cautious when the 1950 production year opened. Olson was desperate for cash and scrambled to get investors—doctors, lumber yard owners, bankers, insurance people, feed people, and a few farmers—to partner with him in turkey production. Several corporations, all operating under different arrangements, were created. This was the only way he could get additional capital to raise more turkeys for the new plant.

In the 1950 marketing season, the independent growers at Swift Falls were unwilling to sell their turkeys to the creamery on the traditional consignment basis under which they received a small initial payment and then partial payments each month until fully paid. The

A 1950 picture of early trucks used to haul turkeys to Plant No. 1. The average trailer carried six hundred 22-pound turkeys, which took six to eight people about 40 minutes to load. At the plant a forklift was connected with the metal hooks on the tops of the cages and toted the cages to the kill line, where the birds were removed.

creamery board was not willing to change its policy and pay its patrons cash for turkeys on delivery. Olson was caught in the middle, for he was still managing the creamery, but at Farmers Produce his policy was cash on delivery. He wanted to keep the Swift Falls producers happy so they would continue selling to him. He wanted to operate the Farmers Produce plant at full capacity, so he accepted turkeys from the patrons of the Swift Falls creamery. This caused hard feelings between Olson and the board.

Fortunately, 1950 was a better turkey year. When the season ended, Olson received $40,000 in lease payments (1¢ a pound plus all expenses on 4 million pounds of turkeys processed at Farmers Produce) from the creamery, plus his salary of $1,200 and commissions of more than $25,000 on gross sales, as well as income from the truck routes. He continued his job at Swift Falls through 1951. Then he turned his full attention to Farmers Produce, EBO Farms, and his turkey raising partnerships. From then on, the turkey grower patrons of Swift Falls Cooperative Creamery sold to Farmers Produce Company, and within two years the creamery was bankrupt.

W HEN the creamery business had expanded, it had become necessary to employ a bookkeeper. Olson had gladly shed the detail work, but with the purchase of All State Supply, he again found himself saddled with management concerns. Day-to-day management was not his forte; he was a builder. In June 1949 Don Handahl, a four-year navy veteran who had just graduated from college with a degree in accounting and economics, was hired as office manager and bookkeeper of Farmers Produce. Handahl's wife was Dorothy Olson's niece. Don Handahl knew that "Earl was a real stem winder, whose nickname was Governor Olson, who liked to drive big cars fast, and whom everyone felt was going places."

For the first time, Olson had a confidant other than his wife, Dorothy, who was a strong supporter of whatever her husband did, and his banker, Russell Hanson. Handahl worked full time at Farmers Produce until June 1952. The turkey business was not good, and Handahl wanted to increase his income, so he took another job that paid considerably more and did the records for Olson evenings and weekends. During the interim, Handahl remained as Olson's "sounding board." In August 1953 Olson offered him a management position with a substantial salary increase. Handahl became Olson's closest business associate and eventually became company executive vice president. Their relationship was still in effect in 1998.

Handahl commented:

> Earl had to learn to delegate, but he always had his finger on what was taking place. He had a knack for asking the key questions. If he had something on his mind, he could not sleep and would call me or the plant managers at any time of the night. He was no respecter of time, but he needed a full eight hours of sleep.

Handahl continued:

> Earl has a clean mouth. I have never heard him use profanity. I can add that I have never really seen him angry. The closest he ever came was one day he came to where Spencer Kostad and I were working and said, "You guys have good heads. Why don't you use them?" He gave us that much freedom. After I was made executive vice president, everyone reported to me.

Handahl offered further insights on Olson's strength of character:

> Earl has an amazing ability to absorb criticism and complaints without getting upset. I remember turkey growers getting really nasty, but he never rebutted them. No matter how upset the growers got, he would always listen to them and cool them down. He could look like a whipped puppy, but it was a front. After all was over, he never held a grudge.

In the early years of his employment, Handahl sat opposite Olson when they were both buying so that they could hear each other's conversations. Handahl noted, "We never confirmed sales or buys by letter. We jotted everything in spiral notebooks. Earl impressed on me from the start that once we made a commitment, we stand by it, even though we might have made a mistake and regardless of what it cost us."

Olson always wanted to make the most of his time, and that got him into the habit of being late. He sometimes tried to make up for being late to meetings by driving fast. Handahl related incidents when he was riding with Olson as they raced down the highway at top speed: "Earl would say, 'Take the wheel. I have to write down some notes.' There I was in the right seat with one hand on the wheel."

By 1952 EBO Farms produced 400,000 turkeys, all of which were raised to be ready for processing after the traditional growers had sold their birds in the four big marketing months starting with October. By stretching out the processing run, the per unit cost of birds going through the plant was significantly reduced. This enabled Farmers Produce to outbid competition, thereby securing more turkeys. Olson stressed:

> Volume is the only way to reduce per unit cost. It is the best way to compete with other U.S. firms and with foreigners. We have to produce a good quality product at a reasonable price. This is how you develop markets worldwide. What helped me most was that I led and changed the industry by creating volume and year-round demand.

In 1953 all other segments of the former All State business were discontinued, and the plant was converted solely to processing turkeys. Then, Farmers Produce became a federally inspected plant and started eviscerating. That enabled the company to secure a contract to produce a frozen turkey roll for the military. Although the contract was of short duration, it was a marketing coup, for it opened the way for large-volume year-round sales.

The Olson enterprises were growing rapidly, creating a greater demand for cash than the business could generate. That caused the Wholesale Dealers Section of the Minnesota Department of Agriculture to question Olson's ability to pay for turkeys delivered and made it difficult for him to get his wholesale dealers bond renewed each year. Olson admitted that the shortage of capital was a real stumbling block.

In the fall of 1953 the cash problem was compounded. Olson said:

> One of the growers, who was a friend, got into my desk drawer and discovered that I had lost a million dollars during the previous six months. He went around and told all the growers to follow their turkeys to the plant and ask for cash. I laid awake nights trying to figure out how to get money. The banks would not lend me any. I decided I had to buy a new 1954

Fleetwood Cadillac to counteract the rumors. As soon as this person saw the car, he turned to the growers and told everyone that we were making too much money. But the ploy worked, because the growers gave me time to pay. That freed up more than a million dollars in capital because I was able to pay the growers about 40 days later when I got money from the chains.

Later, this same grower was in Handahl's office, and when Handahl stepped out, the grower thumbed through checks that Handahl was signing to be mailed to other growers. Handahl returned and took them from the grower's hand. The forgiving nature of Earl Olson is revealed when, at a later date, he and Dorothy went on a vacation trip with that grower and his wife. This writer knew the person involved and was amazed how friendly Olson was to him.

In 1955 Sidney Fox, a New York broker buying for one of the food chains, told Olson that Farmers Produce was an old-fashioned name and that he ought to adopt a name that "had a catch to it and included the word 'foods.'" Olson brought this up to his family gathered at the table for Thanksgiving dinner. Dorothy suggested Jennie-O, after their daughter, Jennie. Olson said that he did not like it, but the next Monday he called Fox and asked how he liked Jennie-O for a brand name. Fox replied, "A h--- of a name. It will sell food all over the world. It's short and catchy." Years later, when the company incorporated, it took the name Jennie-O Foods.

The big event of 1956 proved to be another test for the company. A major fire destroyed the packing department and damaged the blast freezers. Rebuilding included expanding and upgrading the plant, so, despite the misfortune, production rose to 21 million pounds, up from 14 million in 1955. Olson never stopped encouraging more turkey production. He always worked on the assumption that if he could get the turkeys, the company could build or buy more processing facilities and then develop the market.

In 1955 the Litchfield Produce Company, 28 miles from Willmar, went broke. Litchfield Produce was still trying to be a complete poultry facility. It dressed geese, ducks, turkeys, and chickens. It also handled eggs. Olson said, "Just like All State, Litchfield could not cover its bases and was not doing a good job of marketing. I had learned marketing by selling butter to big firms that were also purchasing turkeys. My big buyers were A&P, Safeway, and other major chains that were rapidly expanding at that time."

The longer processing run and full eviscerating facility of the Farmers Produce plant, along with volume selling, gave Olson the ability to outbid Litchfield for the available turkeys, which was part of the reason for Litchfield's failure. In 1957 the Litchfield plant was for sale. Olson learned that Armour was the only known opposing bidder. Armour was so sure it was going to win the bid that it had started to move in supplies. Olson and his aides calculated that Armour was bidding $200,000. The bids were to be opened on the coming Monday, so Olson delayed until late Friday to jump the bid to $200,800 on the assumption that the Armour people could not react fast enough to get a new bid in. He assumed correctly and came out the winner.

The Litchfield plant was put into use immediately. However, some remodeling was required, so it was not until 1959 that it went into full production. The combined production of Plant No. 1 in Willmar and Plant No. 2 in Litchfield reached 60 million pounds. In 1959 Olson was relieved of a major obstacle in turkey production when Al Huisinga, of the Willmar Produce and Egg Company, agreed to specialize and hatch turkeys for EBO Farms. That year the farms produced more than 500,000 turkeys. Willmar was considered the world's leading turkey center. Willmar Grain Terminal, recently acquired by Olson, had a million-bushel storage capacity and manufactured 20,000 tons annually of a complete-ration turkey feed for EBO Farms and other area growers.

The next year, 1960, was not a good year for many turkey growers and processors. However, because of his lower cost of growing and processing, Olson was better prepared than his competitors. His contrarian nature allowed him to act quickly when Donald Sonstegard, a turkey grower who was heavily involved in the Melrose Produce Company, about 50 miles from Willmar, approached him. Melrose was financially sound but had difficulty marketing and wanted to get out of the business. The asking price was only $75,000. As usual, Olson did not have the capital, but he was not about to let the opportunity slip from his grasp. Within a week a deal was consummated, and Olson purchased the Melrose facility (Plant No. 3) for "a penny down and 10 years to pay." A fire in the Melrose plant shortly after it was acquired forced it to shut down, so it was completely remodeled before reopening for the 1961 season. That remodeling and later additions have quadrupled the original plant's capacity.

In 1961 Olson's back was to the wall financially, so he tried to raise money by selling $300,000 worth of debentures. He offered 7 percent interest and guaranteed to pay double the money back in 10 years. This was not a successful floatation, but he plunged ahead and survived.

At the same time he was trying to sell the debentures, he entered into a chicken processing venture that was a disaster for everyone involved. From 1960 through 1962, Jennie-O, the Peavey Company (feeds), Jack Frost, All State, and other hatcheries, plus several broiler growers, developed a vertically integrated organization to hatch, grow, and process broilers. Jennie-O was to guarantee 19¢ a pound for the broilers, but when the crop was ready, the market was only 9¢.

All the partners agreed to split the loss. The combined loss for Jennie-O and EBO Farms was $500,000. The hatcheries and many of the chicken growers went broke. Even though Olson was in a financial bind, EBO Farms immediately took over the defunct broiler farms and converted them to turkey production, increasing capacity by 25 percent. Now Olson was forced to "bite the bullet" and turned to the Peavey Company for help. He sold half interest in Jennie-O for $1 million, with the provision that he could buy back at any time. Within two years Peavey was repaid, including interest. That was the only time Olson ever shared ownership of his company. In retrospect, Olson said that no one could have foreseen the drop in prices, but the volume of broiler production was too low to do the job efficiently, which added to the losses. From that point on, he concentrated on what he knew and did best—growing and processing turkeys.

The peak year for the broiler venture, 1961, was also a soft year in the turkey business, but by then Olson expected periodic downturns and weathered them without being too concerned. However, what he saw in the bigger picture disturbed him. The market was no longer capable of absorbing more whole turkeys, so Jennie-O had to get into further processing.

IN 1962 Charles, the eldest of the Olson sons, joined Jennie-O. Charles's first experience working in the business came while the family was still in Swift Falls. He assembled egg cases for 1¢ each. Charles said his father believed in having the children work and preferred to pay them on a piecework basis. When Charles was 12 he worked for EBO Farms doing everything from mowing the grass, to cleaning around and in the barns, to whatever else was assigned by the farm manager.

After the Olsons moved to Willmar, he had to punch the clock and did virtually every job in the plant. At first he worked chiefly at the loading docks. During summers of his last two years in college, he drove a truck. The work force was still small enough that everyone knew he was the boss's son. He was careful never to tell his father when he saw workers "goofing off," because he did not want them to think he was a stool pigeon. On the other hand, he felt that he had to prove to those workers that he could work just as hard as they could.

Charles could not remember that his parents ever spoke to him about entering the business. He said, "Sometime during my last semester in college [Spring 1962], I spoke to Dad about going into the business. He gave me no encouragement. I began to wonder if he wanted me. He was not making any offers. In fact, he suggested that maybe I should look elsewhere."

Charles was married, so rather than wait around, he took a position with a firm that did claims adjusting for insurance companies. He was sent to Atlanta. He recalled that in December his father contacted him, stating that sales and marketing had a job opening. Charles related, "When I joined the company, I had no idea there was a financial problem. I was there about six months before I realized that we were losing on all the chickens for which we had contracted at 19¢ per pound and the market was 9¢."

Charles went on the road looking for new customers when he was not on the phone taking orders from the company's established market. Suddenly, the two people in charge of the sales and marketing department were hired by a competitor. At age 24, with less than two years of experience, he was the senior person in the department. From his daily contact with the customers, he realized that they wanted more turkey products. But he also sensed a growing resistance to whole turkeys. From the company point of view, he knew that it was his responsibility to make a change. He was much more focused on the bottom line than his father, who loved to concentrate on expansion.

Charles's concern for profits caused him to push for further processing because he realized that that was where the greatest profit potential was. In the early 1960s, about 80 percent of the company business was in the fryer-roaster turkey, a 6- to 8-pound bird. Jennie-O was king of the fryer-roaster business, but Charles felt that the firm should come up with a bigger variety of birds because the customer wanted 14- to 22-pound birds as well as other turkey products. Because Jennie-O had EBO Farms, the company could move ahead and produce what the customer wanted. It took 11 weeks to produce a 6- to 8-pound bird, 16 weeks to produce a 14 pounder, and 23 weeks to produce a 22 pounder. The longer growing season caused resistance from the independent contract producers, so EBO Farms had to prove that it was profitable to produce larger birds.

Charles recalled:

I had the confidence we could sell the products but had no knowledge of how to further process. We were selling larger toms in canner bags to a further processor. I visited that company and learned what it was doing. Several other firms were buying whole turkeys from us, and we saw what they were doing in further processing. We hired several people from those companies, including one who became our director for research and development.

Initially, further processing during the mid-1960s was simple deboning. The first Jennie-O further-processed product was a 2-pound pan roast of half white and half dark meat. It was packed in an aluminum pan that could be placed in an oven and was done in 1 hour and 20 minutes. That was an instant success, and 30 years later, Jennie-O still had 74 percent of the turkey pan roast market.

In 1963 Jennie-O processed more than 3 million turkeys, of which 700,000 were grown on EBO Farms. More than 70 million pounds of finished turkey products were handled. By 1968 further processing enabled the market to expand to a year-round basis, which was the best solution to absorbing the increased turkey supply.

Earl Olson was bent on continued expansion and more integration, which gave him better control of turkey production and at the same time reduced his risk. To accomplish that, he had to process more feed. The Willmar Grain Terminal, acquired in 1963, was operating at full capacity, so he built another mill at Swift Falls to manufacture feed exclusively for EBO Farms concentrated in that area. In 1965 he purchased the Farm Service Elevator in Willmar, followed by a feed mill and chicken growing facilities at Brainerd, Minnesota. The Brainerd growing farm was converted to turkey production, increasing the capacity of EBO Farms to 1 million birds by 1967.

Olson liked associating with people. He enjoyed walking around the plants and farms and talking to the people who made the business function. He liked going to the lunch room and visiting with the workers. Initially, most of the workers were farmers and farm wives who spent four months a year working at the plants to earn extra income. Olson never forgot his humble beginnings and felt that he was one of them. This and the fact that he was so public relations and people minded made him feel that there was no need for personnel services. However, as the company expanded and the culture changed, conflicts between labor and management arose.

When asked about Olson's patience with key people, Handahl replied:

> When Earl employed someone, he wanted to see that person be a winner. If he had a problem with an employee, especially a plant manager, he would take lots of time to get the employee in line. He didn't like to admit he had made a mistake when he hired the person. I am sure many people worked for us longer than they should have because Earl tried to help them by searching for their talents so they could stay on.

The first of 28 union campaigns came in 1951 after the purchase of All State Supply. The company responded by employing its first personnel professional, but the personnel department remained quite loosely structured until 1965, when labor problems intensified. The change bothered Olson, who enjoyed the one-on-one relationship. He admitted, "I personally am not very tough with people and bend over backward to help them succeed. I am certainly not good at firing." But times changed, and what bothered him most was the total freedom union agitators had to talk to the workers and the promises they could make to them while the company was forbidden to defend itself.

Olson said, "We had to pay for their [the agitators'] promises, but when the union lost the campaign, the professional leaders just disappeared." In the early decades there was always a supply of underemployed farm and village people seeking work. In the 1960s and 1970s, because of farm consolidation, surplus farm workers were still available, and the company experienced the good fortune of having a waiting list of 100 to 200 people. That too changed as the area became more industrialized and the demand for labor increased.

During the 1960s the basic inner circle of the Jennie-O family management team was established. In 1965 and 1968, respectively, sons Jeff and Bruce graduated from college and entered the business. Olson, his three sons, and Don Handahl were in administration, with John Jeffords as general plant manager; Spencer Kostad, plant manager at Litchfield; and Bacon White, manager of the largest plant, at Melrose.

At this stage a major change took place in the way the company functioned. Up to this point much of the day-to-day management was "by the seat of the pants." There was no organizational structure. Some people reported to Olson or to Handahl, and some to Charles. Charles recalled, "We needed to define and make an organizational chart. We had no regularly scheduled meetings—only crisis meetings." Charles did not like the haphazard way of operating, and as soon as he was made executive vice president, he pushed for an organizational and planning structure. Charles's appointment was not announced by his father. The people in management found out from Charles when he sent out directives indicating his new title and stating that he was calling for uninterrupted staff meetings from 7 A.M. to noon each Friday. An agenda was sent out regarding topics to be discussed. Charles commented, "Dad never came to the meetings. Once in a while he would pop his head in the door and make a suggestion, but I don't recall that he ever overruled any decision we made at those meetings." He added, "Dad once told a friend, 'Charles has never had a boss.' It was his way of turning over the reins."

The second major transition during the 1960s came with the creation of an external (nonfamily) board when Peavey became 50 percent owner of Jennie-O. Originally, Olson relied on Handahl and Mr. Julie Davis, his attorney from Minneapolis, as his "sounding board." Davis was very respected by Olson and could be very direct with him. After Peavey became involved, Bill Stocks and Gil Geibink (both from Peavey), Davis, Charles, and Olson made up the five-person board.

After Peavey was paid off and no longer part owner, Olson decided to continue the board and asked Stocks and Geibink to stay. The board met quarterly, and its input proved very beneficial in helping to change a one-person–guided venture into a more formal organizational structure of the business world. The board members encouraged Olson to sell some stock to Charles.

In 1968 veterinarian Peter E. Poss joined the company with the combined task of managing EBO Farms and helping partner growers, as well as independent contract growers, operate more efficiently and profitably. Poss, aided by Jim Johnson, D.V.M., also did in-house research on disease control, vaccines, medication, biosecurity, feed formulation, feed additives, management techniques, ventilation, fuel use, litter management, types of equipment, building construction, and other phases of turkey production. Olson wanted all information shared with the growers because he realized that they could not individually afford to do the type of work Poss and Johnson were doing.

Once again the company ran head on into another downturn in the turkey cycle. With the Brainerd addition in 1967, EBO Farms produced 1 million turkeys for the first time. EBO Farms lost nearly $1 on each bird grown, but to make matters worse, the processing business suffered financial reversals. It was time to make some crisis management decisions to improve overall margins. Olson's contrarian instincts paid off in the first decision. He felt that many turkey growers would be discouraged and would quit or cut back production in 1968, so he would increase his production to 1.5 million birds. The price of turkeys rose from 40¢ a pound in 1967 to 60¢ in 1968, and the losses were more than recouped.

Next, the company decided to sell directly to the chain stores to save broker and commission firm fees. This is when it pushed the Jennie-O Foods label for company-processed turkeys in an effort to gain nationwide recognition. Then came a concerted effort to convince the chains to adopt year-round marketing of turkey products instead of following the traditional practice of selling the "holiday bird" during a four-month season. The next logical step was to encourage the chains to buy a full semitrailer or carload at a time. This was a hard sell, but the volume-conscious Olson realized it was the key to improved margins for Jennie-O and the chains.

The combined efforts paid off, for by 1970 Jennie-O products were being sold in full semi loads to chains with as few as 10 stores. Half of all the chain stores in the nation were marketing Jennie-O products. By 1972 products carrying the Jennie-O label made up 70 percent of the sales from the company plants. The other 30 percent of the 105-million-pound production was custom labeled for large retail outlets. Olson enterprises produced and processed 31 percent of the 19 million turkeys grown in Minnesota. Olson changed the buying habits of the consumer, for 55 percent of all turkey products were marketed in the nontraditional turkey months.

The practices of the entire industry were changing, and EBO Farms was no longer the only grower operating on a year-round basis. The same applied to turkey processing plants, which were fewer in number but larger, automated, and far more labor efficient than the Swift Falls creamery and its contemporaries. However, by the 1960s the larger, more mechanized Jennie-O plants cost $15,000 each day they were idle, so it was more important than ever to keep them in production. This necessitated going to commercial year-round turkey production, replacing the historic small-flock operation geared to supplementing the farm income. The result was greater economies for the growers and processors, which benefited the consumer with year-round turkey products at lower cost.

By 1972 Jennie-O employed 1,100 people in its processing plants; another 30 processed 70,000 tons of feed. Generally, about 100 truckers delivered feed and equipment or hauled turkeys from a five-state area to the plants. The 25 EBO Farms covered 4,000 acres, none of which was tilled, and employed 100 workers to raise 1.5 million turkeys. In addition, another million turkeys were grown on 13 farms in which Olson was a major stockholder. Contract and independent growers furnished another 3.5 million. More than 3 million bushels of locally produced grains were required to feed the turkeys grown in the immediate area. The local demand for grain was so strong that the corn market improved from traditionally selling 10¢ under Minneapolis prior to the growth of the turkey industry to 5¢ over. The farmers who once resented Olson's success now appreciated what he had done.

Probably the greatest concern among all the growers was the constant threat of disease. Managers of EBO Farms and supervisors of all partner, contract, and independent operators were on constant alert for any outbreak. Under 1972 technology it was deemed advisable not to have more than 500,000 birds in any five-square-mile area. However, by that date, EBO Farms controlled enough land that it could safely produce 5 million birds.

From the day Olson grew his first turkeys, he was determined to raise about 35 percent of the company's total processing requirements. This was contrary to the industry pattern, but integration gave him greater flexibility and "some immunity to the uncontrollable variations of the industry." Further processing was very labor intensive,

This turkey farm at Wilcox Lake is the largest in Minnesota and probably the largest in the United States at one location. Each of the 12 environmentally controlled barns measures 64 by 1,224 feet, totaling nearly 22 acres under roof. More than 850,000 tom turkeys weighing a total of 30 million pounds are produced here annually. Better health care has made these larger, more efficient units possible.

and even with the most automated equipment, labor cost per bird increased by 300 percent. The surest way to offset higher labor cost was to go to two shifts per day and year-round plant operation. This reduced the fixed capital cost 40 percent per bird.

Charles was the strongest advocate of further processing because it not only aided profitability but it also served to broaden the market for turkey products. From his position in marketing, he saw a change in what the consumer wanted. Jennie-O was not alone in responding, for further processing was being adopted throughout the food industry to suit the needs of the two-worker family. "Besides," Charles added, "I personally liked further processing because it was far more exciting. There were so many ingredients involved and new things to try. It was certainly more profitable than whole-turkey processing."

After the big success with the 2-pound pan roast, the company added a 2-pound all-white-meat boneless roast. Then came a cooked 9-pound all white turkey roll, immediately followed by a cooked 9-

pound white and dark meat turkey roll. The last two products were for commercial customers. Other products, such as turkey ham, baloney, and salami, sliced turkey, and turkey bacon all had marketing success. Probably the most unique product was basted turkeys using a company-developed low-fat broth solution that retains the turkey flavor.

Once further-processed goods were developed, the company was challenged with organizing a sales campaign. One of the first big customers for these products was Freezer Queen Foods, at Buffalo, New York. It had extensive retail accounts, which helped to get Jennie-O products known. Freezer Queen was soon taking nearly 50 percent of the cooked product output, putting the company in a risky position of relying too heavily on a single customer. "We practically held Freezer Queen's hand." The German market gave Jennie-O another experience in being too dependent on a single customer. That country was taking 15 to 20 percent of the output of a product when a political move in the European Community caused the demand to dry up instantly. From that time the company focused on keeping its market base well dispersed so that no customer was taking more than 5 percent of its sales.

By 1970 the gross volume of the 20-year-old business exceeded $30 million. The firm marketed in every state and in 17 foreign countries. Olson grew 3 percent of all turkeys raised in the nation, and his plants processed 7 percent of the total. While his father pushed expansion, Charles stressed the further-processing value-added concept. From his work in sales, Charles was attuned to ever-changing consumer demands. New products had to be developed, and each new product meant more space was needed. Soon Plant No. 1 in Willmar was no longer large enough.

In 1971 computers were brought into the business. They hastened the shift to a more structured organization with tighter cost accounting. The entire company was "revamped" to follow the long-range plans. Consultants were engaged, and specialists from various sectors of the food industry were employed. Human resources, budgeting, forecasting, incentive programs, job definition, and an organizational chart were all put into place. The company was no longer operating "by the seat of the pants" and, almost as if by design, was poised for rapid growth.

To remain a leader in the industry, the firm spent more money on development of new products and advertising. A research and development (R&D) department was started on a small scale to handle the time-consuming and expensive process of developing new products. Jennie-O ran a conservative R&D program, and about 20 percent of all products developed succeeded. This was far better than the food industry nationally, which had only a 1 percent success rate. But, according to Charles, "Value added was the real opportunity to improve the bottom line." It also fit his father's goal of keeping the plants operating more months per year, enabling him to reduce cost, pay the grower more, sell the products cheaper, capture a bigger market share, and improve profits.

In 1972, partly because of his urging, Charles was given the task of overseeing the construction of a new 35,000-square-foot plant. This plant, south of Willmar, was designated Plant No. 4 and was to be strictly for further processing. The fully automated plant cost $12 million, more than all previous plants and additions to that date had cost. Most of the equipment came from the United States, some came from Germany, but a few very specialized pieces were developed in-house. Jennie-O Foods had acquired a sound reputation by this time and was able to bargain for financing. Willmar industrial revenue bonds offered very favorable loans, a far different experience from the trying times of the 1960s.

All the further processing from Plant No. 1 was transferred to Plant No. 4, which immediately allowed for doubling the slaughtering capacity of the original plant. In 1974, Plant No. 4 was brought into production, and Charles was named president. Olson became chairman of the board and left the day-to-day operation to his sons. Jeff continued in marketing, while Bruce was active in accounting and finance. The company was large enough that each family member could function on his own without the conflict so common in family businesses. Olson kept busy in community, charitable, and industry activities and was always at the forefront during any of the union campaigns. He never stopped thinking about the business and kept up his habit of calling Charles or Handahl whenever an idea struck him.

In 1968 whole turkeys made up virtually 100 percent of the market, but by 1974 further-processed products made up 25 percent. Up to that time the demand for dark meat was so weak that the thigh, which made up about 40 percent of the pre-boned turkey, was sold for whatever price the market, chiefly on the West Coast, was willing to pay. Jennie-O marketing personnel observed that Hormel Foods had a 2-pound pork ham called Cure Master that was attractive and popular. Jennie-O decided if it could prepare a 2-pound turkey ham, the product would sell because the company could profitably market it for half the price of a pork ham. The turkey ham was an instant success, which enabled the firm to sell dark meat for 10 times what the unprocessed thighs brought.

The development of turkey ham in 1975 was a major breakthrough, not only for Jennie-O, which was one of the first to market it, but also for the entire industry. It was too successful in the eyes of the leaders of the pork industry, who brought a lawsuit against the turkey industry for the promotion of turkey ham. The ensuing publicity helped the turkey people educate the public about turkey ham. The turkey ham attracted the attention of European buyers, and Jennie-O profited from an exclusive contract with one of the largest distributors in England. As more processed products were developed, Canada, Germany, and Japan invited Jennie-O to market to their distributors.

Soon after the turkey ham was developed, R&D produced a turkey loaf that consisted of ground short-end products. An extensive cooperative advertising campaign was conducted, but the turkey loaf was a major disappointment. A Japanese deboner used in the fish industry made it possible to produce a turkey wiener that was well suited to international marketing because it held up well in shipment. Charles's push for further-processed value-added products was well timed, and by 1978 Jennie-O had more than 120 turkey products on a rapidly growing market. Depending on the year, the company ranked from second to fourth in the nation in production of processed turkey meat.

To keep up with the increased demand, production had to be expanded. The smaller and outdated Litchfield plant was closed in favor of the larger and more automated Melrose and Willmar plants. Melrose was tripled in size, and over the next several years, three additions were made to the Willmar plant. The larger plants brought economy of scale, but the major cost cutter came when the plants adopted two shifts. That did not come without problems. "It was like starting another business. Country people were used to working daytime only." The night shifts were started slowly, employing mostly women, but eventually the second shifts were fully staffed.

Rapid plant expansion geared for year-round operation, plus the increased demand for a greater variety of turkey products, put the squeeze on EBO Farms. The company farms had to expand to maintain their 35 percent of the total supply of turkeys going into the plants. They also had the burden of demonstrating how to produce 28-pound toms on a year-round basis to independent contract growers who were convinced that year-round production was not possible. The bigger birds had leg problems, and winter production created challenges with ventilation and manure that did not concern seasonal growers.

Newly designed buildings were needed, and incentives had to be devised to get the growers to fall into line. EBO Farms had to come up with buildings suitable for winter production. At the same time, it had to be the least-cost producer so that it would know what to pay the independent contract growers, who theoretically should be able to produce for less than the company farms. The challenge was made more difficult because just when funds were needed for expansion, a financial hit caused by the overproduction of amino-protein foods—beef, pork, chicken, and turkeys—took place. But it was imperative to expand turkey production to keep the plants running at capacity, for that was where the biggest investment overhead was.

When asked what was the greatest challenge in keeping Jennie-O running smoothly after 1974, Charles replied:

> Growing a steady supply of turkeys without producing them ourselves. We had to figure out a contract to prevent the growers from taking a loss. Historically, we had agreed to take a grower's birds at market price and charged 12¢ to 14¢ a pound processing fee. The industry was going to base-price contracts whereby the grower received a floor price, and if the market rose, the increase was split between us and the grower. This virtually guaranteed that the grower would not lose money. It was risky for us, but as long as the market for turkey products was growing, we came out O.K.

In 1974, when Charles became president, the company employed 900 people, many of them for only part of the year. In 1986, the last year the Olson family owned the company, there were 2,200 employees, most of whom worked full time. Between those years, the Olsons built more than 100 turkey barns, added a feed mill at Atwater, renovated Koronis Mill and Supply, and made three additions to Plant No. 4 at Willmar, all of which were for further processing. The key to the entire growth was the added profits made possible by further processing, which at the end of 1986 made up 75 percent of all sales. Basically, everything had gone as planned.

After the 1974 industry-wide financial crunch caused by the oversupply of all meats, the company did not face another major financial setback until 1996, when high corn and soybean prices negatively affected the entire feeding industry. The intervening years represented an era of continuous growth and good profits. It was at the beginning of this period that representatives from Hormel Foods, Inc., made their first contact with Jennie-O leaders. Other mergers in the industry, such as the Oscar Mayer purchase of Louis Rich, probably prompted Hormel. The Olsons were not interested in selling at that time.

The door was kept open, and in January 1986, Dick Knowlton, president of Hormel, asked Charles for a meeting. At that point the Olsons had no idea whether they wanted to sell and no idea what Jennie-O was worth, although they had been offered $50 million by another company. Jennie-O had experienced a very profitable decade, and it appeared that leadership was on board for the future. But by June 1986, family members decided they should sell, with Charles probably the one holdout, and he was doing the negotiating.

The deal was consummated in December 1986 for $86 million cash, for Hormel did not want minority stockholders. Charles signed a five-year non-compete agreement and was offered a five-year employment contract. He had managed the company well, and Hormel wanted him to remain on board as president. Don Handahl continued in the same capacity as he had under the Olson ownership. Earl Olson remained as chairman of the Jennie-O board and served two terms on the Hormel board. He was very instrumental in the leadership of the company.

Some of Olson's dreams were still being realized. In 1992 Jennie-O consummated a three-year option to buy West Central Turkeys, of Pelican Rapids, Minnesota, which added 80 million pounds of processing capacity to the existing 350 million pounds of capacity. On October 1, 1997, the company acquired Heartland Foods, of Marshall, Minnesota, with another 130 million pounds of processing capacity. This fulfilled Olson's long-sought goal.

When Handahl, Olson's colleague of 50 years, was asked to describe Olson's key traits, he said:

> When things were the very darkest, he shifted gears. He really has perseverance—he cannot stand to lose. He loves the impossible. He thinks the impossible is an opportunity. He does not give up until he figures his way out of a problem. Earl has no fear of the future. He still wants to charge right into it.

The other two individuals working closest to the 82-year-old Olson were asked to name a single trait to describe him. Rahn Annis, in charge of procuring turkeys and developing grower contracts, said, "Earl B. is the liveliest mind I know."

James Rieth, president and CEO of Jennie-O Foods, shot back, "He has a real intuition on how to do things. Earl is a driving force and comes up with the ideas. . . . He has a marvelous way of leading."

When Olson was asked his plans for the future, his answer was a direct, "I suspect I will die with my boots on. I've always enjoyed the business."

The business he started in 1941 by raising 300 turkeys financed with his disability income insurance money had grown to a $30 million operation by 1970. In 1998, 75 EBO Farms contained 67 brooder barns and 212 grower-finisher barns and employed 225 people producing 12 million turkeys. These workers were joined by 84 independent contract growers, who produced another 24 million turkeys. The total 36 million turkeys required more than 14 million bushels of soybeans and 28 million bushels of corn, which translates to the production from nearly 500,000 acres of land, an area of nearly 800 square miles.

Those turkeys weighed nearly 900 million pounds, which Jennie-O processed into more than 400 products that sold for well over $700

This 1998 trailer is 49 feet long (three times as long as the 1950 rig) and hauls three thousand fifty 14-pound turkeys. One person using automatic loaders can fill the semi in 20 minutes without touching a bird. That is 75 times faster than in 1950.

million. Jennie-O employed 4,500 people working in eight plants in six Minnesota towns. Olson had met his long-sought goal of becoming the nation's largest processor of turkeys. He had changed turkey raising from a part-time "Ma and Pa" farming venture to one of the first fully automated sectors of agriculture.

Bibliography

Interviews

Annis, Rahn. Willmar, MN, August 7, 1997.

Handahl, Don. Willmar, MN, August 6–8, 1997.

Olson, Charles B. Willmar, MN, June 8–9, 1998.

Olson, Earl B. Willmar, MN, April 24, 1972; August 6–8, December 16, 1997.

Rieth, James N. Willmar, MN, August 8, 1997.

Miscellaneous

Franklin, Lynne. "Earl Olson: The Horatio Alger of Turkeys," *Commercial West*, January 3, 1981, 8, 10, 11, 29.

Handahl, Don. Data on production of EBO Farms and processing volume at Jennie-O Foods, December 7, 1997.

"Jennie-O Foods, a Family Turkey Processing Business." Minnesota Department of Economic Development, April 1980.

"Jennie-O Foods, Tops in Turkey Processing," *AgWeek*, November 30, 1998, 27.

"Something to Gobble About at Jennie-O," *Star Tribune*, July 18, 1999, D1, D4.

J. R. Simplot

"I'm a Farmer at Heart"

JOHN RICHARD (J. R.) SIMPLOT was born January 4, 1909, in Dubuque, Iowa, to Charles Richard and Dorothy Haxby Simplot. J. R.'s parents and siblings had contracted tuberculosis and were advised to leave Iowa for a drier climate. In late 1909 Charles and Dorothy Simplot, with their two daughters and infant son, J. R., loaded their belongings into a railroad immigrant car and left for Idaho. They settled first in Boise and then in Burley, where Charles worked as an independent contractor building homes. In his spare time he made improvements on their 80-acre homestead near Declo. The homestead was limited to 80 acres because it lay within an established irrigation district.

Unlike most homesteaders, the Simplots were by no means destitute. Charles had between $5,000 and $6,000 from the Simplot estate, and Dorothy had inherited $20,000. Charles used his money to build a good set of "Iowa type" farm buildings on the homestead, but Dorothy never used her money. J. R. said, "She salted it away." He recalled that at one time she earned only 2 percent interest in postal notes. As soon as Charles finished erecting the buildings on the homestead, he purchased another 40 acres. When not building houses for others, Charles used a fresno (a horse-operated scraper) to level his land in readiness for flood irrigation. He had the entire 120 acres ready for irrigation by the time water became available.

J. R. and his sisters attended a four-room school in Declo. When not in school he learned to do every job on the farm. He was big for his age and liked every challenge his parents gave him. At age nine he became involved in a sheep project. He acquired 22 orphan lambs from his parents' flock that required bottle feeding. He liked sheep and continued raising them, giving the proceeds to his mother for safekeeping.

UNFORTUNATELY, J. R. and his father did not get along. One day his father "got tough with him," so J. R. quit school and left home. He admits that he did not like school, did not do well at it, and never graduated from the eighth grade. He said, "I asked my mother for the money from the sheep project. I remember so clearly. She gave me four $20 gold certificates. I can still see that gold seal." The 14-year-old lad went to the Enyeart Hotel in Declo, where he roomed and boarded for a dollar a day for the next four years. He later purchased the hotel building and still owned it in 1999.

From the day he left school to the present, J. R. Simplot has never been out of work. He did everything—unloading coal and lumber, working on farms, riprapping irrigation ditches. In his spare time he traded in cattle, pigs, horses, junk, "anything I could get my hands on." He added, "I had a knack for trading, and I liked dealing with people."

When he was not busy, he "hoboed on freights" to California and Canada. One year he and a friend, Gibson, hired out as spike pitchers on a threshing run in the Declo area for $2.50 a day. They heard that they could earn $5.00 a day in Canada for the same work, so they "rode the rods to the Canadian harvest fields." When they returned, J. R. got a job as a saw operator in a sawmill at Eugene, Oregon. One of his friends belonged to the local National Guard unit that was preparing to go to camp but was not at full strength. The next thing the husky 15-year-old realized was that he was in the Guard and on his way to camp.

WHILE J. R. was out making his fortune, his parents continued farming. The Charles Simplots never farmed more than 240 acres, but they milked cows, raised beef cattle, sheep, and hogs, and did well. They lived frugally and used farm profits to buy several farms prior to the increase in land prices that came with the farm prosperity of World War I. In 1919 and 1920 they sold those farms and moved to California to operate a chicken ranch. After farm commodity and land prices collapsed in 1921, the purchasers defaulted on their payments, and within a few years the farms reverted to the Simplots. Then Charles contacted J. R. and asked him to return to farm with him. J. R. accepted the challenge and returned to Declo.

The price of hogs dropped sharply, and J. R. decided to capitalize on the opportunity. He had enough money to purchase about 600 feeder pigs at $1 per head, materials for fencing, and a cooker necessary to prepare a ration of horse meat, cull potatoes, and ground barley. His father helped him get "set up."

J. R. had a Model T Ford pickup with a two-speed rear axle, and he had a 22-caliber rifle that he used to hunt wild horses in the Raft River area about 10 miles from the farm. When he needed feed, he shot two horses, skinned them, cut up the carcasses, and hauled them to the cooker to mix with 6,000 pounds of cull potatoes plus ground barley—the 1920s version of a total mixed ration. The horse hides provided him with money for ammunition, gasoline for his Model T, and coal to fire the boiler when he did not have time to gather brush.

When the hogs were ready to market, the price had risen to 7¢ a pound, and the young entrepreneur found himself with about $7,800. He recalled in a 1982 video: "That was big money for a teenager in the

Temporary hog facilities similar to what J. R. used to feed 600 pigs and start his rise in agriculture.

1920s." Armed with profits from his hog venture, he struck out on his own for the final time. He had come to know Lindsay Maggart, a "very sharp" potato producer and dealer. J. R. rented 40 acres of irrigated land from Maggart on a crop-share basis and another 120 acres from Kate Maggart Walker (Lindsay's sister) on the same basis. He operated the acreage as a single unit.

J. R. BOUGHT used machinery and six horses with his hog money and was ready to raise potatoes. Maggart was a progressive potato producer and required his tenant to plant certified seed and use phosphate fertilizer. Neither was standard procedure for that area at that time. The partners raised 20 acres of potatoes, plus beans and clover for seed, as cash crops on half the farm and oats and hay for horse feed on the remainder.

Certified seed was purchased from Bill Axhley. After harvest, Axhley invited Maggart and J. R. to northern Idaho for elk hunting. Axhley's son and J. R. went hunting while Axhley and Maggart "sat in a tent and drank." The two hunters got five bull elk, but because J. R. shot one within the boundaries of Yellowstone National Park, he was fined $50 and had to give the authorities four of their bag. The experience of seeing Axhley and Maggart "dissipate their lives with booze" made a deep and lasting impression on J. R., who from that day on has steadfastly opposed alcohol and nicotine. It was during Prohibition, and he recalled, "I got sick of having to track down booze for Maggart."

While sharecropping with Maggart and his sister, he learned of an electric potato sorter that was being built in Shelly, Idaho. He and Maggart investigated and jointly purchased one for $254. J. R. immediately took the sorter to a blacksmith shop to "beef it up." J. R. and Maggart sorted their own potatoes and then custom sorted for whoever hired them, dividing the income 50-50. One day during spring potato planting, Maggart discovered that J. R. was sorting for a farmer who was selling his potatoes to a broker engaged in a feud with Maggart. Maggart wanted J. R. to quit sorting and start planting, but J. R. wanted to continue sorting. They flipped a coin to settle the dispute.

J. R. won the flip and entered the next stage of his life, the one that would lead him to the top of the potato industry. The 19-year-old employed a crew and contracted with farmers to sort their potatoes and deliver them to the rail siding. He also brokered the potatoes. He commented:

> I was serious about the potato business. I was on the wire service and took orders for potatoes from Kansas City and Chicago. Once I had orders, I had a credit line at the local banks, and everything went my way. I started building potato sheds, and during the next 12 years [by 1940] I had 33 of them in an area from Vale, Oregon, to Idaho Falls, Idaho.

The potato sheds were 60-by-300-foot structures dug about 3 feet into the ground and lined with matched timbers. The potatoes, once in storage, were covered with straw, dirt, and woven wire, which remained in place until time for the potatoes to be sorted and marketed. When J. R.'s sorting crew was not busy sorting or loading potatoes, it worked in the woods cutting and hauling timber to the site of another shed. In the meantime, J. R. solicited more business or encouraged farmers to buy certified seed. He financed and delivered as few as 10 sacks of certified seed to farmers, because he wanted them to learn how much good seed could do for them.

In 1930 he established an office in Declo and employed Birdell Curtis to do bookkeeping, check writing, rail car billing, and whatever else needed to be done to keep the business in order. J. R. said:

> I couldn't do these things because I couldn't spell and was not good at bookkeeping. I give Birdell credit for building the company in those years. He was really great. He knew no time constraints and could do everything. He stayed with me for 15 years and became a partner before he went on his own.

In 1931 J. R. married Ruby Rosevear, from nearby Churchs Ferry. The couple had four children—three sons and one daughter. Richard was born in 1935, Don in 1936, Scott in 1947, and Gay in 1945. Every time the business changed headquarters, the family relocated. The Simplots lived in Declo, Burley, and Caldwell before moving to Boise in 1946. J. R. and Ruby were divorced in 1962.

J. R. often commented that he liked trading and dealt in anything that came along. He was not afraid to take risks. He recalled a land deal in the early 1930s when the Holland Land Bank (he was not positive of the exact name) got caught with 15 ranches ("ranches," not "farms," is the preferred term in Idaho) totaling 1,200 acres. The loan agency wanted to dispose of the land it had repossessed from the homesteaders. J. R. purchased the land on a contract-for-deed and immediately resold it, mostly to the former owners, on contracts and for more than he had paid. Most of the buyers were nearly destitute, so he made those sales attractive by offering to take potatoes delivered to his cellars for 50¢ a hundred. He not only profited on the resale of the farms, but he increased his potato business and made several farmers happy by giving them a second chance.

Shortly after that, he purchased the business of J. O. Sewell, at Nampa, Idaho, which included nine onion/potato sheds, increasing his storage units to 42. He was only in his mid-20s but already operated the largest potato firm in the state.

That deal was no more than finished when he learned that G. Pace, the sheriff of Cassia County, Idaho, was planning a sale of tax-forfeited land in the Raft River Valley. The 18,000 acres, virtually in one block, came chiefly from homesteaders who had gone broke or had given up on farming. J. R. bid 50¢ an acre, with 5 percent down and the balance to be paid over 20 years at 4 percent interest. The fact that he was able to make a deal with such favorable financial terms indicates how desperate conditions were. First, the county was in dire need of getting the land back on the tax rolls. Second, even with such generous terms, no one else was willing to compete for the land.

At that point J. R. encouraged his brother, Robert, to join him, and the two immediately set to work drilling wells for irrigation. These were the first irrigation wells in the Raft River area. The partners had to drill only about 60 feet to get an adequate supply of water. The partnership lasted until Robert "fell to drinking," but J. R. still retains the land (1999).

J. R. LAID THE GROUNDWORK for his empire early, for he expanded nearly every year during the 1930s, almost as if he had a vision of better things ahead. In that decade he started growing potatoes in the Grand View area, a beautiful flat block of land along the Snake River south of Mountain Home. He recounted: "The land was not so costly then. At first I could buy Grand View land with water rights for $40 to

$50 an acre." Grand View's favorable climate was important to him later when he was forced to get into livestock feeding.

By 1940 he owned more than 30,000 acres of farm land along the Snake River and shipped more than 10,000 carloads of potatoes annually, in addition to onions. But he was not processing yet. At this time he had the foresight to purchase from the government for $5 an acre 2,200 acres of land bordering Boise. "I wish I had bought twice as much as I did." That land is where his home overlooking Boise was built.

DURING THE 1930S J. R. started raising onions. "I got big in the onion business, up to 300 acres." He also brokered onions, which unexpectedly opened doors to another venture. On January 23, 1941, he attended a meeting of the Dehydrated Food Manufacturers of America in Chicago. The chief purpose of the meeting was to discuss ways the industry could serve the national defense effort. At the meeting he met John Sokol, head of a spice business in Chicago, and Colonel Paul Logan, Assistant Quartermaster General.

Through his brokerage business J. R. came to know Milton Grosz, of Burbank, California, who purchased all the onions that did not make table grade that J. R. could provide. Grosz ran up a bill of $8,500 and would not respond to requests for payment. In mid-July 1941, J. R. made a trip to Burbank. Grosz was not in the office, so J. R. waited. Soon John Sokol arrived. Sokol was purchasing flaked and powdered onion products from Grosz and was there to discuss some problems. J. R. and Sokol got into a discussion that ended in Sokol's giving J. R. a contract, reputedly written on the back of an envelope, for 300,000 pounds of onion powder at 21¢ a pound and 200,000 pounds of flake at 31¢ a pound.

Neither J. R. nor Sokol saw Grosz that day, but after J. R. finished with Sokol, he followed a truckload of onions to where they were being dried. He observed that a prune drier was being used to dry the onions and saw how simple the process was, so he copied the name and address of the machine's maker from its nameplate. Knipchild Manufacturing Company, Santa Maria, California, was his next stop. He reasoned that if Grosz could buy onions from him and pay 65¢ a hundred to ship them to California, he could dry them in Idaho and save that cost. He ordered a six-tunnel Knipchild drier and prepared to go into the onion drying business.

He returned to Idaho to seek a site for an onion processing plant. The city of Caldwell offered five acres on a rail spur. J. R. sensed that he was on the verge of something big, so he focused on getting a plant erected. "It was a real stroke of luck. I was there when it [the $750,000 structure] was being built, and I stayed right there to make it work." By October 1941 the plant was finished, and 80 people were soon drying 2½ carloads of onions daily. This was an especially profitable business because onions that could not make table grade were no longer sold as culls. Instead, they could be dried at Caldwell. J. R. reminisced: "In the first year with the drying machine, I made [netted] a half million dollars. Sokol & Co. paid cash when the product arrived."

On October 17, 1941, soon after the drying plant opened, Colonel Logan and Sokol stopped at Caldwell. To prepare for anticipated wartime demands, they were touring the country, looking for operating vegetable dehydrating plants and for locations where plants might be built. According to J. R., only five vegetable dehydrating plants were in existence, so the government was eager for him to join the effort. On February 18, 1942, he signed a contract to dehydrate potatoes for the military and entered a new phase in his career.

The dehydrating industry was still in its infancy and faced many technical difficulties. The company tried dehydrating several different vegetables, each of which presented a unique problem. The best success was achieved drying onions and potatoes. Because J. R. hoped to garner as much business as possible, he used the same drier for both products, but the aroma of the onions carried over into the potato flakes. Until more driers could be built, the firm discontinued onions and concentrated on potatoes, which were in greatest demand by the military.

Soon J. R. faced another challenge—how to peel potatoes with as little loss as possible. The company experimented with several methods of peeling: grinding the peeling off; using hot water under pressure to force the peeling off; and, finally, using a super-heated lye wash that softened the skin so it would come off with steam from jets and gentle rubbing by slowly rotating brushes. Once the peeling process was perfected, several driers were built in-house, and it was full steam ahead.

The labor force rose from about 100 to 2,000 by the end of 1942 and increased steadily during the war years. The plant became the largest in the world, requiring 640,000 pounds of raw potatoes daily. It produced an average of 20 million pounds of dehydrated potatoes annually, about one-third of the total U.S. military consumption.

In 1943 total shipments of table-stock potatoes dropped to 7,000 carloads because 160 million pounds of potatoes were used to produce 20 million pounds of dehydrated product. Another 8 million pounds of onions were dried into 1 million pounds of onion powder and/or flake. Gross income for the 34-year-old entrepreneur from the dehydrating facility alone was $6.5 million. The contracts for 1944 called for 25 million pounds of dehydrated potatoes and 2 million pounds of onion products, from which J. R. grossed about $7.75 million. That year his plants produced about 25 percent of the total military demand for dried potatoes and onions. The Army/Navy Award for Excellence was bestowed on the facility because production of dehydrated potatoes per worker hour was 4 pounds more than the average of all other plants producing for the military. J. R. was also named one of Idaho's 10 outstanding young men that year.

The dehydrated products were put into 5-gallon lightweight metal cans that had to be placed in wooden crates for shipping. By 1943 wartime demand for lumber products was so great that the company could not secure wood needed for the crates. J. R. said, "I needed the boxes, so I had to get into the sawmill business to assure myself of a supply." The Caldwell Box Manufacturing Company was organized, and a plant was built adjacent to the dehydration plant. Demand outgrew the available sources of lumber, so the Caldwell Lumber Company was formed to purchase a sawmill in northern Idaho. Then J. R. purchased half interest in Cal-Ida Lumber Company in California. Before long he had half interest in two sawmills in California and owned two outright in Idaho. After the war the lumber division became a major producer of finished lumber products, turning out up to 100 million board feet of lumber annually and netting a million dollars per year. However, profits dwindled after the slowdown in housing activities in the 1960s, and the business was sold in 1969.

L AND has always been dear to J. R., so in 1943, when profits started rolling in, he purchased the Bruneau Sheep Company, which had a large block of land and 12,000 sheep. In Simplot fashion, he soon expanded the flock to 30,000 head. Contrary to J. R.'s boyhood experience, sheep were not profitable, so after several years he converted the ranch to cattle. He also acquired the Grand View Farms, adjacent to land he already owned along the Snake River, and Jamieson Farms, at Jamieson, Oregon. Each of those farms contained 1,200 acres of irrigated land, and the Jamieson unit also had 5,000 acres of range land.

A new headache occurred. The potato plants at Caldwell and at Burley produced increasing amounts of slurry, which had to be disposed of. This was before the environmental movement, but because of the tremendous volume involved, everyone recognized that the slurry had to be disposed of in an environmentally friendly manner. In some cases it had to be hauled up to 150 miles. "We had to get rid of the slurry." Feeding it was the logical answer.

It was assumed that hogs might be the best animals to consume the potato slurry. Historically, potatoes were a major part of hog rations in other countries and on some U.S. farms, and J. R. had used them in his early hog feeding venture. In typical Simplot style, he amassed 2,000 sows into one unit to raise pigs on potato waste. This did not work because the operation "could not keep the hogs alive." Then, he turned to cattle. The slurry has more feed value for ruminant animals, but, except for rejected french fries or hash browns, it is 80 percent or more water. This makes it expensive to transport any great distance. Attempts to reduce the water content by spinning did not prove practical.

Fortunately, the region has very favorable weather for cattle feeding and is near ranch country with a large supply of feeder animals. A pipeline was installed under Highway 19 west of Caldwell, extending from the potato plant to the feedlot on the opposite side of the highway. This was a very efficient method of disposing of the slurry. The hog operation was shut down, and Caldwell Feeders was organized with partner Bill Richardson. A feedlot with 1,000-head capacity was established. The profitability of the feedlot encouraged everyone and solved the crucial problem of disposing of the slurry. Other feedlots were opened, and by 1955, when John Basabe took over the farming operations, 4,500 cattle were on feed at any one time. The ranch and the cow herd were expanded and crop farming was increased to supply feeder cattle and to provide the forage needed to blend with the ever-increasing volume of slurry.

J. R. NEVER FORGOT the lessons of Lindsay Maggart about using adequate phosphate to increase the potato yields. He recalled an event in the early 1930s when he got top-quality red potato seed from the Red River Valley of North Dakota, phosphate fertilizer from California, and a fertilizer spreader, also from the Red River Valley, to attach to the Iron Age planter. However, the sprockets that controlled the fertilizer flow were the wrong size, and all the phosphate was used on the first 60 percent of the field. In spite of that error, J. R. gloated over the results:

> We got 350 bags an acre on that part of the field and 100 bags on the rest. The university was not recommending fertilizer. I could not get them [the extension people] to agree to the need for fertilizer at that time, but that experience was very valuable to me. The fertilized field also yielded a much nicer quality potato, but what we did not know was that the potato solids also increased 2 to 3 percent.

The additional solids were of little concern for table-stock production, but they were very important in processing because they reduced the pounds of potatoes that had to be processed.

In 1943, when the demand for phosphate climbed rapidly, J. R. received notice from the Anaconda Company, his major supplier, that it would not be able to make any future deliveries. However, the head of sales for Anaconda encouraged him to build a plant to process phosphate. That was all the challenge he needed. "I had people out looking for mines, particularly phosphate. Initially I was limited by [law] as to how many acres of phosphate leases I could own."

J. R. steamed ahead and, after a feasibility study, acquired 140 acres for a plant site near Pocatello. Because of the need for fertilizer to grow potatoes, he had little difficulty securing funds from the Reconstruction Finance Corporation (RFC) for a plant with an annual capacity of 120,000 tons. A second RFC loan was negotiated to finance a plant to produce sulfuric acid needed in the production of superphosphate. When J. R. learned that he could not secure phosphate rock from Anaconda, he leased jointly from the Shoshone-Bannock tribes on the Fort Hall Indian Reservation and the U.S. government 2,500 acres of phosphate reserves that his people had located. Once he controlled a source of ore, he built a crusher to supply the fertilizer processing plant, and the Simplot Fertilizer Company was ready for operation.

Expectations were high, and a 1948 article in a major national publication reported that the company hoped to net a million dollars from phosphate sales. But a series of roadblocks prevented the fertilizer business from blossoming until 1955, when the Pocatello plant was finally producing 100,000 tons of superphosphate annually.

Most American farmers were not using much commercial fertilizer. However, J. R. hoped that would change. Fortunately, Ralph Nyblad and Joe McCollum joined the company and conceived the idea of a chain of retail fertilizer dealerships. Farmers needed to be educated about the value of using fertilizer. The first Simplot Soilbuilder outlet that offered fertilizer, pesticides, and expertise on their use was opened at Twin Falls. A second store was opened in the same town after the war, and the race was on. By 1968, with the processing plants at Pocatello and at Brandon, Manitoba, in operation, 70 Simplot Soilbuilder retail fertilizer outlets were in business and served virtually every major agricultural region of the country.

Looking back on the obstacles that had to be overcome to get the fertilizer business going, J. R. reminisced:

> I think I was lucky. I hung on when things were tough. We did a better job of storing and selling fertilizer than our competition. Fertilizer has been a consistent money maker. I was offered $150 million for the fertilizer plants but did not sell. The smartest thing I have done to this day is that I did not sell.

J. R. WAS PROBABLY THE BEST SALESPERSON in the company, particularly of phosphate, so when the demand for potatoes grew during World War II, he encouraged farmers to increase the amount of phosphate they used. He financed much of his sales to farmers because he was so convinced that they would profit by using it. He had more than one reason for encouraging them, but the farmers who worked with him also profited. He enjoyed telling how he financed his sales.

> I bought 5 percent of the Burley National Bank from Maggart for $2,400 when the bank was in trouble. It was a risk, but the bank rebounded when it was merged into the Idaho Bank and Trust, the largest bank in the state. For a while it was a great source of credit, as was the First Security, but neither

would lend more than a dollar for each dollar of security I could give it.

About 1943 I started with U.S. Bank in Ontario, Oregon. I told the bank that I had $1.5 million in securities. I wanted to sell fertilizer and seed potatoes to farmers and take their notes. It was my intention to cosign the notes and sell them to the bank. The Ontario bank agreed with the plan and gave me a line of credit in excess of $15 million. I made money on sales to the farmers. I felt I had little risk doing so. All I sold was certified potatoes and single superphosphate [0-46-0], but it got the income stream really rolling.

I T IS NO SECRET that World War II propelled J. R.'s good fortunes. However, like most people, he hoped for the end, even though he was overwhelmed by the uncertainties ahead. Reality struck immediately after V-J Day when orders for dehydrated potato products dropped to 20 percent of wartime level. The company had 4,000 workers, mostly in the potato plants, and J. R. was determined to keep them employed. Many of those people lived in the large apartment complexes he owned in Boise and in Pocatello.

By late 1944 the Research and Development (R&D) people in the plant, led by chemist Ray Dunlap, were running tests on canning and freezing vegetables, fruits, and berries and were intensifying the search for new ways to process potatoes. By December 1945 the plant expansion was underway for canning and freezing vegetables and fruits. A 10,000-ton cold-storage unit, a 60-ton-per-day ice maker, and internally made automatic dipping machines were included.

The R&D crew had already developed a much superior dehydrated potato product—the potato granule—which replaced the dehydrated product made during the war. The granule resulted in a superior mashed potato product that quickly gained acceptance in the marketplace. But Dunlap never gave up on his pet product—a frozen french fry. During 1946 a french fry was developed to the point that a commercial production line was built. The major remaining challenges were overcoming the batch process for applying and recirculating the oil and perfecting a continuously run fryer for mass production. It was not until 1953 that the last hurdles were overcome.

After the war, the potato industry faced a dilemma because the demand for the traditional table-stock potato started to decline. The number of two-worker families was increasing rapidly, and to save time these families wanted some sort of processed potatoes. The industry had to refocus. During the war, J. R. continued marketing table-stock potatoes and was one of the first to realize the changing consumption pattern. That did not concern him, because he was well established in processing. However, he quickly learned that his company was not well known by either the institutional buyers in the food industry or the public.

A nationwide potato shortage in 1950–1951 provided an unexpected opportunity. Because the company was one of the nation's major processors and had a large inventory, it was able to penetrate the national institutional market, which was searching for potato products. The Simplot name became better known. Leon Jones, head of the food division, and J. R. both felt that the frozen french fry had greater market potential at the institutional level than at the retail level, so they directed their attention to the institutional buyers.

In the meantime, the company had perfected other new products and was ready to test the market. In 1950 R. Starr Farish became the firm's first professional sales manager and was directed to launch a national marketing effort exclusively for frozen french fries. Other companies were doing the same, and the combined campaign was successful. From 1951 to 1961 the total national production of frozen potato products rose from 25 to 484 million pounds. Per capita consumption of processed potatoes rose from 6.5 pounds in 1950 to 36 pounds in 1965, to 66 pounds in 1973, and to 94 pounds in 1997. Fresh table-stock potatoes, which initially represented nearly 100 percent of J. R.'s business, declined correspondingly, and by 1998 they were only 5 percent of the total volume handled.

By the early 1950s the company was poised for the new demands and quickly became the number one processor in the nation. R&D learned more about blanching and compressing with freezing to produce a constantly improved french fry. The consumer demanded processed potatoes and kept pushing the industry, and in 1953 the first viable commercial french fry was produced. Some credit the company as "essentially being the inventor of the frozen french fry business." In any case, it was certainly in the front ranks.

By the mid-1950s the flagship Caldwell plant was consuming more than half the potatoes grown in surrounding Ada, Canyon, Gem, and

Payette counties. It led the nation in dehydrated and frozen potato products. In addition, the company had plants in three other Idaho cities that turned out 600,000 cases of non-potato canned vegetables annually plus 5 million pounds of dehydrated and 10 million pounds of frozen products.

IN 1955 J. R. headed 26 companies in 10 states. These grossed more than $30 million annually. However, expansion opportunities were growing faster than capital could be generated. Like other entrepreneurs, he was risk oriented, but, like his counterparts, he understood a calculated risk as opposed to an outright gamble. When asked what his most limiting factor had been, he said:

> Money has been a constant problem. You cannot do everything you want, but you have to keep expanding. I never went public because when I wanted to go, they wouldn't accept me. Now I don't sell out because if it [the business] is worth that to them, it is worth more to me. [However,] I was never really close to going broke. I always kept my head above water.

> Continental Bank of Chicago was good to me and for years provided me with $20 million in operating capital. In 1960, when we were getting ready to make several expansions, John Hancock [Life Insurance Company] sent a loan officer, John Quincy Adams III, out here for a month and then loaned me $15 million. That was the biggest long-term loan I had. Earlier I had military backing to get money for the fertilizer business.

ONCE financing was in order, J. R. made several rapid expansions. He completed the plant at Burley, Idaho, and constructed another at Heyburn. The three plants used the equivalent of about 12,000 hundred-ton cars (the crop from more than 75,000 acres) annually, or 16 percent of all potatoes produced in Idaho. In the meantime, he started a plant at Carbury, Manitoba. He was excited about the prospect of growing potatoes there "because we could drive down a sand point and get water. That could become our biggest plant." Today, the company partners that plant with Nestlé and ships

500 million pounds of processed potato products annually direct to McDonald's.

In 1960 J. R. partnered with Jules Salzbank, of New York, to build a processing plant at Presque Isle, Maine. He and Salzbank also grew and contracted for potatoes there. By 1968 that plant had $25 million in sales, and the three Idaho plants grossed $60 million. The partners built a second processing plant in Maine, but, unfortunately, could not solve the slurry problem. The weather was not conducive to fattening cattle, and they encountered other problems unique to the area. After 15 years they discontinued operations in Maine.

During the 1960s a sixth plant was established at Taber, Alberta. The six plants processed 2.5 million pounds of potatoes daily, which resulted in 1.6 billion pounds of finished product annually. Those plants came on line none too soon, for in 1959 J. R. had come to know Ray Kroc, founder and head of the McDonald's fast-food chain. The fast-food industry, started in the 1920s, grew slowly until the late 1940s. From 1950 to 1990 the fast-food industry increased its share of national food sales from 10 to 40 percent. Although McDonald's was not a pioneer in the industry, after the 1950s it became a major player in the rapid acceptance of fast food. J. R. proudly stated that for the first 10 years of doing business with McDonald's, he was the chain's sole supplier of processed potatoes. This helped the Simplot company maintain its position as the leading potato processor.

After being involved with McDonald's, J. R. realized the importance of what Dunlap had done and patented the frozen french fry process. He added, "We never had a written contract. New stores were added [by McDonald's] as fast as we could furnish the french fries. Kroc had me on his board for 15 years."

The company was encouraged to follow McDonald's around the world wherever it started new outlets. By 1968 the Simplot french fry output reached 550 million pounds, and sales totaled $100 million on that product alone. With a solid, reliable customer and growing demand, J. R. convinced First Security Corporation and Continental Illinois, of Chicago, to increase his credit line to cover inventory and receivables to $50 million.

Following the McDonald's team in its worldwide search for sites for new restaurants got the Simplot company involved in potato production and processing in several countries. J. R. soon learned that U.S. farmers were not the only ones with problems. He found in most countries that fields were too small to gain economies of scale and that there were more problems controlling diseases. The company consistently had better yields in Canada and in the United States and was able to keep quality problems to a minimum. Much of the time, J. R. found it more feasible to supply processed potatoes to McDonald's stores in other countries from the North American plants. When asked why he did not follow McDonald's into Russia, J. R. replied that everybody he met to discuss starting a processing plant there "seemed to have a problem with alcohol. But maybe I should have gotten involved there. McDonald's has done well there, and the Canadian partner [of the Simplot company] has started in Russia."

Following McDonald's to Australia, Mexico, China, Japan, several countries in Europe, and other nations made Simplot an international company. J. R. recalled that he had "bad" partners in Germany, Holland, Sweden, and Turkey. His mistake was relying on foreign contractors and management instead of employing American nationals who spoke the language of the country and having them provide the leadership. "I lost $30 million in Germany. [We] built the plant first and then, because of poor planning, got into environmental troubles. There was nothing we could do, so we just walked away from it."

Despite problems generally caused by language or cultural differences, the company continues to look internationally for future growth. When asked what he thought of the North American Free Trade Agreement (NAFTA), J. R. replied: "NAFTA is a challenge, but we are in Canada and Mexico, and we manage. It will take time but eventually will work to everyone's benefit."

J. R. HAS NEVER DENIED that even though he was "backed into the business," the phosphate fertilizer division has proven a virtual cash cow in his empire. During an interview he stated that "fertilizer has been our biggy since 1948." In 1965 a privately employed consultant who had formerly worked with the U.S. Geological Survey Service placed a value of $500 million on phosphate reserves in the ground. At the same time, J. R. received royalties on taconite ore reserves in Wyoming. True to his entrepreneurial spirit, he did not sit back and wait for the cash to come in. He used those assets to leverage future ventures.

In 1965 he set out to erect a fertilizer plant at Brandon, Manitoba. Working through the Manitoba Development Fund, he learned that the province was interested in promoting greater agricultural production and that such a plant was important to that goal. The Development Fund agreed to back the project with $40 million if he put up the first $5 million. That plant has been very successful, and in 1998 the company was in the process of building a $300 million addition.

In 1968 the 70 Simplot Soilbuilder stores had sales of $65 million in fertilizers and chemicals and showed a profit despite the fact that the industry nationwide was losing money. The fertilizer business expanded rapidly during the 1970s and was extremely competitive. By the 1980s the chain of Simplot Soilbuilder stores numbered nearly 100 and was still growing. By the 1990s the 115 stores sold about 25 percent of all Simplot fertilizer products. The other 75 percent of production was sold in bulk to independent dealers.

At various times, J. R. owned half interest in three dredges used near Cascade, Idaho, for mining monoazite, a mineral that yielded thorium needed in producing nuclear energy. He was also involved with mining barite and feldspar in Idaho and iron ore in Montana, Wyoming, Nevada, and Alaska. At another time, he was involved in a copper mine in the Dominican Republic, which, besides copper, yielded zinc, sulfides, and gold. That mine paid for itself the first year, when the government took only 20 percent of the proceeds. The second year, the government took 45 percent. The third year, when it grossed about a million dollars a day, the government took over the operation and paid the owners $75 million.

The company eventually established the Minerals and Chemical (M&C) Group, which now has facilities in 16 states. The M&C Group has phosphate holdings in Utah, a silica sand mine in Nevada, and natural gas reserves in Alberta. It concentrates on agricultural fertilizers, specialty fertilizers for turf and consumer markets, industrial chemicals, animal feed phosphates, and silica sand. It is a major distributor of crop protection and nutrition products. By 1998 the Pocatello plant produced about a million tons of phosphate and nitrogen fertilizers (liquid and dry) annually. That plant is fed by an underground pipeline that slurries crushed phosphate rock from a mine 87 miles away.

Simplot and Farmland Industries joint ventured SF Phosphates Limited. This operation has a 96-mile pipeline to transport slurried phosphate ore from the mine at Vernal, Utah, to the processing plant at Rock Springs, Wyoming. This plant produces about 460,000 tons of finished fertilizer annually.

The M&C Group possesses the largest deposit of high-grade silica sand in the western United States. The ore runs 99.4 percent pure and is marketed to glass manufacturers and foundries. Another silica operation produces silicate, which is used for laundry detergent and ceramic tile mortar. Both mines have extensive reserves.

The Lathrop plant, near Stockton, California, produces phosphate fertilizers, livestock feed formulas, about 50 bulk and packaged products labeled Simplot's BEST, specialty fertilizer products, sulfuric acid, ammonia hydroxide, and ammonium sulfate. The Helm facility, near Fresno, produces nitric acid, used in the computer, airplane, construction, and dairy industries. Simplot Canada Limited, at Brandon, uses natural gas from Alberta to produce 1,200 metric tons of anhydrous ammonia daily in its new $300 million addition.

ONE THOUSAND POTATO GROWERS raise nearly 110,000 acres under contract to provide for the fresh and processing markets. The plant at Grand Forks, North Dakota, alone requires 28,000 acres of potatoes grown under contract. This represents 15 percent of the production of Minnesota and North Dakota. In addition, 8,000 acres of company land are used for potatoes. The 120 varieties of french fries and other processed potatoes sold under the Blue Ribbon and Golden Seal labels represent the major product of Simplot plants. Because crop rotation is necessary, other vegetables are grown on the land and utilize the same plants, so the company expanded into other lines. Probably the best examples are the two sweet-corn canneries at Pasco and Quincy, Washington, that process sweet corn (both cob and kernel) in season and then shift to potatoes for the remainder of the year. The by-products of sweet-corn processing, just as in the case of potato slurry, are also fed to cattle. Plants at Caldwell and Heyburn, Idaho, also use potato by-products to produce ethanol.

Broccoli, carrots, peas, lima beans, asparagus, avocados, cauliflower, strawberries, and tropical fruits are also processed into many finished products under several private label names. The Simplot Classic line, for example, provides more than 200 frozen fruit and vegetable products, in addition to pasta and rice blends. Four jointly owned plants in Mexico produce under the Mar Bran name. Besides Mexico, the company has three plants in Australia, a combined pea and potato plant in Tasmania, another in China, one in Canada, and five potato and several vegetable and fruit packaging plants in the United States.

J. R. was quick to point out that not all ventures were successful. He stated that the company tried operating a 1,500-cow dairy near Burley for seven years and "lost money every year." It also had big turkey and poultry operations "for several years but could not make them profitable." He continued: "I didn't get into the cheese business. One of our presidents got us into that line in [Arpin,] Wisconsin. We lost money at first but have since gotten profitable and big in cheese." The other cheese plants are at Lacey, Washington, and Salmon and Nampa, Idaho.

J. R. LOVES LAND, and he likes being the biggest in whatever venture he tries. Because of that, the cattle operations are one of his great satisfactions—they are large. The Agricultural Group contains several enterprises. The ranch unit is made up of 12 ranches in 6 western states, encompasses nearly 3 million acres, and is populated by 31,000 cows. This makes the company the number two cow/calf operation in the nation. One 35,000-acre ranch containing deep peat soil produces a heavy crop of grass that is cut by swathers and then formed into large round bales without the use of twine. This gets the grass off the ground, enables the cattle to eat it without too much waste, and avoids the danger of the cattle consuming any twine.

When asked about the profitability of the cow/calf operation, J. R. replied:

> The 31,000 brood cows are not really profitable, but they utilize the ranch land, and we wanted to experiment with cattle to develop a more uniform, leaner calf for leaner beef. We have [the most] potential for more uniform, leaner beef in the entire world. Furthermore, the ranches contain water rights, and I think have great agriculture potential.

J. R. understands the long-range implications of the project for leaner beef because the company owns a livestock packing plant where uniformity of carcasses is a major economic consideration. More beneficial for the future of the cattle industry is the need to produce a more consumer-friendly and lower-cost product. Fortunately, entrepreneurs like J. R. are willing to devote their resources to the industry when most of the smaller ranchers either cannot afford to upgrade their product or do not see the need to do so. In any case, he is happy with his ranch operations. He says, "I'm a farmer at heart. I like owning land. I cannot believe I bought some of the land for the price I did."

As the potato processing business grew, the volume of slurry increased, and by 1968 the company had three feedlots feeding 150,000 cattle per year to consume it. At that time the slurry reportedly reduced the cost of finishing cattle by 4¢ to 6¢ per pound. Plans were made to increase the one-time capacity to 90,000 head so that 270,000 head could be finished per year.

According to Tom Basabe, president of the Agricultural Group, the company feedlots at Boardman, Oregon, and at Grand View and Caldwell, Idaho, plus a Simplot Feeders Limited Partnership feedlot

The Grand View lot, with its one-time holding capacity of 150,000 head, is the largest facility of its kind in the country.

at Pasco, Washington, have an annual capacity of 600,000 head. This makes the company among the top six largest fed-cattle providers in the nation. J. R. wants to become the number one U.S. cattle feeder. To reach that goal, he visualizes increasing the Fair View lot to a 250,000-head capacity and expanding the other lots.

Besides the feedlots, Basabe oversees 3 million acres of agricultural land, of which 88,000 acres are crop land. In 1998 the crop land was used for 8,000 acres of potatoes, plus alfalfa, corn, asparagus, onions, sugar beets, wheat, and barley. Fair View contains 17,000 acres, all of which are irrigated. Most of the corn is used for silage and, with the alfalfa, provides much of the roughage for the ration.

To reduce the cost of acquiring corn, the company has erected a terminal consisting of six bins of 7,000 tons each, for a capacity of 1.5 million bushels. It has a spur large enough for a 108-car unitrain. A complete trainload of corn (more than 400,000 bushels) is purchased weekly at Omaha and delivered to Mountain Home. There it is unloaded nonstop, and the train immediately returns to Omaha. The corn is then trucked 30 miles to the Fair View feedlot.

A 108-car unitrain slowly moves, without stopping, while unloading 400,000 bushels of corn into an underground pit. The Agricultural Group financed the more-than-two-mile spur and the terminal to give it access to midwestern corn.

With the closing of the original Caldwell feedlot, it became more important to find some method of reducing the moisture content of the slurry to make it more economical to transport and at the same time make it more viable as a ration component. The feedlots have 3 trained veterinarians, and each day 10 people mount horses to ride the pens to detect any animals needing treatment. Fair View has 25 semitrailer rigs for hauling cattle to market and for picking up feeders from local ranches.

The company has a beef packing plant at Nampa, Idaho, that handles cattle processed exclusively for the Japanese market. Those cattle are fed about three times longer than most cattle finished by the company and weigh 1,800 to 2,000 pounds at slaughter. They have the marbling preferred by the Japanese.

J. R. is not directly involved in the other packing plants but is a "very large" stockholder in Iowa Beef Processors, which has a plant at Pasco, Washington, to slaughter the other cattle fattened by the company. Simplot, via its subsidiary SSI Food Services, partners with Cargill, at Wilder, Idaho, and ConAgra Fresh Meats Company, at Montgomery, Alabama. Those plants provide cooked, fresh, and fresh-frozen patties, precooked taco meat, beef and pork crumbles, fajitas, and burrito filling for Wendy's, Burger King, Taco Bell, and other restaurant chains in the United States and for the Pacific Rim.

THE RIDE TO THE TOP did not come without a few bumps, sometimes self-inflicted. J. R. started with meager resources, but once he learned how to generate capital, he disliked having it taken from him by taxation. He never ceased to have new ventures in mind, and every one required cash. Obviously, every means to delay taxation had to be examined.

It appears that his first bout with the Internal Revenue Service (IRS) came in 1948, when he transferred commercial properties to a trust for his children so the properties would not go into his estate and be taxed as inheritance. By then, he possessed property in 5 states, owned 34 corporations, and was a stockholder in many more.

He was also challenged by the IRS because during World War II, when confronted by a 90 percent excess profits tax, he formed 52 partnerships. J. R. said that he got the idea from two bank auditors. "I financed the 52 partnerships to get management help. I could not depreciate fast enough to make money. I paid the partners well, and they reinvested their income in the business, but the initial money was all mine. After the war the easy money stopped, and I had to get out of the partnerships." The IRS contended that partnerships were formed purely to avoid the excess profits tax, and the agency construed the moves as "schemes of tax avoidance, not evasion."

Another time, the IRS had 28 auditors on his account for four years. Part of the problem was that he had moved money between accounts as needed. He recalled that he needed money to build plants in Canada and that by moving money between his various divisions, he had access to internal funds. "I was fined $1.5 million, but that was cheaper than fighting them." National magazines reported that in 1971, 1973, and 1977, he paid fines for underpaying his income tax.

J. R. has never left any doubt about his willingness to take a gamble. In the mid-1970s he and P. J. "Pete" Taggeras gambled on nearly a thousand futures contracts involving 50 million pounds of potatoes worth about $4.2 million. They lost. The price of potatoes did not go in the direction the two had hoped, and they failed to deliver by the expiration date. It was the biggest potato default in the 104-year history of the New York Mercantile Exchange. The incident became known nationally when it was published in *Time*, June 7, 1976. In 1978 J. R. was fined $50,000 by the Commodity Futures Trading Corporation, ordered to pay $1.4 million in damages, and suspended from trading for six years as a result of a civil suit.

JOHN RICHARD SIMPLOT stepped down from the presidency of his companies in 1973 at age 64 but remained as chairman of the board until 1994. He credits the extension service and "university" people with significant help in developing new seed potatoes. In his estimation, the Idaho legislature was a "great help" to the state's potato industry when it prohibited the export of anything but No. 1 and No. 2 potatoes. He was active in the movement that brought about that legislation. But he added, "We have 60 people in seed potato and fertilizer [and food processing] labs working on ways to improve our products. We find that we can develop products faster ourselves, but we rely on them [universities] to fine-tune."

By 1982 J. R. Simplot Company was the twentieth largest private industrial company in the nation. In 1998 it was still one of the fastest-growing privately held companies. It had annual revenues of $2.8 billion, with 12,000 employees, and was looking global for expansion. In 1996 it made $210.5 million, and in 1997 it had $241.1 million in capital expenditures. Today the company has a fleet of about 400 trucks and has 600 owned or leased tank or hopper cars, including 150 cryogenic rail cars that use carbon dioxide for cooling instead of mechanical refrigeration.

Western Stockmen's Inc. (WSI) produces a full line of feed and animal health products for beef, dairy, horses, poultry, and pets. Its Windy Hill label represents one of the largest pet food manufacturers in the nation. Vet Direct distributes instruments and pharmaceuticals to veterinarians in the United States and abroad. Simplot AgriService handles a large line of agricultural products for domestic and foreign markets. The Jacklin Seed division is a leader in grass seed research and a major producer of grass seed and has conditioning facilities in Arizona, Idaho, Oregon, and Washington. All the divisions are free-standing.

When asked to what he attributed his success, J. R. replied:

I stayed on top of my business. I called the shots. We got big in potatoes and land early. I have relied on my judgment for employing topnotch people. I saw what alcohol and tobacco did to people, and it paid off. The only surprises I have had with [management-level] employees have been alcohol related. I learned to take care of myself, have been a great exerciser. I still love to ski [at age 89].

His son Don says:

Dad is a risk taker—that's what keeps him getting up each morning. He could forget about business when he was away. He was [not] a manager, but he was a leader of people. He picked good

J. R. Simplot in his 80s. When interviewed at age 89, he was not using glasses and still driving. His voice was strong, and his hearing good.

people—this was a strong knack of his. He would tell them their job and then let go. He never bothered about the little things. At the same time, he had great difficulty getting rid of people. He could build loyalty, and he was loyal to them.

Don recalled that his father always had need for money, so he had to spend lots of time with bankers. He was a great charmer with bankers. Bankers liked him because he was full of creative ideas, and they were quick to put him on their boards. But the freewheeling J. R. was reluctant to get too closely tied to any institution because he "did not like to answer all the 'don'ts' the big finance firms required of him." In a 1968 *Fortune* article, he commented that his business could have been larger but that he wanted to own what he built and he preferred to work alone. At that date the estimated income of all enterprises was $200 million, and his personal worth was probably $200 million.

At age 85, after he was "free from the company," he continued to speculate, especially in the cattle business, and parlayed those investments into another fortune. One of his favorite ventures was Micron Technology, in which he reportedly owns 20 percent of the stock. This person who did not receive a diploma from the eighth grade was a true entrepreneur, yet he "enjoyed being teased about his reluctance to spend money on a new car or a jet."

Bibliography

Interviews

Basabe, Tom. Fair View, ID, November 10, 1998; telephone conversation, December 16, 1998.

Simplot, Donald. Boise, ID, November 10, 1998.

Simplot, John Richard. Boise, ID, November 9, 10, 11, 1998.

Miscellaneous

Brandt, R. "J. R. Simplot: Still Hustling, After All These Years," *Business Week*, No. 3176, September 3, 1990, 60, 64, 65.

"The Great Potato Bust," *Time*, June 7, 1976, 70.

Hadley, C. J. "Mr. Spud," *Range Magazine*, Summer 1998, 30–35.

"Idaho's Henry Kaiser," *Fortune*, July 1944, 248.

"King of the Hill: John Richard Simplot." May 19, 1982. Video.

"The Magic of 'Mr. Spud,'" *Newsweek*, November 27, 1989, 63.

Murphy, C. J. V. "Jack Simplot and His Private Conglomerate," *Fortune*, August 1968, 122–126, 166, 168, 171.

Neuberger, Richard L. "Idaho's Fantastic Millionaire," *Saturday Evening Post* 220, June 19, 1948, 20, 21, 110, 113, 114, 116.

Origins of the J. R. Simplot Company, 1909–1955. Published by J. R. Simplot Company, 1997.

Serwer, Andrew E. "The Simplot Saga: How America's French Fry King Made Billions More in Semiconductors," *Fortune*, November 27, 1995, 69–86.

"Simplot: Bringing Earth's Resources to Life." Published by J. R. Simplot Company, 1998.

"Simplot Covering Costs to Compete," *AgWeek*, April 7, 1997.

Keith and
Richard Walden

Ready to Tackle the Global Market

A RIZONA'S 7,800 farms make it the fortieth-ranking state in number of farms, so it might seem an unlikely place to look for a large-scale, entrepreneurial, commercial-type farm. Its $2.145 billion in gross sales place it thirty-first among the states in that category, while its 28 million acres of land in farms place it fifteenth in that respect. However, its 14,300 acres of irrigated pecans make Arizona the fourth-ranking state in pecan production. Pima County, which encompasses the area around Tucson, has 419 farms covering 2,913,607 acres of land that generate only $46.9 million in gross sales, a mere $16 per acre. However, one farm of slightly more than 6,000 acres is responsible for nearly one-fourth of the county's agricultural income; and if the proceeds of its value-added enterprise are included, its income matches the total income of the other 418 farms. This farm contains 42 percent of the state's total acreage of irrigated pecans and reputedly is the nation's largest irrigated pecan grove. What is surprising is that the grove did not come about by a deliberate plan but as the result of a disease that ruined the farm's major crop, cotton. Fortunately, years of on-the-farm experimentation conducted by a curious entrepreneurial managing partner, who was always searching for something better, came to the rescue.

R . KEITH WALDEN was born July 4, 1913, the eldest of five children. Keith's father was a banker at Santa Paula, California, who also owned a 50-acre citrus farm and partnered with his brother on a 160-acre farm that contained 40 acres of oranges, 40 acres of peaches, 40 acres of apricots, and the balance in melons, cucumbers, and other vegetables. Walden's parents were comfortable financially, so there was no need for the children to earn income, but his parents believed that the discipline associated with labor was necessary. Walden was expected to help on the farm as soon as he was capable of doing so. He reminisced about how proud he was when at age 12 he got up at 3:00 A.M. to take a truckload of melons and vegetables to the market at Ninth and Central in downtown Los Angeles. His uncle, realizing that buyers might try to take advantage of a youth and attempt to pay less, gave him strict orders not to "budge on price." His uncle told him that if he could not get full price, he should bring the produce back to feed the hogs. Walden made no sales on Tuesday, his first day, or Wednesday and sold about half on Thursday. By Friday the buyers realized there was no bargaining at the Walden stand, and everything was sold at the asking price.

Walden liked the action of the market as much as work on the farm. It was no secret that his parents wanted him to become a banker, but each year he liked agriculture more. During Walden's junior year at college, his father suffered a heart attack, and Walden was asked to leave college to work with his uncle on the farms. This delayed his graduation from Pamona College until 1936. He graduated with a degree in economics and took a job with the Limoneira Farming Company, at Santa Paula, which was then reputed to be the lemon capital of the world. He got his first taste of large-scale agriculture, for that 2,200-acre irrigated citrus ranch was a very large operation for those days. The Teague and the Blanchard families, who owned Limoneira, also individually owned large separate operations. Teague was the first president of Sunkist Growers, Inc., and was later chairman of the Western Region Federal Farm Board.

THROUGH most of his years in late grade school and then high school, Walden had chicken and hog projects. During his final year in college, he laid plans for a citrus, avocado, and vegetable nursery and used the earnings from his projects to purchase and plant seed stock on 10 acres of leased land. His moonlighting on the nursery while working at Limoneira proved to be very profitable. Ironically, Limoneira was one of his biggest customers, and by 1948, when he sold the nursery, it was among the largest wholesale citrus nurseries in California. One of his employees in 1943 was his brother-in-law, Warren Culbertson, who later spent 50 years as one of his associates.

Shortly after Walden was employed by Limoneira, the company pathologist was killed in an accident. Because of his nursery experience and education, Walden was given a crash course in plant diseases, after which the manager asked him to assume the task of leading a crew of 40 to search for diseases. He remained at Limoneira until 1940, when he learned that the Craig Ranch was looking for a manager. The owner's son had been in the Reserve Officers Training Corps (ROTC) while in college and was called into service. Craig Ranch was a thousand-acre combination citrus and vegetable operation. He remained there until 1946, when the son returned.

At that point Walden was ready to branch out on his own. He found 960 acres west of Corcoran, in the Tulare Lake Basin. It had wells that were a thousand feet deep, but core samples from drilling for water produced redwood and silt left by the Merced River, giving a good indication of the fertility of the soil. Walden made good use of that information and during his first year of ownership sold 480 acres at the price he had paid for the entire 960 acres. This left him with 480 debt-free acres. His entrepreneurial mind sensed an opportunity to apply this "unused" equity. That came when he located 800 acres of undeveloped land at Wheeler Ridge, south of Bakersfield, which was about 75 miles south of his Corcoran farm.

As a matter of pride, Walden never borrowed money from his father's bank, preferring to do things on his own, and financed with Bank of America. He prepared a proposal for a loan for the Wheeler Ridge land. The loan officer informed him about a water witcher (sometimes referred to as a diviner) and told him if he could find water on the property, the bank would lend the needed funds. The witcher found water. A 450-foot-deep well produced 2,500 gallons a minute.

Mechanical cotton harvesters were just coming on the market, and Walden sensed that the time was ripe to apply for a cotton allotment. Historically, sodium nitrate from Chile and manure were applied to fertilize cotton, but he decided to use anhydrous ammonia (NH_3) instead. NH_3 was in its initial stages of production, and it seemed like a more economical source of fertilizer. By rotating sugar beets, potatoes, and other vegetables plus small grain with cotton, he produced an average of 2.8 bales per acre for the 14 years he owned that farm. He quickly became a recognized leader in the cotton industry and rose

When R. Keith Walden started producing cotton in the late 1940s in California, hand-hoeing to control weeds was still required. Within a couple of years, chemicals were developed that controlled weeds cheaper and more effectively.

to several leadership positions while he remained an active cotton producer. Even more important, he became acquainted with Paul Greenig, a contact in the anhydrous industry. This would serve him well in the future.

In 1947 he paid $50 an acre for a farm in the Tulare Lake Basin, which was being drained but still had 4 feet of water in it. The first fall after it was dry, Walden did minimum tillage and planted wheat without the benefit of any fertilizer or irrigation and harvested 4,200 pounds (70 bushels) per acre. He said, "That was almost pure profit." The next year, for $85 an acre, he purchased a nearby farm with a crop of edible beans on it. He had another stroke of luck, for the bean crop "just about paid for the farm." However, instead of paying off on the mortgage, he kept the cash in hand.

Walden realized that he was living in a fast-changing agricultural era and that he was benefiting from it. As a boy he grew up using horses, but when he started farming, he used only tractors. He saw threshers replaced by combines, and buggies by pickups, and pickups by airplanes. The first crop he raised in the Corcoran area was harvested by leased combines pulled by Holt track-type tractors. Those rigs were replaced by smaller but faster self-propelled machines that were far more efficient. He attributes some of his good fortune to the agricultural prosperity of the postwar years and to his good standing with Bank of America. Part of the good bank relations came about because of a lesson he learned in college. He was taught that when applying for a loan, one should always include details about how expenses were being budgeted and a plan for repayment. Walden did so throughout his career, and the practice served him in good stead when things did not turn out well.

In 1947, at the suggestion of his liaison at Bank of America, he formed Farmers Investment Company (FICO) to interest private investors in his business. Throughout the 1940s and the 1950s, he was in partnership on 70 acres of lemons, which provided a good cash flow while he expanded into other farm ventures. In 1947, to better enable him to manage his widely separated farms, he took flying lessons and upon graduation purchased his first airplane. He was one of the pioneer flying farmers, who were very active nationally in the postwar years. The airplane became a big part of his farming business for the next decades.

IN HIS BUSINESS TRAVELS, Walden became acquainted with Kemper Marley, an Arizona realtor and investor. Although Walden had done well in California, he felt that land prices were rising too rapidly and that the state was becoming too crowded. He told Marley he was interested in looking for land out of state. Marley took him to Yuma, Arizona, because that area was starting to develop irrigated vegetable farming. The Yuma area did not appeal to him, so both continued to search. Marley found what interested Walden—the 10,000-acre Continental Farm at Sahuarita, just south of Tucson. Continental had an illustrious but short history, and it was large enough to attract his attention.

Continental Farm was founded in 1915 by Bernard Baruch, Joseph Kennedy, and J. P. Morgan, three prominent national figures of that era. The government reputedly sensed the danger of a rubber shortage if German submarines cut off the latex producing islands of the East Indies from exporting their goods. Guayule, which grew naturally in the Santa Cruz Valley, was a known source of rubber, and the government was interested in increasing its production, just in case a shortage occurred. The three above-named men were well connected in high circles and were given the opportunity to venture. They developed irrigation on 4,000 acres to produce guayule, and a plant was erected to process the crop for wartime needs. The project was discontinued at the end of the war.

In 1922 Continental Farm was sold to Queen Wilhelmina of the Netherlands, who rented it to cotton farmers through 1945. The Queen was a friend of the McCormick family of International Harvester Company fame and learned that the company was developing a mechanical cotton harvester. This would make cotton a more economical crop to produce, which should make it more profitable. The Queen was interested in the potential for cotton and purchased five pickers. Continental was operated by the Queen for the crop years 1946 through 1948. Whether the Queen's death in 1948 had any bearing on what happened next is not known, but in any case Continental was placed on the market.

When Walden looked at it in late 1948, the land was heavily infested with Johnson grass, a vigorous weed that grows high enough to smother cotton. He learned that yields had declined to one bale per acre and that $300,000 in losses had accumulated from 1946 through 1948. Fortunately, he met the former head of the Department of Plant Pathology at the University of Arizona, who informed him that a new chemical with the trade name Dalapon was being developed that would control Johnson grass. With that news, Walden saw a possibility that he could make Continental pay. However, before he made a final decision, he wanted to know if he could secure anhydrous ammonia to increase the yields, because using NH_3 was far more economical than importing guano from Chile.

Walden knew that NH_3 was in short supply, so he contacted his friend Paul Greenig, who was the distributor for NH_3 in Arizona and California. Fortunately, Continental had a rail siding, so the NH_3 could be shipped directly to the farm, and Greenig promised what Walden requested for a three-year period.

The original Continental owners had built nine adobe houses and one larger home, all of which were rundown but were very repairable. The deal also involved 17 tractors, 8 mechanical cotton pickers, and 4 pickups, plus the tillage equipment. None of the machines were in good condition, and only one pickup would start. But Walden saw the potential, knew that the problems could be overcome, and was prepared to buy.

Walden had money because he had sold one farm in the Tulare Lake Basin for $100 an acre. It had cost him $35 an acre. For the 800-acre Wheeler Ridge farm, he had paid $10 an acre plus the cost of developing the irrigation system. After 14 profitable years, he sold that for $400 an acre. He also had cash from the farm he purchased in 1947, from which the first bean crop yielded nearly the cost of the farm. In addition, he retained ownership in two farms totaling 1,280 acres, on which he was growing cotton and vegetables, as well as other property in California. But in 1948 the $300,000 price for Continental was still a handsome sum. Once again connections paid off.

HENRY CROWN owned a prosperous sand and gravel business in Chicago prior to World War II. During the war he served as a colonel and was known for "buying more parts for less money than any other officer in the country." While in the service, Colonel Crown learned of Colonel Fred Sherrill, the head of procurement for spare parts for the military. On a trip to California after the war, Crown wanted to meet Sherrill, who was a member of the California State Water Board, so he stopped in for a visit. Crown let it be known that he was interested in investing in agriculture. Sherrill knew Walden and realized that he was capable of farming more land than he currently operated and told Crown the story of Continental Farm. A few weeks later Sherrill received $300,000 from Crown for Walden to purchase Continental. Walden was not sure that he wanted to accept the offer, so Sherrill held the check while Walden pondered his decision. He decided that for $500,000 he would sell half interest in FICO and that he would be the managing partner of the company. This is exactly what Crown wanted, and on January 9, 1949, Continental Farm was purchased. Since then, neither partner has invested any additional funds in FICO, and as of May 2000, the partnership was still thriving.

WITH Crown's financial strength to back him, Walden, the entrepreneur, was ready to roll. As soon as Continental was purchased, Walden called Warren Culbertson and said that he needed an accountant and the local school needed a teacher. Culbertson was working for General Electric and was taking accounting courses at night school. The Culbertsons were a young and eager couple ready for a challenge, so they accepted the offer. They moved into one of the renovated adobes near the school, where Mrs. Culbertson taught. The Continental office was located between the school and the Culbertson's new home. When not occupied with office work, Culbertson was expected to fill in where needed on the farm. He made regular trips with a truck to downtown Los Angeles to pick up workers (mostly winos) to hoe weeds and pick cotton. Out of every 30 who accepted work, only about 10 stayed as long as three months, which meant he had to make regular trips to solicit more workers. That practice continued until the truck loaded with men was involved in an accident.

In the 1949 crop year, Walden put all his efforts into Continental. It was time well spent, for he harvested nearly two bales of cotton per acre on 3,000 acres. Initially, the remainder of the land was used for grazing cattle. Gradually, it was developed for other purposes. With the aid of Dalapon for weed control and far less cultivating for weeds, Continental produced 2.5 bales of cotton per acre in 1950 and had the same results in 1951. In the first three years, on just the 3,000 acres of cotton, the net profit equaled the purchase price of the entire 10,000 acres. By 1950 Walden was comfortable with the potential of Continental and was sure that with proper irrigation, adequate NH_3, and Dalapon, he had the keys to success in Arizona.

He recalled that when he came to Continental, all the ditches were dirt lined; there was not 1 foot of cement ditching. He put flow meters at the pumps and then in the flow boxes and measured a 43 to 50 percent seepage in the mile run to the field. That convinced him to reshape and to cement all the ditches at once. The savings in pumping water repaid the investment in cementing the ditches in just over one year.

As soon as he realized that Culbertson had the makings of a good manager, Walden shifted him away from his accounting role and decided to employ a professional controller. In 1950 he engaged Robert Stuart, a graduate of the University of Chicago who had become well established in the Chicago area. However, Stuart asked to be transferred to Tucson because Mrs. Stuart was advised to live in Arizona for health reasons. Walden, a stickler for good records because "they were the key to running a good business," had to convince Stuart that joining FICO would be worthwhile. It was, for Stuart remained with the company for more than 20 years as controller and

treasurer until he developed a health problem and had to pursue less stressful work. In the 1970s Bob Hutchinson, who had worked with FICO's external auditors, was employed as Stuart's understudy and took over as controller. Hutchinson was skilled in computers and used that knowledge to enhance the FICO records.

With everything under control at Continental, Walden was ready to expand and purchased 4,000 acres at Eloy, 65 miles northwest of Tucson. This was a totally irrigated farm on which cotton was grown with a small grain rotation. Culbertson was sent there to clean up and develop Eloy to fit into FICO's pattern of operation.

In 1954 a 10,000-head cattle feedlot was erected at Continental. The original plan was to integrate the farm further by feeding the grain and the alfalfa produced as rotation crops with cotton. In 1955 Kemper Marley informed Walden that Jack Harris wanted to sell 5,000 acres of irrigated land and 3,000 acres of pasture land at Sahuarita. Harris wanted to devote full time to his farming in California. This is the Jack Harris of the Harris Farms chapter. The price was $5 million, but the land nearly adjoined Continental, which made it attractive. In a few years it became the site of FICO headquarters.

IN 1957 Walden found a 138,000-acre ranch at Aguila, 60 miles northwest of Phoenix. This was an undeveloped piece of "really poor land." In five out of six years, it could not support more than 400 brood cows. However, when 8 to 10 inches of rain came in late November or early December followed by warm weather, it produced a stand of grass about 6 inches tall that could support 8,000 head of cattle long enough for each to gain 250 pounds. This was a virtual bonanza if that many cattle could be secured quickly enough to utilize the grass. After the grass was gone, the cattle were ready for the feedlot at Continental utilizing all the synergies that Walden had conceived.

But he had bigger plans for Aguila. To determine if water was under any of those 138,000 acres, he immediately called upon the water witcher he had used at Wheeler Ridge. Water was located, and 40 wells were dug. These were capable of producing 80,000 gallons a minute, adequate water to irrigate about 16,000 acres. With the assurance of an ample water supply, Walden started construction on a rail siding and a vegetable packing facility, as well as on a mess hall and dormitories for the 1,600 workers needed for about seven months each year.

After a few years Walden became concerned about the potential supply of water at Eloy and decided to sell that farm. He immediately purchased a 1,000-acre farm at Picacho, near Eloy, which was converted to a citrus program. That proved to be an excellent move. In 1960 he secured 1,500 acres at Maricopa, near Phoenix. This was choice property, and after eight wells were completed, it was used for cotton and vegetables. Culbertson was transferred from Eloy to manage Picacho and Maricopa. All the above capital expenditures were made on FICO's operating line of credit rather than by seeking long-term financing. This occurred at a time when cotton prices were low, which caused cash flow to decline. Bank of America became concerned and put pressure on FICO because the company's operating loan was getting too large. For the second time in his career, Walden had to use long-term, real estate–based financing and secured a loan from Connecticut Mutual Life Insurance Company.

As soon as water was available at Aguila, a cotton, small grain, melon, and lettuce rotation was established. Lettuce quickly became a major crop, with 2,500 acres being double-cropped each year. Walden was in his glory, for he was acclaimed the largest lettuce producer in the nation. On an average day, 400 carloads of lettuce were loaded nationally, of which 100 came from Aguila. One day Aguila loaded 119 carloads.

Everything was going perfectly until Walden sensed that he was headed for trouble. He called Gene Mezillier, his liaison with Bank of America, to report that the price of lettuce was down and he was going to lose "lots of money." He wanted Mezillier to come out and survey the situation. Bank of America financed 90 percent of the lettuce production in Arizona and California at the time and was fully aware of what was happening.

Mezillier flew to Arizona and was shown around the farm. After he made his assessment, he reminded Walden of the 80-page outline he had presented when he applied for financing, which showed that over a 20-year period, 8 out of 10 crops grown in a 5-year period would be profitable. Then he said, "You've had three average years and one bonanza. Now you have your first disaster. What's the matter, Walden? Haven't you any guts?" He explained that that was why the bank always insisted that before anyone secured financing for lettuce production, the farmer had to agree to grow the crop at least three years. Mezillier returned to California and left Walden with the assurance that he had financing for the next crop. The crop that caused the worry resulted in a $1.3 million loss, the first in seven crops. However, the fall crop was good enough that the year ended with a net profit of $17,000.

BY THE EARLY 1960S, FICO operated every irrigated acre, 16,000 in all, in the Santa Cruz Valley, from Tucson to within eight miles of the Mexican border. Much of this was leased land, but a rotation was established to provide for the maximum acreage in cotton. Once again the water witcher was called in. He succeeded in finding sites where 41 wells were later drilled, 40 of which were good. Before he witched for the forty-first well, he stated that it was a slim chance, which led Walden to give him a 100 percent success rating.

In 1962 FICO erected a cotton gin at Sahuarita, and Culbertson was assigned to secure contracts with independent growers to produce for the gin. He was also instructed to offer crop financing for any contractor with the FICO gin. According to Walden, one reason for the success of the gin was that everyone produced the same variety of cotton, which enabled the company to do an excellent job of ginning. This resulted in the best end product, which helped FICO obtain a higher price from the mills. Now cotton seed hulls from the gin were available to supplement the ration at the feedlot, which was only about six miles down the road. This provided another synergy to the business.

IN THE LATE 1940S, after becoming a large-scale cotton producer, Walden became active in the industry's organizations. By 1954 he was an officer in the Arizona Cotton Growers' Association. In 1956 he was a trustee of the National Cotton Council of America. He then became a board member of the Cotton Producers' Institute and of Cotton, Incorporated. In 1961 he was elected president of Cotton Council International. In addition to being active in the cotton industry, he served on the Advisory Committee of Stanford Research Institute's Agricultural Research Center during the 1960s. He was also chair of the Arizona Oil and Gas Commission, a trustee of Pamona College, a director of First Interstate Bank of Arizona, and a director of Arizona Feeds Company.

Walden was a very talented person and was sought after to serve, but he ruffled many because he was an ardent free enterpriser and strongly defended his beliefs. In 1958, addressing the American Cotton Congress in a speech entitled "What Is the Future of the American Cotton Producer?," he stated his views:

> The national cotton policy, which, in my judgment, if permitted to continue another 10 years, perhaps even as short a period as 5 years, will bring about the complete destruction of the raw cotton business in America. Our policy has been one of high prices, which has encouraged competition from synthetics at home and abroad and from cotton producers in other lands. From World War II to 1958, world cotton consumption has increased 15 million bales (1.25 million a year), while U.S. exports have dropped from 5 million to 2 million bales annually.
>
> Through our system of high price supports and acreage controls, we have:
>
> • Kept high-cost producers in business (with both good- and bad-quality cotton)
>
> • Kept farms inefficient by not permitting economy of scale
>
> • Kept the efficient producer at a low volume, which forced the textile industry to look for a substitute
>
> • Forced the government to accumulate cotton that cannot be sold except at a huge discount

Walden argued that the subsidy for cotton should be dropped so that the low-cost producer would be free to enter the market against the competition. He insisted that a high price for cotton was not the solution for the small-scale operator. He encouraged farmers to sign up for the Soil Bank by proposing a permanent cotton retirement program for the half million farms with fewer than 49.9 acres of cotton. An acreage that small or smaller was not enough to support a family. Walden suggested that these farmers should receive payments for five years, similar to Soil Bank payments, which would come jointly from federal Soil Bank funds and the sale of their allotments to the larger growers. He realized that such a proposal presented a socioeconomic problem in which the "high-cost 525,000 farmers" would outvote the efficient, large-scale, low-cost farmer by more than 12 to 1. He closed his speech by declaring, "The industry needs some fresh thinking."

He continued to expound his views throughout the 1960s and into the 1970s, after which he turned his attention elsewhere. He maintained that there were too many varieties of cotton, causing difficulties for the mills in turning out a uniform product. At the same time, the high subsidized price kept the mills from buying cotton. When cotton prices dropped sharply in the 1970s, he argued that the set-aside programs had perpetuated the problem. He pushed Cotton Council International to assess all growers $1 a bale to improve marketing of cotton products.

Walden reaffirmed conclusions he made in the 1950s from his world travels during which he learned of producers in other countries, particularly Egypt and Turkey, who grew cotton for $10^{1}/_{2}$¢ a pound and sold it on the world market for 36¢, while the American established price was 34¢. He said, "Cotton production has not bounced back in the United States because so many other nations learned to make a profit and sell for less than our protected product. . . . Our program encouraged other countries to increase their production."

His strongly worded messages and his personal involvement in selling some of his cotton abroad while on organization business "drew lots of heat." This probably caused him not to be elected president of the American Cotton Council. On the other hand, he gained from the experience.

In the late 1960s, when verticillium wilt and rust infested cotton fields in the Santa Cruz Valley, he reacted quickly. Cotton yields dropped from more than two bales per acre to one bale, the lowest yield he had ever experienced, and there was no known cure for the diseases. In 1968 he made 13 trips to Washington, D.C., and secured permission to move his cotton allotment from the Santa Cruz Valley to his land at Aguila.

In the early 1970s FICO was exporting a big portion of the cotton production from Aguila to Achiele Roncoroni, who owned five textile mills in Italy. Roncoroni visited Aguila and proposed to buy the farm from FICO if the company would agree to operate it for two years. For the next two years FICO operated it, making a profit of $500,000 the first year and $1.25 million the second year for each of the partners. At that point the sale was made for $15 million for Aguila real property plus $1.6 million for the machinery. A water shortage had developed on the eastern portion of the farm, and Walden suggested that irrigated crops be discontinued there. The new owners followed his suggestion, and the irrigated portion of Aguila was reduced to 10,000 acres for vegetable production.

In the 1970s there was much concern over a possible world food shortage, and Walden had an answer to that problem. His opinion was that farmers had to be allowed to produce to the maximum, that a worldwide distribution system should be established to get food where it was needed, and that a credit system had to be created to finance the distribution of food. He closed his commentary by stating that "the full power of private enterprise must be released to improve the lot [of humanity]."

RICHARD (DICK) WALDEN was born in 1942 in California. While the family was still located there, he had 30 laying hens and sold eggs to the neighbors. He was permitted to put all the income into a savings account. He was 7 years old when the family arrived at Continental. At age 8 he was assigned to sweep the shop on Saturdays, and at age 10 he was taught to put "hard face" on cultivator sweeps. When Dick was 11 he mowed alfalfa, using a Farmall H with a sickle mower. The following year he drove a Farmall M with a four-row cultivator for cultivating cotton. He and his brother, Tom, had sows and raised pigs. For a couple years they also had steers that they raised on Johnson grass, which thrived on the tail water (the water on the headland at the end of rows of cotton or later pecan trees). The pigs and the steers were their 4-H projects. During Dick's junior high years, he attended a boarding school in Tucson and was home only on weekends, so the projects came to an end. At age 14 he drove silage trucks for corn, milo, and millet harvest and then graduated to operating combines and herbicide spray rigs.

In the summer of 1962 Dick worked on the crew constructing the cotton gin and then, with other employees, attended a seminar on how to operate it. The following year he had his first lessons in cattle buying for the 1,300-cow ranch FICO owned in Osage County, Oklahoma. In his first year after college, he purchased cattle for the feedlot at Continental and bought drought-stricken cows, which were placed on grass at Aguila and in Oklahoma.

Dick had taken ROTC while at college, and in January 1965 he was called to service in the Army, where he flew planes for surveillance purposes. In January 1968, after completing his duty in Viet Nam, he was named assistant manager of the expanded 12,000-head feedlot, where his duties were to buy cattle, grain, hay, and supplements. At the same time, the manager taught him the art of selling cattle. He also flew to locations in Texas to check on FICO cattle that were being backgrounded in preparation for moving them to the feedlot.

While the feedlot was receiving grain and hay grown on the farm in rotation with cotton, plus getting cotton seed hulls from the gin, the synergies worked to make it quite profitable. However, when the cotton allotment and the gin were transferred to Aguila and then that property was sold, cotton seed hulls no longer were available, nor was alfalfa or grain. Then, virtually 100 percent of the inputs had to be purchased and trucked in, and the synergies were lost. Because of this, making a profit in the feedlot became a real challenge.

A 2000 photo of Dick Walden, Chairman, President, and CEO of FICO, with his father, R. Keith Walden, Chairman Emeritus. FICO is the largest integrated grower-producer of pecans in the nation. (Photo by Tamietti Photography 2000)

Another problem arose. The population of Sahuarita and the retirement community of Green Valley had grown considerably, and FICO received many complaints about feedlot odors. Mountain breezes swept through the lot on their way into the valley, and even though every precaution was taken at great cost, the odors could not be eliminated. This led to the decision in 1976 to close the feedlot after 22 years of successful operation.

WALDEN'S BACK was to the wall when cotton production was doomed in the Santa Cruz Valley. He commented on his plight: "I bought this land [Sahuarita] and Continental to grow cotton, but when disease came, we were driven out of that business. I never dreamed I would be growing pecans." Fortunately for FICO, every year that Walden farmed, he had an experimental plot. In 1950, the second crop year at Continental, he planted an experimental plot that contained 100 each of peach, apricot, pear, plum, apple, nectarine, walnut, almond, and pecan trees, in addition to 8 to 10 varieties of grapes. By the mid-1960s he realized that he could be competitive in two varieties of grapes and was offered a 15-year contract to produce them. But grape production had too many negatives to contend with—summer rains, mildew, labor intensiveness, perishability, and the fact that the nearest winery was 700 miles away. Pecans were native to the area, and, equally important, Walden knew that nuts could be mechanically harvested, which meant lower labor cost.

He also had noticed pecan trees on several residential lots in the Valley, where they appeared to thrive. His interest was whetted from his observation of the impressive 12-acre pecan grove on the Midvale Farms that FICO managed and operated in the 1960s. Midvale was owned by Mrs. Charles Wiman, of John Deere fame, who knew that the grove had been planted in 1936. George Parker, a friend, had four pecan trees in his yard, and Walden asked if he could put a fence around them to keep squirrels and humans away so he could get a more accurate yield potential.

Walden had become acquainted with Dean Stahman, of Las Cruces, New Mexico. Stahman was a well-known cotton producer and a pioneer large-scale pecan grower. He had grown pecans for 25 years for commercial food production in addition to having a nursery. Over the years the two men became close friends, for both had an intense curiosity about agriculture, read a great deal, and freely shared their information. Walden was a "fantastic researcher," and as soon as he realized that cotton was doomed, he read all he could about pecans. He convinced himself that pecans were the answer to survival in Santa Cruz agriculture.

He liked doing things on a big scale and decided that the entire 5,000 acres of irrigated land at Sahuarita and Continental and another 1,000 acres at Maricopa should be converted to pecans. Walden presented his case to his partners, the Crowns, who gave him the "go ahead." In the 1960s it cost about $2,500 to prepare an acre of

land for pecans over and above the cost of owning the land, drilling wells, and constructing cement ditches. The wells had already been drilled and the ditches already built for raising cotton.

Even with greatly increased technology, the cost of preparing a pecan grove in 2000 has risen to $5,000 per acre. The cost increases each year after the trees are planted because they and the land have to be cared for, including applying 6 acre-feet of water annually at a pumping cost of $20 per acre-foot. As soon as the trees are planted, one person is needed for irrigation per 250 acres on a year-round basis. Intense weed control is mandatory in a new grove, and during the first years, people are needed to carry shields to protect the infant trees as the sprayer travels down the tree lanes. This expense has to be financed because the trees do not yield fruit for several years.

As was his custom, Walden prepared a complete proposal for Bank of America to secure increased operating capital. Next, he and Connecticut Mutual worked out an arrangement for a unique "agricultural step-up long-term loan." Connecticut Mutual agreed to lend 70 percent of the appraised value of the land to start the project, and each year it increased the amount of the loan to pay the ongoing cost. The lender's reasoning was that with the passing of each season, the pecan grove became more valuable and its security was improved. It is reputed that this was the first time the company had made such a loan.

STAHMAN came to Sahuarita to give instructions on how to establish the grove. All the soil had to be tested to indicate how quickly or how slowly it would absorb water, for this determined the degree of slope. Fortunately, the land lay in a slope stretching along the Santa Cruz River for about 20 miles. It was made perfectly flat, but sloped for water flow. (In those days, this work was done with transits, but now it is accomplished with lasers.) Then, the land was marked off for placement of a tree every 30 feet in each direction. Originally, 48 trees were planted per acre, which provided maximum sunlight for the trees. Trees cost $2.10 each in the 1960s, in contrast to about $12.00 in 2000. When trees are about 15 years old, they start shading each other. Then, half of them are removed, leaving 24 per acre in a spacing pattern of 30 by 60 feet. In another 5 to 10 years, shading again becomes a problem. Half the remaining trees are then removed, leaving 12 per acre with a 60-foot spacing in each direction. The removed trees are cut into proper sized chunks and sold directly to consumers as a fireplace wood.

In late 1999 FICO developed a concept for an automated pruner needed to trim trees to gain the maximum exposure of fruit-bearing wood to the sun. After the company was convinced its concept was right, it engaged an engineer/inventor to produce the machine. The finished product is a long boom, with several circular saws attached, installed on a Caterpillar excavator unit. The machine shapes the trees on all sides. It is hoped that this will delay the need for thinning out more trees. The project cost more than $250,000, but farm manager Layne Brandt said, "We have no idea how much it will be worth to us."

IN OCTOBER 1983 the pecan grove suffered its first setback when a hundred-year flood put 1,000 acres under water at harvest time. After the waters receded, the shakers were mounted on Caterpillars. Then, canvas tarps were placed around the trees before shaking them to prevent the pecans from falling onto the wet ground. Then the tarps had to be picked up by a large crew to pour the pecans into the carts for transportation to storage. The cost of that operation was about $5 million in addition to the expense of putting the land back in shape.

Other than flooding, which is FICO's greatest natural risk, lightning regularly destroys trees, causing the need for replanting. Texas root rot, a fungus hosted by the native Mesquite tree, can kill the pecan tree. If leaves are caught at the first sign, the canopy of the tree is cut off to relieve the stress on the roots. Then the roots can be treated to preserve the tree. Fortunately, the fungus does not travel in the soil, so Texas root rot can be controlled by eliminating Mesquite trees in the vicinity of the grove. Yellow and black aphids spread a honeydew deposit on the leaves that prevents them from producing carbohydrates needed to fill out the nut. A severe infestation can cause a 40 to 60 percent crop loss. DDT was very effective against the aphids. When that was banned, the cost of insect control rose significantly for all citrus, fruit, nut, vegetable, cotton, and floral producers. A chemical to control the aphids that has recently become available is placed in the soil and follows the root system into the tree, but the results are not finalized at this writing.

THE FIRST PECANS appear in about five years, and the yield is enough to pay the harvest cost. By the eighth year, the trees produce a profit. They reach peak yield in about 15 years. The oldest pecan tree in the nation is believed to date to about 2000 B.C., but it is not known how long the more recent strains of hybrid pecan trees will produce at a profit, because commercial production is still that new.

FICO started planting pecan trees in 1965, had its peak plantings during 1966 through 1968, and completed the 5,000 acres at Sahuarita in 1969. At that time it was reputedly the largest irrigated pecan grove in the nation, not including the 1,000-acre grove at Maricopa. Later, FICO purchased a 1,000-acre pecan grove near Albany, Georgia, "just to get an idea of what the competition was doing," but with a 50-inch rainfall, higher humidity, and other condi-

Part of the FICO pecan grove that stretches for nearly 20 miles along the Santa Cruz River near Sahuarita. The area has an ideal climate for pecans when adequate irrigation is provided.

tions that were so different from Arizona, it was difficult to make a valid comparison. Nevertheless, that property has proven to be an excellent investment.

In 1964 FICO, under Walden's management, was the largest grower, packer, and shipper of lettuce in the United States. In 1967 it was the largest grower, ginner, and grower-seller of cotton in Arizona, and by 1969 it had the largest pecan grove. The new venture became profitable, and everyone at FICO was pleased, particularly Walden, who had been forced into a corner and for the third time had created a new niche for himself. From 1971 to 2000 FICO averaged 2,200 pounds of pecans per acre, with a peak yield of 2,900 pounds. Depending on price, it takes about 1,200 pounds per acre to pay for cash operating costs.

WALDEN has always farmed where irrigation was necessary with the exception of his first farm in the Tulare Basin, where the water-holding capacity of the soil was sometimes high enough to grow a crop without supplemental water. In Arizona, the sandy soil always made it impossible to produce a crop without irrigation. This made water the number one concern of FICO. Presently the company is prohibited from drilling any additional wells in the Santa Cruz Valley because of the limited water supply, but it can replace an existing well. The only chance for an increase in available water is the possible discontinuation of operations at a nearby copper mine. Should that occur, authorities would determine how the water would be allocated. Current pumping cost is $29 an acre-foot, which brings water expense to $174 annually per acre. Walden commented, "It's always been a challenge to make the use of water more efficient." Ongoing work with fertilizer applications and the computer to determine how to get the maximum use of water is essential to any chance of increasing acres at the Sahuarita-Continental location.

In 1978, in an effort to cut costs, FICO management implemented a program of purchasing only used pickups, tractors, and trucks. At one time, three shops with 10 mechanics were required at Sahuarita-Continental to keep equipment in repair. But because most of the land has been converted to pecan groves, the tillage requirement is greatly reduced, and so is the need for repairs. For a short time after the switch to pecans, the shop workload continued at an intense pace because of the constant repair needed on the diesel-pump engines. However, after the adoption of electric motors, the demand dropped. Today, one full-service shop at Sahuarita does all the work, including complete engine overhauling. When the company decided to buy only used pickups, tractors, and trucks, it also decided to standardize the equipment to reduce the parts inventory. A reliable source of used equipment keeps FICO's current fleet of 30 pickups and trucks and 25 tractors operating efficiently and economically.

IN 1971 the first pecans were harvested and delivered to Stahman's processing plant at Santa Cruz as Walden and Stahman had previously agreed. Each year, as more trees came into production, the volume increased, and FICO continued deliveries to the Stahman plant. Because of the friendship between Walden and Dean Stahman, it was assumed that as the FICO supply increased, Stahman would expand to handle it. Then the unexpected happened. Stahman died, and his son Bill called Walden on September 10, 1974, saying that he could not sell any more pecans than what he was already processing and would not take FICO's production. Walden had never thought of building a plant and "challenged Bill on a moral basis," saying that his father and he had a handshake agreement. Walden got Bill Stahman to process half the crop, about 2.5 million pounds. The rest of the crop went into storage.

Walden immediately traveled to St. Louis to meet with people at the Funston Nut Company, a division of Pet Milk Company. They entered into a joint venture to build a pecan processing plant. The Santa Cruz Pecan Company was organized, and a plant with a capacity of 12 million pounds was completed in time for the 1975 harvest. As part of the agreement, Funston was responsible for managing the plant, but because of a difference in philosophy, that arrangement did not endure.

FICO was willing to sacrifice short-term profits to build long-term relationships with retailers. For FICO, the processing plant was an extension of the grove, which was the company's major interest. Dick Walden commented, "I would not want to have this processing plant without the grove, nor would I want to have the grove without the processing plant." He was alluding to the synergies involved in the long run. In 1979 FICO purchased Funston's interest in the Santa Cruz Pecan Company.

At first the processing plant handled only FICO-produced pecans, but in 1981 the cold-storage facilities were enlarged so pecans could be stored for year-round processing. Without refrigeration, pecans can be held for only about six months without their quality being affected, but if they are dried to 4.5 percent and placed under refrigeration at 26 degrees, they can be held safely for several years. At that point FICO started purchasing pecans from other producers to keep the plant going full time.

Initially, the plant required 250 people working in three shifts to process 12 million pounds of pecans. This was a large enough group to make the plant a target for the unions. After a period of union activity, a vote was taken, and the union won the right to organize the plant. But when the union merged with another, the company challenged the new union. After a strike was called, only about 15 percent of the workers walked out, and the others voted the union out. FICO determined that the best way to avoid future labor turmoil was to increase the benefits. The company decided to adopt a total program covering health, dental, and vision, because these were the greatest needs of the workers who were new to the country.

However, before the company could improve the benefit and wage package, it needed to upgrade the technology in the plant to increase productivity. In 1989 FICO enlarged the plant and the processing line using state-of-the-art technology, which enabled it to double the capacity to 25 million pounds. Prior to the renovation, it cost 70¢ a pound to shell and sort pecans. Following the refurbishing, it cost 40¢. After new machines were installed, the work force was reduced from 250 on three shifts to only 70. This meant that wages and benefits could be increased. Since then, turnover in the labor force has been reduced to "almost none," and many employees are second-generation workers. Many of them live in the 60 company-owned houses. All the plant supervisors are bilingual, which greatly minimizes misunderstandings on the line.

FICO management philosophy is that people are the organization's greatest capital, whether in the plant or in the field. That creates a positive atmosphere. Because of the company's location between the border and Tucson, the nearest large city, FICO feels that it is a training ground for people in the food industry. It also profits by being near

An early morning view of the Green Valley Pecan Company processing plant. Its 120,000-square-foot refrigerated storage (on the left) holds pecans in condition for year-round processing. Pecan grove and Santa Rita Mountains in the background.

the University of Arizona, which has been a good feeder for beginning management personnel.

After the break with Pet Milk, FICO had to send salespeople on the road to make "cold turkey" calls on bakeries, ice cream companies, and candy manufacturers to sell the company's production. It soon learned that large retail chains would take pecans in bags under private label. Today private-labeled pecans make up a major portion of the company's sales. Cosco alone takes 25 percent of FICO's total output under Cosco's private label. FICO is the sole supplier for Nestlé.

An in-house retail store sells about 2.5 percent of the total output, much of which is sold to people who come to tour the plant. Plans for enlargement of the store and for e-mail marketing are in process. About 20 percent of the plant's production is exported to Europe, Mexico, Canada, Japan, and Israel.

One of the biggest unsolved problems with processing pecans is that no one has developed a product that makes use of the shells. This keeps the market price for shells at a mere $5 a ton at the plant. The oil fields are the current largest users of ground shells.

FICO management is strongly in favor of the North American Free Trade Agreement (NAFTA) because it has helped the company's relations with Mexico. FICO purchases pecans from growers in Mexico who are able to deliver through customs without delay. Pecans are native to Mexico, which produces about 20 percent of the world's supply. The United States produces most of the rest, except for Australia and China, two minor producers.

In 1998 the pecan subsidiary name was changed to Green Valley Pecan Company, which improved the image. In 2000 Green Valley produced about 10 percent of the nation's total processed production. The company is sixth ranking among the 10 largest processors but is the only major one that has its own source of production. From 40 to 60 percent comes from its 7,000 acres. Walden's closing remark about acquiring pecans reaffirms his faith in free enterprise: "The shellers [processors] buy for cash, and there is no government program to foul up the supply chain."

A FTER selling the Aguila operation to the Mortori family in 1978, Walden had to look for a place to invest the $15 million from that sale. He recalled that that probably was the biggest investment mistake of his career. In part, this may have been because the time allotted was insufficient to locate a comparably large agricultural unit. He purchased 28,000 acres near Wewahitchka, in the panhandle of Florida. Culbertson, who had been onsite manager at Aguila after Tom Walden was killed in an airplane crash, was sent to Florida to manage that operation. He moved his family to Mexico Beach, where living conditions were more satisfactory than at Wewahitchka.

Except for 8,500 acres that had been cleared, the remainder of the land was in trees, so the plan was to continue removing timber and put the land into other crops. By 1988 about 12,000 acres had been cleared, and the timber was sold to a mill at nearby Panama City. Walden had hoped to raise soybeans there, but the land was too sandy. Instead, the company planted 300 acres of blueberries, which it cooled, graded, packed, and sold under the Blue King label, and 800 acres of pecans. An assortment of speciality crops was tried, but Culbertson said nothing but trees really seemed to have potential in that rainy, humid climate with all the insects and plant diseases. After five years of battling the elements, he contacted Walden and said that he could not see a future there unless it was in trees, so he wanted to retire.

The Culbertsons returned to Sahuarita, but Walden would not let his long-time trusted friend and ally retire. Soon Culbertson was managing Farmers Water Company, a subsidiary that FICO founded in 1958 originally to serve its shops, warehouse, cotton gin, and the 60 company homes, plus the private homes of employees. In 1984 the company supplied water to 110 businesses and residents of Sahuarita and Green Valley. By 2000 it had more than 1,100 accounts.

Dick's father asked him to take charge of the Florida operation. FICO purchased a Cessna 421 with a pressurized cabin so he could cover "all his bases," which now included managing the Florida property, the pecan grove in Georgia, and the grove at Maricopa, as well as taking regular trips to Mexico in search of pecans. The company stabilized the Florida operation at 12,000 acres of crop land, of which 5,000 were double-cropped, and bided its time waiting for a buyer.

FICO was in no financial bind, because in the late 1970s, Walden sold 5,000 acres at Continental for $10 million. The land was to be used for an upscale development community. Fortunately, he had enough influence with the governor to get the highway department to alter its original plans on Interstate 19 from Tucson through the Santa Cruz Valley. FICO donated land for the La Pasada Retirement Home and gave land in Green Valley to the University of Arizona College of Medicine and other land in Sahuarita to the community. Where the new route went was much more beneficial to all concerned.

Culbertson commented that the biggest disappointment of his management career was that he could not make Florida profitable. He was comfortable knowing that his brother-in-law had never placed himself in a position where he had to sell land. True to his track record, Walden held on until 1997, when the Florida property was sold for its purchase price to a pension fund that was looking for a long-term investment. Florida produced operating losses, but few entrepreneurs go through life without some setbacks. Their ability to face those challenges is what sets them apart from the crowd.

When asked what part of his career he liked best, Walden replied: "I really liked cotton because it was such a responsive crop. I grew it from 1949 to the early 1970s. I grew an average of 6,000 acres, with a yield of more than two bales and up to 1,480 pounds many years." He pointed out that cotton was responsive to fertilizer, pest control, irrigation, weather, and weed control. Most of the time water was the biggest challenge, and leveling the fields for good distribution was an ongoing problem.

Those 20-plus years of cotton production were exceedingly profitable for FICO, though much of the time, subsidy payments played a role in the success. To his credit, Walden lambasted the government program and advocated that growers be freed from the shackles and be allowed to compete on the global market.

Bibliography

Interviews

Brandt, Layne A. Sahuarita, AZ, March 29, 2000.

Culbertson, Warren E. Sahuarita, AZ, March 29, 2000.

Walden, R. Keith. Sahuarita, AZ, March 27–29, 2000; telephone conversation, May 10, 2000.

Walden, Richard S. Sahuarita, AZ, March 28, 2000.

Miscellaneous

Goff, John S. *Arizona: An Illustrated History of the Grand Canyon State*. Windsor Publishing Co., 1988, 176–177.

Petrakis, Harry Mark, and David B. Weber. *Henry Crown: The Life and Times of the Colonel*. Henry Crown & Company, 1998.

Riddick, John. "Clouds on the Farming Horizon," *Today's Business: Arizona's Leadership Magazine*, January 1979, 19–22.

Walden, R. Keith. "History of Farmers Investment Company." Manuscript in company files.

———. "The Pecan—A Distinguished Nut," December 7, 1994. Speech manuscript in company files.

———. Speech given at the Arizona Junior Achievement Award banquet, Phoenix, October 10, 1996. Manuscript in company files.

———. Vita dated April 6, 1993. In company files.

———. "What Is the Future of the American Cotton Producer?," June 7, 1958. Speech manuscript in company files.

Index